11th Asia Pacific Conference on Plasma Science and Technology (APCPST) and 25th Symposium on Plasma Science for Materials 2012 (SPSM)

Journal of Physics: Conference Series Volume 441

Kyoto, Japan
2-5 October 2012

Editors:

Takayuki Watanabe
Toshio Kaneko
Makoto Sekine
Yasunori Tanaka

ISBN: 978-1-62748-673-6
ISSN: 1742-6588

Printed from e-media with permission by:

Curran Associates, Inc.
57 Morehouse Lane
Red Hook, NY 12571

Some format issues inherent in the e-media version may also appear in this print version.

Printed by Curran Associates, Inc. (2013)

For permission requests, please contact the Institute of Physics
at the address below.

Institute of Physics
Dirac House, Temple Back
Bristol BS1 6BE UK

Phone: 44 1 17 929 7481
Fax: 44 1 17 920 0979

techtracking@iop.org

Additional copies of this publication are available from:

Curran Associates, Inc.
57 Morehouse Lane
Red Hook, NY 12571 USA
Phone: 845-758-0400
Fax: 845-758-2634
Email: curran@proceedings.com
Web: www.proceedings.com

TABLE OF CONTENTS

PAPERS

LIFE INNOVATION

GREEN INNOVATION

BASIC PLASMA PHYSICS

PLASMA GENERATION

APPLICATIONS

MATERIALS

PREFACE

The 11th Asia–Pacific Conference on Plasma Science and Technology (APCPST-11) was held in Kyoto, Japan on 2–5 October 2012 with the 25th Symposium on Plasma Science for Materials (SPSM-25).

SPSM has been held annually since 1988 under the sponsorship of The 153rd Committee on Plasma Materials Science, Japan Society for the Promotion of Science (JSPS). This symposium is one of the major activities of the Committee, which is organized by researchers in academia and industry for the purpose of advancing intersectional scientific information exchange and discussion of science and technology of plasma materials processing.

APCPST and SPSM are jointly held biennially to survey the current status of low temperature and thermal plasma physics and chemistry for industrial applications. The whole area of plasma processing was covered from fundamentals to applications. Previous meetings were held in China, Japan, Korea, and Australia, attended by scientists from the Asia–Pacific and other countries.

The joint conference was organized in plenary lectures, invited, contributed oral presentations and poster sessions. At this meeting, we had 386 participants from 10 countries and 398 presentations, including 26 invited presentations. This year, we arranged special topical sessions that covered green innovation, life innovation, and technical reports from industry. This conference seeks to bring the plasma community together and to create a forum for discussing the latest developments and issues, the challenges ahead in the field of plasma research and applications among engineers and scientists in Asia, the Pacific Rim, as well as Europe.

This volume presents 44 papers that were selected via a strict peer-review process from full papers submitted for the proceedings of the conference. The topics range from the basic physics and chemistry of plasma processing to a broad variety of materials processing and environmental applications. This volume offers an overview of recent advances in thermal and non-equilibrium plasmas as well as on more new and innovative developments in the field of life innovation, green innovation and a technical report session.

The editors hope that this volume will be useful and helpful for deepening our understanding of science and technology of plasma materials processing and also for stimulating further development of the plasma technology.

Finally, we would like to thank the conference chairmen, the members of the organizing committee, the advisory committee, the executive committee, the program committee, the publication committee, organizing secretariat and financial support from The 153rd Committee on Plasma Materials Science, JSPS.

Sponsors and Supporting Organization: The 153rd Committee on Plasma Materials Science, Japan Society for the Promotion of Science

Organizing Committee Chairperson: Osamu Tsuji, SAMCO Corporation, Japan
Advisory Committee Chairperson: Akihisa Matsuda, Osaka University, Japan
Executive Committee Chairperson: Masaru Hori, Nagoya University, Japan
Program Committee Chairperson: Takamasa Ishigaki, Hosei University, Japan
Publication Committee Chairperson: Takayuki Watanabe, Kyushu University

Editors of APCPST-11 and SPMS-25

Professor Takayuki Watanabe, Kyushu University, Japan
Professor Toshio Kaneko, Tohoku University, Japan
Professor Makoto Sekine, Nagoya University, Japan
Professor Yasunori Tanaka, Kanazawa University, Japan

Peer review statement

All papers published in this volume of *Journal of Physics: Conference Series* have been peer reviewed through processes administered by the proceedings Editors. Reviews were conducted by expert referees to the professional and scientific standards expected of a proceedings journal published by IOP Publishing.

Investigations on Plasma-Biomolecules Interactions as Fundamental Process for Plasma Medicine

Kosuke Takenaka[1], Ken Cho[2], Yuichi Setsuhara[1], Masaharu Shiratani[3], Makoto Sekine[4], Masaru Hori[4]

[1] Joining and Welding Research Institute, Osaka University, 11-1 Mihogaoka, Ibaraki, Osaka 567-0047, Japan
[2] Institute for Materials Research, Tohoku University, 2-1-1 Katahira, Aoba-ku, Sendai 980-8577, Japan
[3] Department of Electronics, Kyushu University, 744 Motooka, Nishi-Ku, Fukuoka 819-0395, Japan
[4] Plasma Nanotechnology Research Center, Nagoya University, Furo-cho, Chikusa-ku, Nagoya 464-8603, Japan

E-mail: k_takenaka@jwri.osaka-u.ac.jp

Abstract. Investigations on plasma-biomolecules interactions have been carried out as fundamental process for plasma applications in medical science and biological treatments. In this study, plasma interactions with L-alanine which are amino acids: the building blocks of proteins have been investigated in terms of physical effects. Effect of ion and photon irradiation on degradation of L-alanine has been studied via x-ray photoelectron spectroscopy (XPS) analysis of chemical bonding states. The XPS results showed that the decrease in the -CH_3 bond, >$C(NH_2)COOH$ bond, and -COOH bond peaks of main component of the L-alanine with increasing ion energy, ion dose and photon energy. The L-alanine degraded by ions and photons with energy over 6eV.

1. Introduction

Plasma applications in medical science and biological treatments recently have shown remarkable progress with worldwide attentions [1-11]. Considering that biological molecules (biomolecules) such as DNA, amino acids and proteins, which are the building blocks of cells and organs in a hierarchy manner, are basically composed of organic chemical compounds, it is significantly essential to perform fundamental studies in molecular and/or atomic levels for understanding plasma interactions with biomolecules during exposure with plasmas, which contain ions, radicals, electrons and photons. In the plasma interactions with organic materials, it is of great importance to study fundamental processes involved in plasma exposure, which may contribute to modification and/or degradation of the organic materials in complex synergetic manners via ions, radicals, electrons and photons. In our previous works, plasma interactions with soft materials have been performed on the basis of surface analysis with x-ray photoelectron spectroscopy (XPS), in terms of physical and chemical interactions using plasmas sustained with low-inductance antenna (LIA) modules. [12-14] Based on these scientific knowledge, the preset study extends the investigations further on plasma interactions with amino acids, which are the building blocks of proteins. Experimental results obtained from plasma exposures to amino acids will be presented in terms of physical and chemical effects via XPS analysis of chemical bonding states.

2. Experimental

The Experiments have been performed by the discharge chamber installed with a set of 8 units of LIA. The LIA units were mounted on the top flange of the discharge chamber and were coupled to a RF power generator at 13.56 MHz via a matching network. Each LIA unit was connected in parallel to the matching network. The LIA unit consists of a U-shaped antenna conductor (with a 80 mm height and a 70 mm width), which was fully covered with dielectric tubing for complete isolation from the plasma [12]. The chamber had a 500 mm inner diameter and a 200 mm height, which was connected to diffusion chamber made of stainless-steel vessel with a 500 mm inner diameter and a 400 mm height [15]. Kinetic energy distributions impinging onto the ground potential (ion energy at the sheath edge) from the plasmas were measured using a mass separated ion energy analyzer (Hiden, EQP500), which was mounted beside the substrate holder. The Ar plasma sustained at an RF power of 1000 W. Optical emission spectra in UV and VUV regions were measured with VUV monochrometer (McPHERSON, 234/302) mounted on sidewall of the chamber.

Ion energy distributions of the argon ions impinging to the ground potential through the sheath edge of the argon plasmas were measured using the mass-separated ion-energy analyzer (the spectra not shown). Peak values of ion energy distributions decreased from15 eV to 6 eV with increasing Ar pressure from 0.67 Pa to 26.6 Pa (15 eV at 0.67 Pa, 13 eV at 1.3 Pa, 9 eVat 6.5 Pa, 7 eVat 13 Pa,

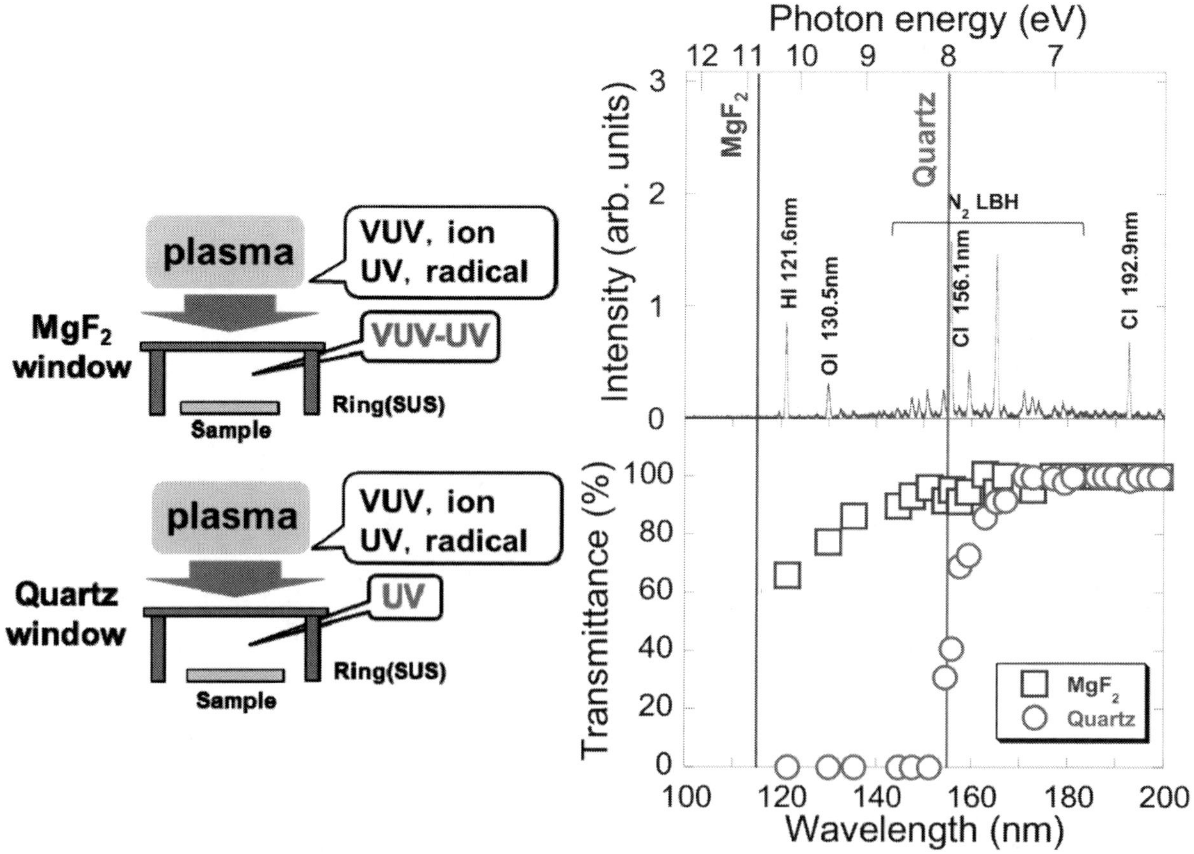

Figure 1. Schematic illustration of the apparatus used in photoirradiation and typical UV and VUV emission spectrum from the Ar plasma sustained at an RF power of 1000W together with the optical transmission properties of the MgF_2 and the quartz windows.

and 6 eVat 26 Pa) [17]. Furthermore, the ion-bombardment energy onto non-conductive substrate (the PET film in the present investigation) is almost equivalent to potential drop from plasma potential to floating potential. As reported in the previous work, the potential drop of argon plasma was in the range of 3–15 eV for Ar pressures of 0.26–13 Pa (15 eV at 0.26 Pa, 8 eV at 1.3 Pa, and 3 eV at 13 Pa) [17].

For evaluating the interactions of the polymer surface with UV and VUV photons from the L-alanine were exposed to photons from the Ar plasmas by covering the L-alanine with MgF_2 and quartz windows as shown in Fig. 1. Here it is noted that thickness of the MgF_2 and quartz windows is 2 mm. The MgF_2 window is transparent in VUV+UV regions (wavelength longer than 115 nm corresponding to photon energy less than 11eV), whereas the quartz window is transparent in UV regions (wavelength longer than 155 nm corresponding to photon energy less than 8 eV). Therefore, it is possible to evaluate the effects of UV and VUV radiations by comparing results of the two samples with plasma exposed samples. [16]

On the bases of these results for ion energy and optical emission, in this investigation, the chemical bonding states of L-alanine exposed to argon plasmas with a varying ion energy and photon energy were investigated.

The analysis of chemical bonding states was carried out by x-ray photoelectron spectroscopy (XPS) (AXIS-165x SHIMADZU Corp., Japan). The XPS analysis was performed with non-monochromatized Mg Kα radiation (photon energy of 1253.6 eV).

3. Results

Figure 2 shows a typical XPS C 1s spectrum for the raw L-alanine together with chemical structure. The C 1s photoelectron spectrum from the L-alanine was composed of the following major components: $-CH_3$ bond peak at a binding energy of 285.4 eV, $>C(NH_2)COOH$ bond peak at 286.7 eV, and -COOH bond peak at 288.8 eV. [18]

In order to investigate effect of ion irradiation on biomaterials, chemical bonding states of the L-alanine exposed to argon plasmas were examined by XPS. Figure 3 (a) shows XPS C 1s spectra of L-alanine exposed to argon plasma sustained at an RF power of 1000Wand argon pressures of 0.67 Pa-26 Pa corresponding to ion energy range of 6 eV-15 eV with an ion dose of 4.3×10^{18} ions/cm^2. The spectra show the decrease in the $-CH_3$ bond, $>C(NH_2)COOH$ bond, and -COOH bond peaks of main component of the L-alanine. With increasing ion energy, the peaks of $>C(NH_2)COOH$ bond, and -

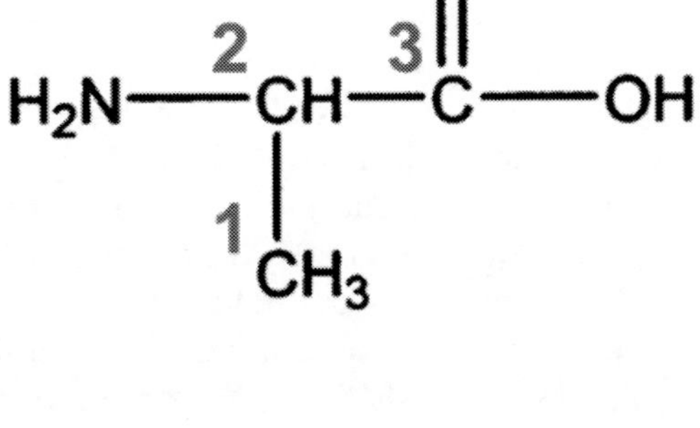

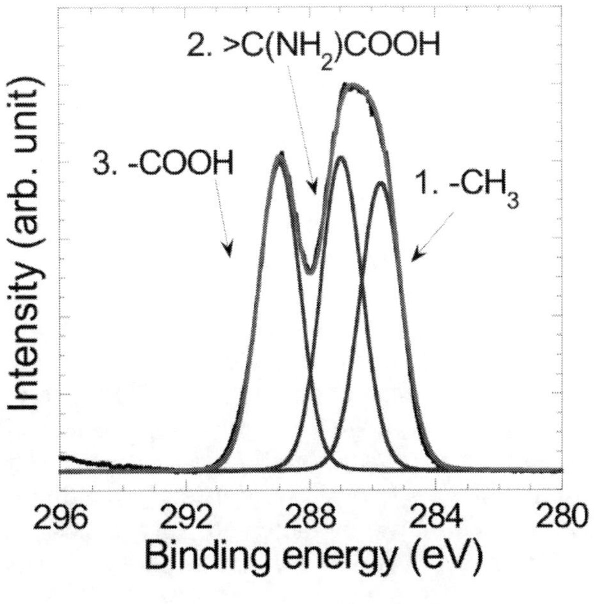

Figure 2. Chemical structure and typical XPS C 1s spectrum of raw L-alanine.

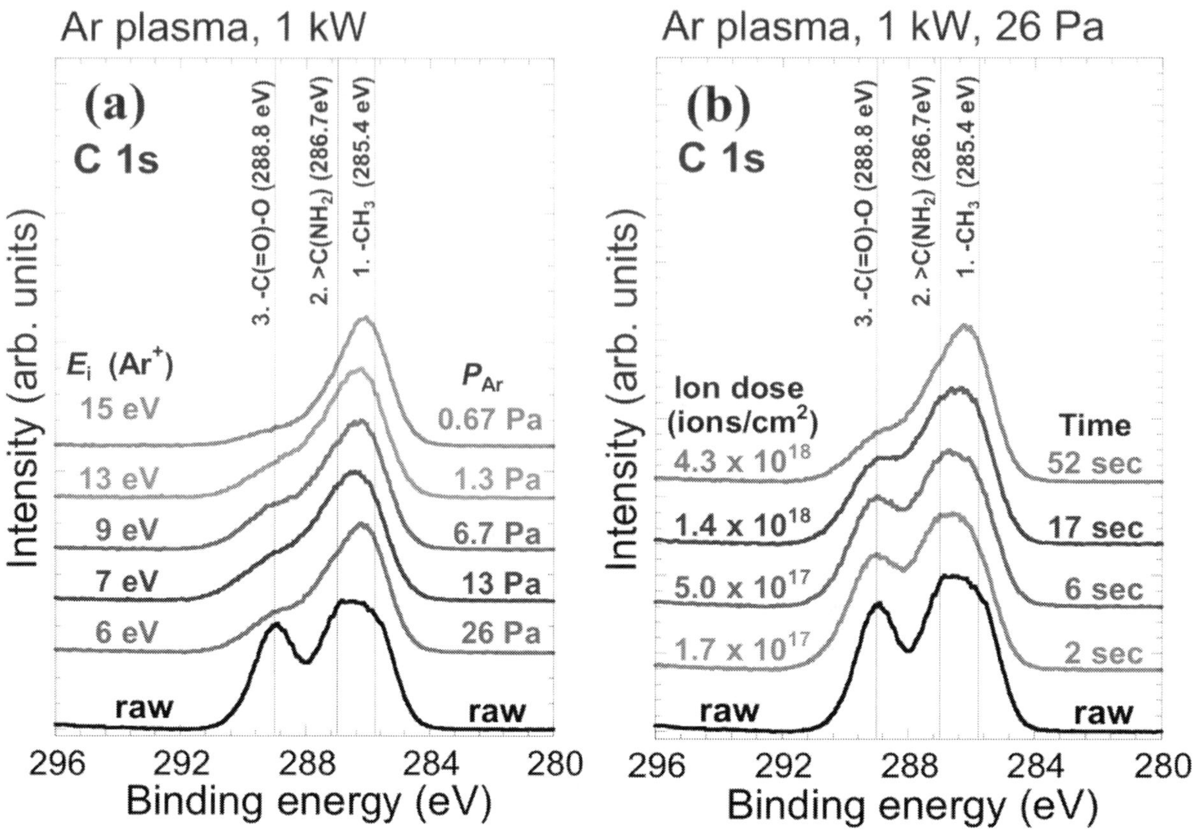

Figure 3. The XPS C1s spectra for L-alanine without and with exposure to argon plasmas with a variation of argon pressures corresponding to ion bombardment energy and ion dose.

COOH bond drastically decreased. Figure 3 (b) shows XPS C 1s spectra of L-alanine exposed to argon plasma sustained at an RF power of 1000W and argon pressures of 26 Pa in an ion dose range of 1.7×10^{17} ions/cm^2 - 4.3×10^{18} ions/cm^2. With increasing ion dose, the peaks of >C(NH_2)COOH bond, and -COOH bond drastically decreased as is the case in the results of ion energy dependence. In the case with polymer materials including polyethylene terephthalate (PET), chemical bonding states of the PET films exposed to argon plasmas with ion energy of 6eV showed no degradation. [19] The results exhibit that L-alanine degraded at low energy compared with polymer materials including polyethylene terephthalate (PET).

Figure 4 shows the conventional XPS C1s spectra of L-alanine; (a) after direct exposure to the plasma, (b) after irradiation with the VUV+UV photons from the plasma (through the MgF_2 window), (c) after irradiation with the UV photons from the plasma (through the quartz window) (d) without photon irradiations (with only radials) and (d)before irradiation (raw). Here the spectra shown in Fig. 4 correspond to probing depth of about 10 nm from the surface. The samples after exposure to the Ar plasma and after the photoirradiation in the VUV+UV, the UV regions and without photon irradiations clearly exhibit significant decrease in the peaks of >C(NH_2)COOH bond, and -COOH bond. In particular, the XPS analyses show that the degradation of L-alanine after the photoirratiation in the UV + VUV regions as well as plasma irradiation is localized in the vicinity of the surface regions as shallow as 10 nm. Substantial degradation is found for the spectrum for irradiation of VUV+UV region corresponding to the spectrum for irradiation of UV region and without photon irradiations (with only radials). These results suggest that the photoirradiation in the VUV regions over 8eV may cause photodecomposition of the chemical bonds to create free radicals.

Journal of Physics: Conference Series **441** (2013) 012001 doi:10.1088/1742-6596/441/1/012001

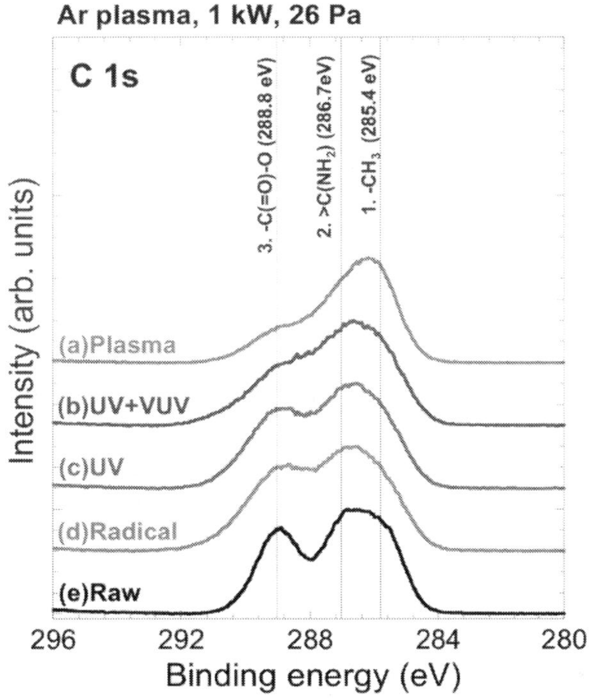

Figure 4. XPS C1s spectra for L-alanine; (a)after direct exposure to the Ar plasma, (b)after irradiation with the VUV+UV photons from the plasma (through the MgF_2 window), (c)after irradiation with the UV photons from the plasma (through the quartz window), (d)without photon irradiations (radical irradiation) and (e)before irradiation (raw).

These results for ion and photon irradiations exhibit that the degradation of L-alanine was occurred for low energy compared with polymer materials including polyethylene terephthalate (PET). It is considered that the degradation of L-alanine was occurred by direct fragmentation due to impinging of ions and photons since L-alanine is monomer.

4. Summary

Investigations on plasma-biomolecules interactions have been carried out as fundamental process for plasma applications in medical science and biological treatments. In this study, plasma interactions with L-alanine which are amino acids: the building blocks of proteins have been investigated in terms of physical effects. Effect of ion and photon irradiation on degradation of L-alanine has been studied via x-ray photoelectron spectroscopy (XPS) analysis of chemical bonding states. The XPS results showed that the decrease in the $-CH_3$ bond, $>C(NH_2)COOH$ bond, and -COOH bond peaks of main component of the L-alanine with increasing ion energy, ion dose and photon energy and the L-alanine degraded by ions with energy over 6eV and photons with energy over 8eV. It is considered that the degradation of L-alanine may be occurred by direct fragmentation of L-alanine due to impinging of ions and photons.

Acknowledgments

This work was partly supported by The Grant-in-Aid for Scientific Research on Innovative Areas "Plasma Medical Innovation" (24108003) from The Ministry of Education, Culture, Sports, Science and Technology (MEXT).

References

[1] Moisan M, Barbeau J, Moreau S, Pelletier J, Tabrizian M and Yahia L H 2001 *Int. J. Pharm.* **226**, 1.

[2] Laroussi M 2005 *Plasma Process. Polym.* **2** 391.
[3] Boudam M K, Moisan M, Saoudi B, Popovici C, Gherardi N and Massines F 2006 *J. Phys. D: Appl. Phys.,* **29** 3494.
[4] Weltmann K -D, Brandenburg R, Woedtke T, Ehlbeck J, Foest R, Stieber M and Kindel E 2008 *J. Phys D: Appl. Phys.,* **41** 194008.
[5] Raiser J and Zenker M 2006 *J. Phys. D: Appl. Phys.,* **39** 3520.
[6] Stalder K R and Woloszko J 2007 *Contrib. Plasma Phys.,* **47** 64.
[7] Yonson S, Coulombe S, Lveill V, Leask R L 2006 *J. Phys. D: Appl. Phys.,* **39** 3508.
[8] Coulombe S, Lveill V, Yonson S and Leask R L 2006 *Pure Appl. Chem.,* **78** 1147.
[9] Stoffels E, 2007 *Contrib. Plasma Phys.*, **47** 40.
[10] Fridman G, Friedman G, Gutsol A, Shekhter A B, Vasilets V N and Fridman A 2008 *Plasma Process. Polym.,* **5** 503.
[11] Morfill1 G E, Kong M G and Zimmermann J L 2009 *New J. Phys.* **11** 115011.
[12] Setsuhara Y, Shoji T, Ebe A, Baba S, Yamamoto N, Takahashi K, Ono K and Miyake S 2003 *Surf. Coatings. Tehcnol.,* **174-175** 33.
[13] Setsuhara Y, Takenaka K, Ebe A, Nishisaka K 2007 *Plasma Process. Polym.,* **4** S628.
[14] Setsuhara Y, Cho K, Takenaka K, Shiratani M, Sekine M and Hori M 2008 *Surf. Coat. Technol.,* **202** 5225.
[15] Takenaka K, Setsuhara Y, Nishisaka K and Ebe A 2006 *Jpn. J. Appl. Phys.,* **45** 8046.
[16] Uchida S, Takashima S, Hori M, Fukasawa M, Ohshima K, Nagahata K, Tatsumi T 2008 *J. Appl. Phys.*, **103** 073303.
[17] Takenaka K, Setsuhara Y, Nishisaka K and Ebe A 2006 *Jpn. J. Appl. Phys.,* **47** 6900.
[18] Gao Y K, Traeger F, Shekhah O, Idriss H, Wöll C 2009 *J. Colloid Interface Sci.,* **338** 16.
[19] Setsuhara Y, Cho K, Takenaka K, Shiratani M, Sekine M and Hori M 2011 *Thin Solid Films* **519** 6721.

Chemical analysis of plasma-assisted antimicrobial treatment on cotton

C W Kan[1*], Y L Lam[1], C W M Yuen[1], A Luximon[1], K W Lau[1] and K S Chen[2]

[1]Institute of Textiles and Clothing, The Hong Kong Polytechnic University, Hung Hom, Kowloon, Hong Kong
[2]Department of Materials Engineering, Tatung University, 40 Zhongshan North Road, 3rd Section Taipei 104, Taiwan

E-mail: tccwk@inet.polyu.edu.hk or tccwk@polyu.edu.hk

Abstract. This paper explores the use of plasma treatment as a pretreatment process to assist the application of antimicrobial process on cotton fabric with good functional effect. In this paper, antimicrobial finishing agent, Microfresh Liquid Formulation 9200-200 (MF), and a binder (polyurethane dispersion, Microban Liquid Formulation R10800-0, MB) will be used for treating the cotton fabric for improving the antimicrobial property and pre-treatment of cotton fabric by plasma under atmospheric pressure will be employed to improve loading of chemical agents. The chemical analysis of the treated cotton fabric will be conducted by Fourier transform Infrared Spectroscopy.

1. Introduction

Cotton fabric is highly popular due to their excellent properties such as regeneration, bio-degradation, softness, affinity to skin and sweat-absorbing property [1, 2]. However, cotton fabric is very prone to be attacked by certain microorganisms, insects and fungi causing functional, hygienic and aesthetic difficulties [3-5]. Extensive efforts have been made to control emerging disease infections. Therefore, a large demand exists for anti-microbial finished textiles [6, 7] and there are three methods of applying anti-microbial agents to textiles, i.e. (i) the anti-microbial agents are directly adsorbed onto fibres; (ii) the anti-microbial agents are confined to the network structure of a reactive synthetic resin formed on the fibre surface; and (iii) the anti-microbial agents are covalently bound to cellulosic fibres [8]. Bacteria have different membrane structures which are classified as Gram negative or Gram positive [9]. Gram negative bacteria exhibit only a thin peptidoglycan layer of around 2–3 nm between the cytoplasmic membrane and the outer membrane, while Gram positive bacteria lack the outer membrane but have a peptidoglycan layer of about 30 nm thick [9]. The anti-microbial finishing can inhibit the growth or kill the invading bacteria by several ways including (i) cell wall damage and/or inhibition of cell wall synthesis, (ii) changing the chemical or physical state of proteins and nucleic acid inside the cell, (iii) inhibition of enzymes inside the cell and consequently retarding the normal biological activities and metabolism of bacteria, and (iv) inhibition of synthesis of protein or nucleic acids thereby interrupting the ability of bacteria to grow and reproduce [10].

---

* To whom any correspondence should be addressed

Journal of Physics: Conference Series **441** (2013) 012002 doi:10.1088/1742-6596/441/1/012002

In this paper, antimicrobial finishing agent will be used for treating the cotton fabric for improving the antimicrobial property and pre-treatment of cotton fabric by plasma under atmospheric pressure will be employed to improve loading of finishing agent.

2. Experimental

2.1. Materials
100% desized, scoured and bleached cotton fabric was used. The fabric was further cleaned with 2% non-ionic detergent at pH 7 and 60°C for 30 minutes and then rinsed thoroughly with deionised water for 15 minutes. The cleaned fabric was conditioned at 65±2% relative humidity and 21±1°C for 24 hours prior to all experiments.

2.2. Atmospheric pressure plasma (APP) treatment
APP treatment was conducted by an atmospheric pressure plasma jet (APPJ, Surfx Technologies LLC, CA) with a rectangular nozzle. The substrate was exposed at a constant speed of 2mm/s to the afterglow plasma generated by a radio frequency of 13.56 MHz. Helium (He) and oxygen (O_2) were used as carrier and reactive gas respectively. Plasma operation parameters were set as: jet distance = 3mm, oxygen concentration = 1%, substrate speed = 10 mm/s and ignition power = 120W. Schematic diagram of the experimental setup was shown in Figure 1.

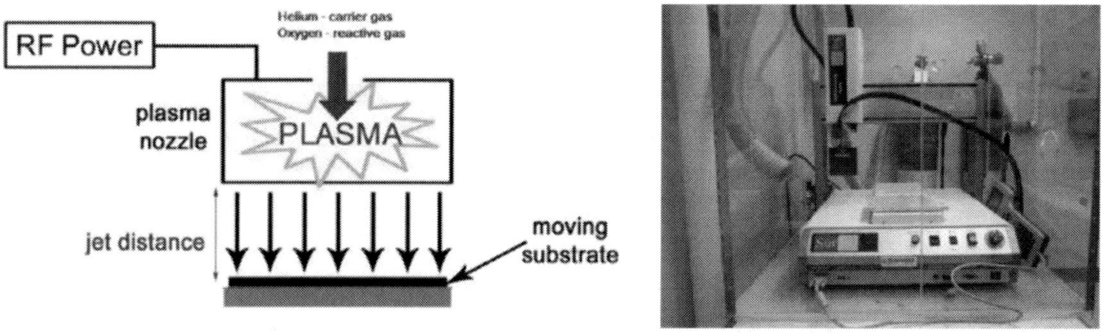

Figure 1. Schematic diagram of APP treatment

2.3. Antimicrobial finishing
The anti-microbial finishing agent (halogenated phenoxy compound, Microfresh Liquid Formulation 9200-200) compound (MF, 0.25%) and a binder (polyurethane dispersion, Microban Liquid Formulation R10800-0) (MB, 0.5%) were used together with a melamine resin. They were applied to the fabric by pad-dry-cure process with an 80% wet pick up. The fabric was dried and cured at 100°C and at 140°C for 5 minutes respectively. Finally, the fabrics were conditioned at 21±1°C and 65±2% relative humidity for 24 hours prior to any further treatment.

2.4. Parallel streak method
The anti-microbial function of the treated cotton materials (five samples) was evaluated by AATCC 147-2004 (Using *S. aureus*, American Type Culture Collection No. 6538 as received) in which the size of specimens was modified to 10mm x 50mm. The preparation and maintenance of the culture medium were conducted according to AATCC 147-2004.

2.4. Fourier transform infrared spectroscopy
Surface chemical composition of the cotton specimens was studied by Fourier Transform Infrared spectrophotometer (Perkin Elmer Spectrum 100) with the scanning range between 4000 cm^{-1} and 700 cm^{-1} and the resolution of 4nm^{-1}, using the attenuated total reflection with 256 scans.

3. Results and Discussion

3.1. Antimicrobial activity

The parallel streak method determines the anti-microbial activity of the treated textile materials against Gram positive bacteria, *S. aureus*. During the test, specimens were placed in contact with nutrient agar which had been previously streaked with an inoculum of *S. aureus*. After incubation, a clear area, i.e. the zone of inhibition, of the interrupted growth underneath and along the sides of the test material indicated the anti-microbial activity of the specimen. Figures 2(a) and 1(b) illustrate the growth inhibition of *S. aureus* conducted on the untreated (control) and anti-microbial-treated fabric specimens (M1) respectively. Figures 1(c) and 1(d) illustrate the growth inhibition of *S. aureus* conducted on the plasma-treated alone (PM0) and plasma pretreated follow by anti-microbial-treated (PM1) specimens respectively. Table 1 summarises the mean clearance distance of the bacteria obtained from the specimens.

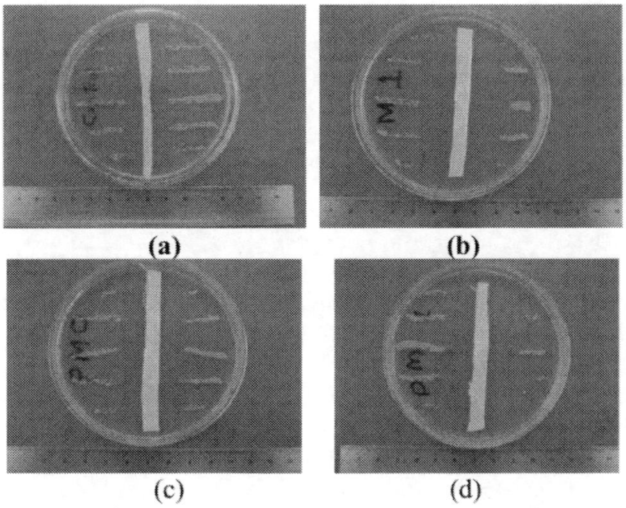

Figure 2. Growth inhibition of *S. aureus* conducted on (a) control, (b) M1, (c) PM0 and (d) PM1

Table 1. The mean clearance distance of the bacteria obtained from the specimens

Sample Symbol	Mean Clearance Distance (cm)
Control	1.2
M1	1.5
PM0	1.3
PM1	1.6

In general, *S. aureus* has a relatively thick wall of 20-80 nm [11], which is composed of many layers of peptidoglycan polymer and one plasma membrane. These layers of peptidoglycan are composed of a fairly open network polymer of N-acetylmuramic acid and N-acetylglucosamine polysaccharide chains with peptide bridges [12]. Hence, the Gram positive bacteria are less resistant to many chemical agents than Gram negative cells [13]. Figure 2(a) shows that the control fabric inhibited the growth of *S. aureus* with an average of 1.2cm ZOC in the agar plate. This was due to the fact that the semi-bleached control cotton fabrics might contain some bleach residues, i.e. H_2O_2, which was used to remove the natural coloration and the remaining trace impurities obtained from the cotton in the pre-finishing stage. Although H_2O_2 was less harmful as compared to other reactive oxygen species such as hydroxyl radicals and superoxide ions, it could still enter the cell exhibiting anti-microbial activity [11]. At a very low concentration of MF, the anti-microbial activity of M1 specimen was enhanced as reflected by a slightly larger zone of inhibition when compared with the control specimen. The result proved that the MF agent does exhibit anti-microbial activity against *S. aureus*.

From Figure 2(a), it was showed that the control fabric (averaged 1.2cm ZOC) inhibited the growth of *S. aureus* because of the bleach residues on fabric surface which could enter the cell exhibiting anti-microbial activity [13]. Moreover, the anti-microbial activity of M1 specimen (averaged 1.5cm ZOC) was enhanced providing a slightly larger zone of inhibition when compared to control specimen. Table 1 shows that plasma treatment may provide anti-microbial activity, i.e. fabric specimen PM0, when compared with the control fabric. The hydrophilic nature of carbonyl groups present in the oxygen plasma pretreated fabric specimen also increases the anti-microbial activity after treatment with MF-MB (PM1) [15-17]. Therefore, PM1 specimen had an average of 1.6cm ZOC which was slightly higher than that of M1 specimen.

2.4. Fourier transform infrared spectroscopy (FTIR)

FTIR was used to characterise the bonding present in the anti-microbial-treated cotton fabric in order to confirm the reactions presented in Figure 3. With regard to M1 specimen, it was observed that there were strong hydroxyl stretching bands at 3420-3250cm^{-1} [18] representing the presence of triclosan on the fabrics. However, the terminal hydroxyl group of triclosan molecule might react with cellulose. Hence, there was no new characteristics peak being formed as shown in Figure 3. On the other hand, Figure 4 (FTIR spectra of control and M1 specimen focused at 2000-1400cm^{-1}) shows two medium sharp bands at 1670 and 1510 cm^{-1} respectively which were due to the presence of benzene ring in aromatic compounds [18]. Therefore, it was confirmed that the MF agent applied was successfully bonded with cotton fabrics.

Figures 5 and 6 show the FTIR spectra of the MF-MB-treated fabric with or without plasma pre-treatment (M1 and PM1) at 4000-700cm^{-1} and 2000-1400cm^{-1} respectively. It was found that the characteristic bands related to the benzene rings at 1670 and 1510 cm^{-1} were also observed in the PM1 specimen as illustrated in Figure 6 [18]. In general, FTIR can be used for quantitative analysis because the strength of the absorption is proportional to the concentration. It was confirmed that the PM1 specimen contained more MF-MB chemical agents than that of the M1 specimen, i.e. PM1 had higher % absorption as shown in Figure 5.

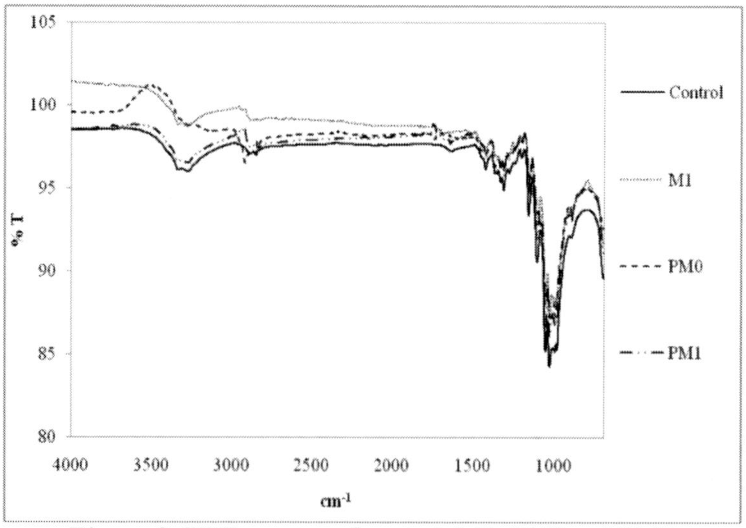

Figure 3. FTIR spectra of control, M1, PM0 and PM1

Figure 4. FTIR spectra of control and M1 at 1400-2000cm^{-1}

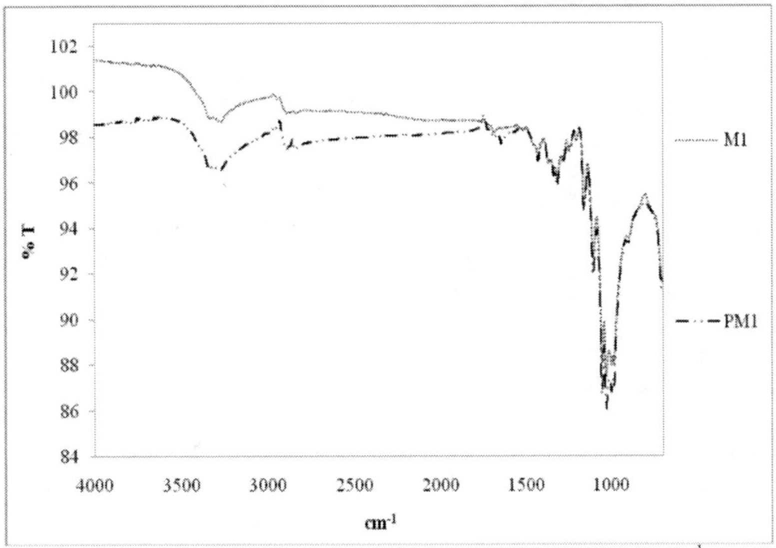

Figure 5. FTIR spectra of M1 and PM1at 4000-700cm^{-1}

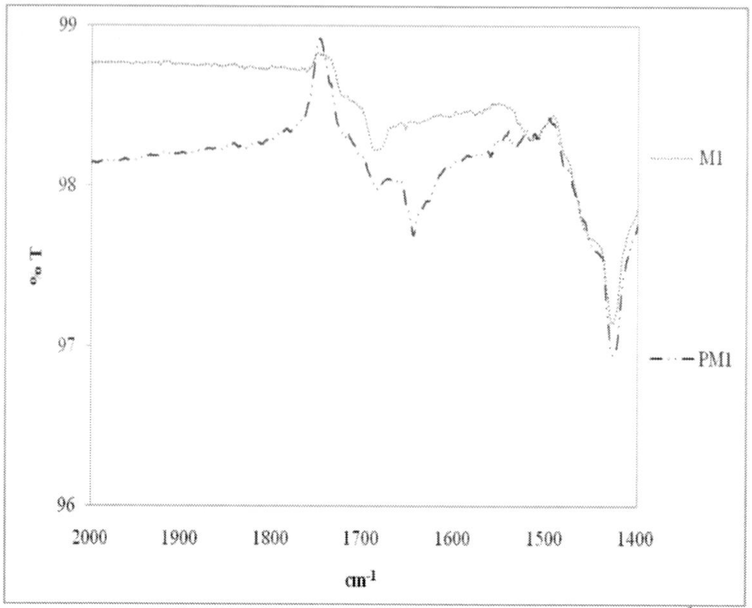

Figure 6. FTIR spectra of M1 and PM1at 2000-1400cm^{-1}

4. Conclusion

In the present study, the control fabric slightly inhibited the growth of *S. aureus* because of the bleach residues left on fabric surface, while the anti-microbial activity of the MF-MB-treated specimen was enhanced as reflected by a slightly larger zone of inhibition. Moreover, the fabrics pre-treated with oxygen plasma only could further enhance the anti-microbial property of the fabric because the plasma gas containing reactive oxygen species which could enter the cell and eventually cause the death of cell. The hydrophilic nature of carbonyl groups present in the oxygen plasma pre-treated specimens also increased the anti-microbial activity when treating it with MF-MB.

FTIR was used for analysing the chemical composition of the fabrics. Two strong hydroxyl stretching bands at 3700-3600cm^{-1} and 3420-3250cm^{-1} respectively were observed in the MF-MB-treated specimen which was representing the presence of triclosan on the fabrics. Based on the FTIR results, it was noted that plasma pre-treated cotton fabric more MF-MB chemical agents than that of one without plasma treatment. Therefore, it could conclude that plasma pre-treatment could enhance the absorption of antimicrobial finishing agent by the cotton fabric and hence increased cotton antimicrobial ability.

Acknowledgements

Authors would like to thank the financial support from the Research Grants Council of The Hong Kong Special Administrative Region, China (Project No. PolyU 5173/11E) and The Hong Kong Polytechnic University for this work.

References

[1] Zhang F, Wu X L, Chen Y Y and Lin H 2009 *Fibers and Polymers.* **10**(4) 496.
[2] Chen C Y and Chiang C L 2008 *Materials Letters* **62**(21-22) 3607.
[3] Chen H L and Burns L D 2006 *Clothing and Textile Research Journal* **24**(3) 248.
[4] Gao Y and Cranston R 2008 *Textile Research Journal* **78**(1) 60.
[5] Daoud W A, Xin J H and Zhang Y H 2005 *Surface Science* **599** 69.
[6] Parthasarathi K and Borkar S P 2007 *Colourage* **54**(7) 43.
[7] Mahltig B, Haufe H and Böttcher H 2005 *Journal of Materials Chemistry* **15** 4385.

[8] Nakashima T, Sakagami Y, Ito H and Matsuo M 2001 *Textile Research Journal* **71**(8) 688
[9] Morones J R, Elechiguerra J L, Camacho A, Holt K, Kouri J B, Ramirez J T and Yacaman M J 2005 *Nanotechnology* **16**(10) 2346.
[10] Abo-Shosha M H, El-Hosamy M B, Hashem A M and El-Nagar A H 2007 *Journal of Industrial Textiles* **37**(1) 55.
[11] Blake D M, Maness P C, Huang Z, Wolfrum E J and Huang J 1999 *Separation and Purification Methods* **28**(1) 1.
[12] Page K, Palgrave R G, Parkin I P, Wilson M, Savinc S L P and Chadwick A V 2007 *Journal of Materials Chemistry* **17** 95.
[13] Fu G F, Vary P S and Lin C T 2005 *Journal of Physical Chemistry B* **109**(18) 8889.
[14] Rajpreet K V and Gita N R 2004 *Textile Research Journal* **74**(12) 1073.
[15] Radetic M, Llic V, Vodnik V, Dimitrijevic S, Jovancic P, Saponjic Z and Nedeljkovic J M 2008 *Polymers for Advanced Technologies* **19**(12) 1816.
[16] Vaideki K, Jayakumar S, Thilagavathi G and Rajendran R 2007 *Applied Surface Science* **253**(17) 7323.
[17] Chen C C, Chen J C and Yao W H 2010 *Textile Research Journal* **80**(8) 675
[18] Chalmers J M and Griffiths P R 2002 *Handbook of Vibrational Spectroscopy, Vol. 4*, (New York: Wiley).

Effect of hydrogen radical on decomposition of chlorosilane source gases

Masatomo Sumiya[1,*], Tomohiro Akizuki[1,2], Kenji Itaka[3], Makoto Kubota[1,2], Kenta Tsubouchi[4], Takamasa Ishigaki[2,5], and Hideomi Koinuma[4,6]

[1] Widegap Materials group, National Institute for Materials Science, Tsukuba, 305-0044, Japan
[2] Department of Materials Chemistry, Hosei University, Koganei Tokyo, 184-8584, Japan
[3] North Japan Research Institute for Sustainable Energy, Hirosaki University, 2-1-3, Matsubara, Aomori, 030-0813, Japan
[4] Graduate School of Frontier Sciences, The University of Tokyo, Kashiwa, 277-8568, Japan.
[5] Department of Chemical Science and Technology, Hosei University, Koganei Tokyo, 184-8584, Japan
[6] Department of Cogno-Mechatronics Engineering, Pusan National University, Miryang 627-706, South Korea

*Email: SUMIYA.Masatomo@nims.go.jp

Abstract. The effect of hydrogen radical on production of Si from chlorosilane sources has been studied. We used hydrogen radical generated from pulsed thermal plasma to decompose $SiHCl_3$ and $SiCl_4$. Hydrogen radical was effective for lowering the temperature to produce Si from $SiHCl_3$. $SiCl_4$ source, which was chemically stable and by-product in Siemens process, was decomposed effectively by hydrogen radical. The decomposition of $SiCl_4$ was consistent with the thermo-dynamical calculation predicting that the use of hydrogen radical could drastically enhance the yield of Si production rather than case of H_2 gas.

1. Introduction

Photovoltaic cells have attracted more attentions and demand of solar grade silicon (SOG-Si) has rapidly increased. SOG-Si is purified either by metallurgical or by chemical process. In metallurgical process, re-solidification and other various processes are used for obtain "six nines (6N)" grade silicon. In chemical process, metal-grade silicon (~99%) is converted to volatile compound such as $SiHCl_3$ for making purification by distillation, which produces highly pure semiconductor-grade Si (SEG-Si) with ~12N orders of magnitudes.

Siemens process[1] has been widely used for producing SEG-Si. In Siemens process, $SiHCl_3$ decomposes on heated high-purity Si rod in reducing ambient of H_2 gas under pressure of ~1atm. Drawbacks of Siemens process are considered to be (1) batch process and (2) low yield of ~25 %, because by-product of $SiCl_4$ can be more favorably generated. To overcome these issues, fluidized bed reactor[2] and vapor liquid reaction[3] processes have been developed for enhancing its production rate and for realizing continuous production. However, there are still difficulties in these processes with respect to the unavoidable impurity incorporation from reactor and low yield. Because reducing reaction of Si production depends on thermal decomposition of source gas in H_2 ambient.

It is necessary to develop process for producing SOG-Si with much higher yield, which must result in reduction of its cost. We have proposed the use of hydrogen radical instead of H_2 gas. Thermodynamic calculation has been carried out to predict the effect of hydrogen radical on the decomposition of chlorosilane sources. Based on this calculation, we have experimentally investigated Si production from $SiHCl_3$ and $SiCl_4$ reacting with hydrogen radical generated from thermal pulsed plasma in this study.

2. Experiments

The schematic of the plasma torch is shown in Fig. 1, of which structures are described in detail elsewhere[4,5]. The solid state amplifier (MP-22CY, Denki Kogyo Co.,Ltd.) was employed for the pulsing plasma generation. The invertor-type power source continuously supplies the electric power of 22 kW with a nominal frequency of 1 MHz, and has a high electric energy efficiency up to 90%. The electric matching with a load is attained by the phase locked loop with variable frequency and an LC matching circuit. The rf power was pulse modulated by imposing the external pulsed signal to switch a static induction transistor. The pulsing rise time of power source is less than 1 msec. The external signal was generated by a pulse generator (HP 8116A) with a response time of 6 ns.

The rf power was modulated with the pulse of on-time of 10 msec and off-time of 5 msec as the optimized condition [6]. The sheath gas was composed of Ar and H_2 at the flow rate of 98 SLM and/or 6 SLM, respectively. The flow rate of H_2 gas was maximum to generate the pulsed thermal plasma. Hydrogen atom density was numerically simulated to be in range of $2.1\sim4.5\times10^{16}$cm^{-3} throughout a pulse cycle [7]. The plasma was first generated at the continuous power level of 10 kW and the pulse mode plasma was generated at low power level of 7.8 kW. By applying pulsed mode in the plasma, a high flux (order of 10^{17}cm^{-3} of hydrogen in ZnO film) of chemically active species can be achieved as reported in ZnO treatment.[8,9]

The reactor pressure was controlled at 200 Torr. $SiHCl_3$ or $SiCl_4$ source in cylinder was kept at 10 °C and -10 °C, and it was introduced into the chamber from side port by Ar carrier gas in the range from 50 to 75 sccm corresponding to amount of supplying Si from 340 to 510 μmol/min. Sapphire and silicon substrates were placed on the position at 10 cm below the bottom of rf coil. Temperature near the substrate was ~1000°C detected by thermo couple. In order to supply more H_2 gas, it was also introduced from side port, not through the plasma region. The plasma was characterized by optical emission spectroscopy (Stellarnet Bluewave UVNb). Materials deposited on the substrates were characterized by Raman spectroscopy (Nihon Bunko, NR-1800) and x-ray diffraction (XRD).

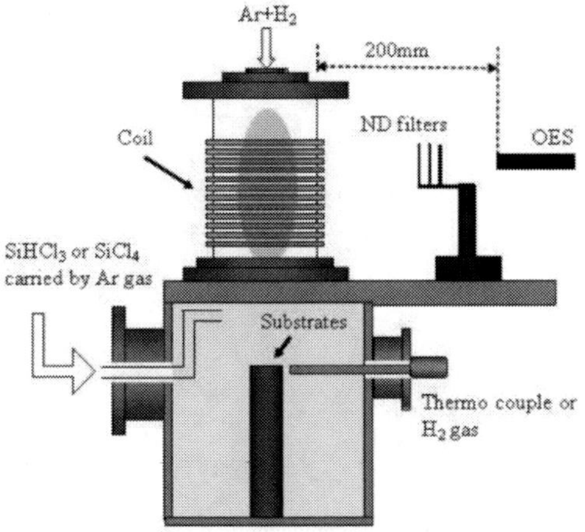

Figure 1. Schematic illustration of the pulsed thermal plasma.

3. Results and discussion

3.1 Effect of hydrogen radical on decomposition of $SiHCl_3$ source

Figure 2 shows Raman spectra for materials deposited on sapphire substrates in the thermal plasma under three kinds of sheath gas condition, when $SiHCl_3$ source gas was introduced for 5 min. Under the continuous plasma of only Ar sheath gas, H_α and H_β emissions are not observed, and Raman peak at around 520 cm^{-1} corresponding to crystalline silicon is not detected (Fig. 2 (a)). When H_2 gas of 5 SLM was introduced from the side port (not through the plasma), H_α and H_β peaks are observed, implying that hydrogen radical may be generated by the Ar plasma (Fig. 2(b)).

The Raman peak at 520 cm^{-1} is clearly detected. Temperature near the substrate was 700 °C under the continuous Ar sheath gas plasma. Although Si was hardly generated at this temperature in Siemens process, Si was confirmed to be produced under hydrogen radical. This indicates that hydrogen radical must enhance the reducing reaction of $SiHCl_3$ source at lower temperature. Furthermore, H_2 gas was added into Ar sheath gas to increase hydrogen radical as shown in Fig. 2(c). Si production was confirmed not only from Raman spectrum but also from XRD as shown in the inset of Fig. 2(c).

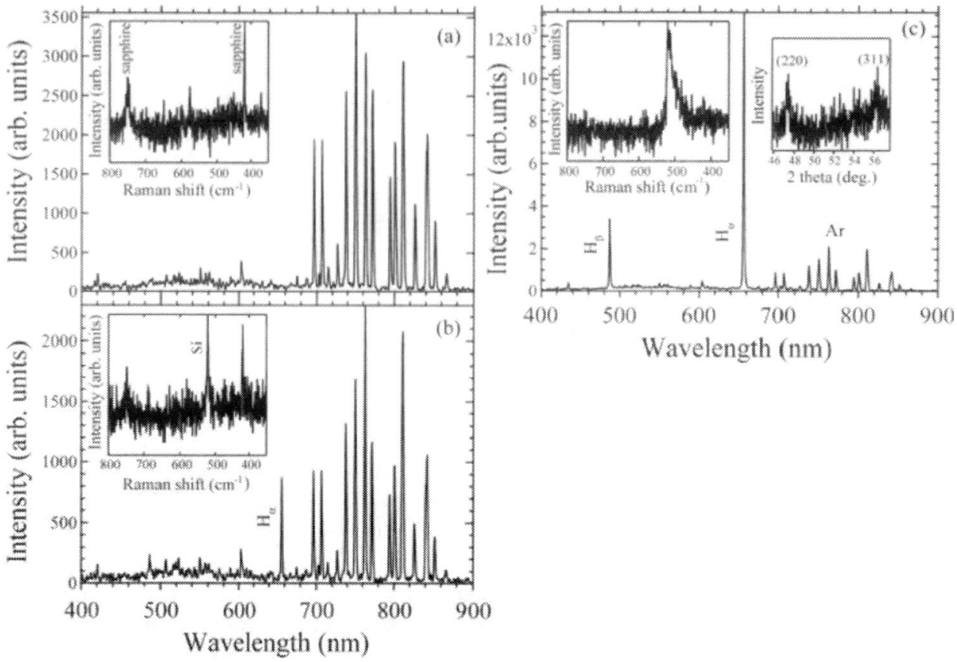

Figure 2. Optical spectroscopy of (a) continuous thermal plasma in Ar sheath gas, (b) H_2 gas from the side port in the Ar continuous plasma, and (c) Ar and H_2 sheath gas in the pulsed thermal plasma. The inset shows Raman spectra of sample deposited under each condition.

3.3 Comparison between $SiHCl_3$ and $SiCl_4$

Alternatively, $SiCl_4$ source was introduced into the pulsed thermal plasma. Raman spectra of materials grown on sapphire substrate exhibits a sharp peak at 520 cm^{-1} as shown in Fig. 3, compared to that from $SiHCl_3$ supplied with the same amount. Si material from $SiCl_4$ shows better crystalline quality than that from $SiHCl_3$.

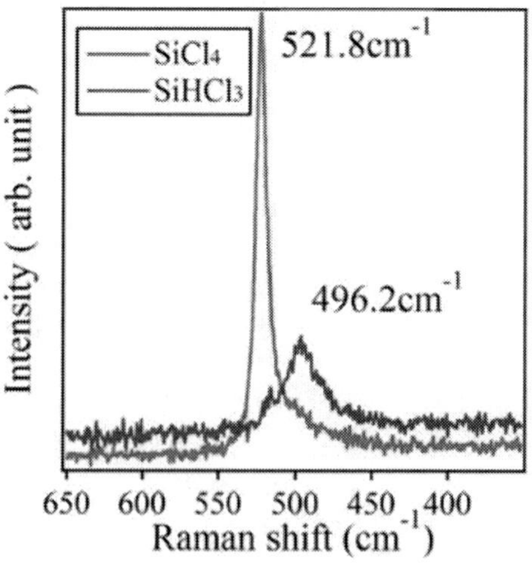

Figure 3 Raman spectra for samples prepared on c-plane sapphire substrates from $SiHCl_3$ (blue) and $SiCl_4$ (red) source supplied with the same amount.

In order to understand the effect of hydrogen radical qualitatively, thermodynamic calculations were performed for the reactions of reducing chlorosilane gases in H_2 and hydrogen radical in follow.

$$H_2(g) + SiHCl_3(g) \rightarrow Si(s) + 3HCl(g) \quad (1)$$
$$2H_2(g) + SiCl_4(g) \rightarrow 4HCl(g) + Si(s) \quad (2)$$

$$2H(g) + SiHCl_3(g) \rightarrow Si(s) + 3HCl(g) \quad (3)$$
$$4H(g) + SiCl_4(g) \rightarrow Si(s) + 4HCl(g) \quad (4)$$

For thermodynamic calculation, standard formation enthalpy ($\Delta Hf°$) and standard entropy (S°) of reaction species are necessary. Temperature dependence of Gibbs Energy was calculated using stored data in the program, MALT-2 [10]. The calculation parameters for partial pressure of H_2, H, HCl, $SiHCl_3$, and $SiCl_4$ were determined from the actual experimental condition of the pulsed thermal plasma. It was assumed that hydrogen radical might be generated at rate of ~2%.

Open and closed circles in Fig. 4 are calculated ΔG of the reaction (1) and (2), respectively. Since the value of ΔG is negative at temperatures higher than 600 °C, it suggests the possibility of Si production from $SiHCl_3$ and $SiCl_4$ under H_2 partial pressure of 5.7×10^{-2} atm. Since $\Delta H_f°$ of hydrogen radical at 298.15 K is 218.0 kJ/mol, which is much larger than that of H_2 gas, it can be expected that reduction of $SiHCl_3$ can be accelerated by using hydrogen radical. ΔG of reaction (3) shows much lower than those of reaction (2). This tendency is remarkable for $SiCl_4$ source gas, because ΔG becomes much further lower. This indication is qualitatively consistent with the experimental results using hydrogen radical as reduction gas.

In order to explain better crystalline quality of Si from $SiCl_4$, the intensities of H_α and Cl at 542.5nm emissions in the plasma were measured. Figure 5 shows the relationship between H_α and Cl intensities emitting from the pulsed plasma for 10 min. Both H_α and Cl are generated more for $SiCl_4$ rather than $SiCl_3$ source. This implies that $SiCl_4$ must react more intensively with hydrogen radical, which is consistent with the thermodynamic calculation. Through reacting with hydrogen radical, chlorosilane can deposit Si containing partially hydrogenated amorphous Si on substrate. Assuming that Cl^- ion could etch the amorphous Si, better crystalline quality was obtained for the case of $SiCl_4$ source.

3.4 Mass analysis in $SiHCl_3$ and $SiCl_4$ plasma

Since the amount of chlorosilne gas was less by four orders magnitudes than that of Ar gas in the pulsed thermal plasma, it was difficult to detect the mass numbers related to the source. We used typical capacitance-coupled plasma to investigate the reaction with hydrogen radical. $SiHCl_3$ or $SiCl_4$ source was introduced by using Ar or H_2 carrier gas of 30 sccm. RF (13.56 MHz) power was applied with 0~90 W under 20 mTorr of pressure. By using quadrapole mass spectroscopy from 1 to 200 amu (MKS e-Vision$^-$), the reducing reaction in the plasma were analyzed. Simultaneously, the optical

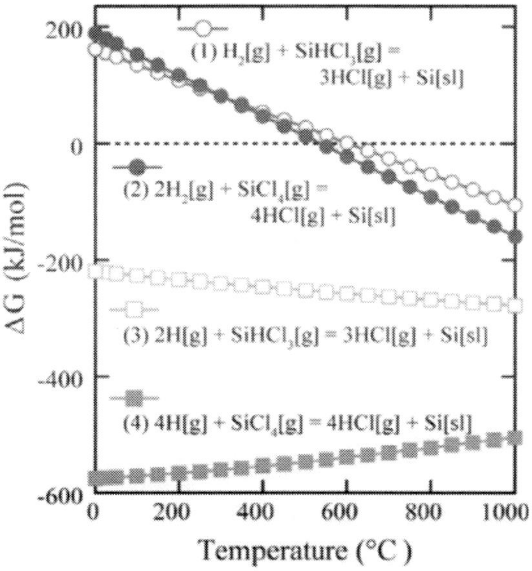

Figure 4. Thermodynamic calculation for $SiHCl_3$ and $SiCl_4$ reacting with H_2 and hydrogen radical, taken partial pressures of H_2 (5.7×10^{-2} atm), H (1.2×10^{-3} atm), HCl $SiHCl_3$, and $SiCl_4$(1.2×10^{-4} atm) under conditions closed to the pulsed thermal plasma.

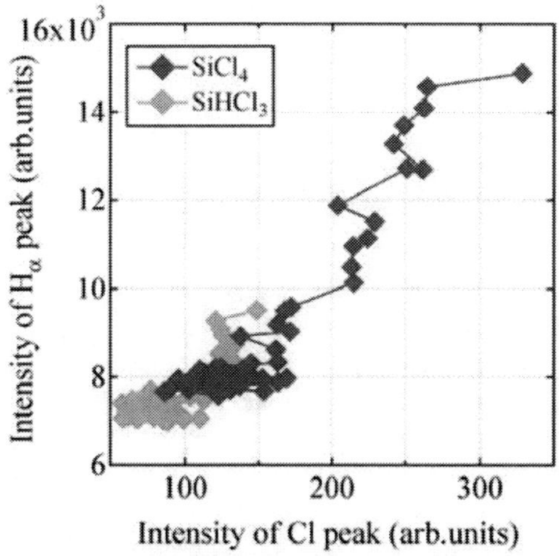

Figure 5 Relationship between H_α and Cl intensity in optical spectroscopy detected from the pulsed thermal plasma under $SiHCl_3$ and $SiCl_4$ source. The detecting time was 10 min after the source was introduced. Plasma lower power was 7.8 kW, and on-time of 10 msec and off-time of 5 msec.

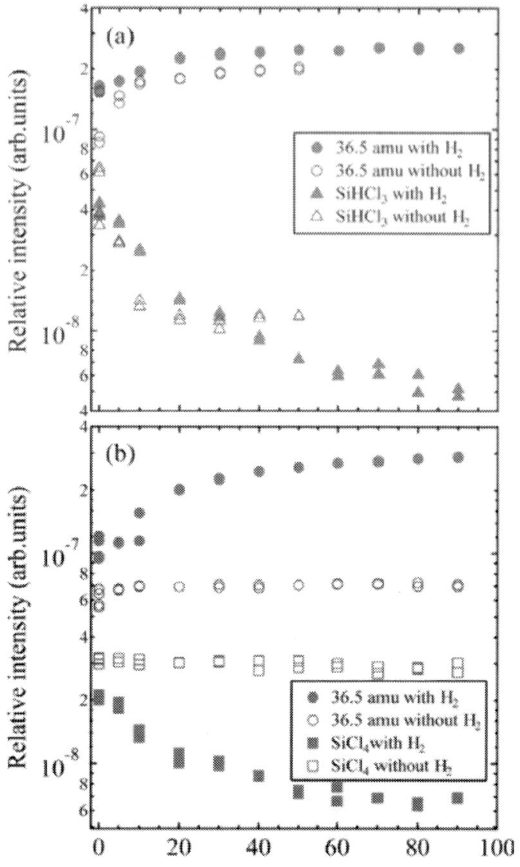

Figure 6. Variation of mass numbers corresponding to HCl and chlorosilane source of (a) $SiHCl_3$ and (b) $SiCl_4$ as a function of rf power applied to capacitance coupled plasma generated in 20 mTorr with and without H_2 gas.

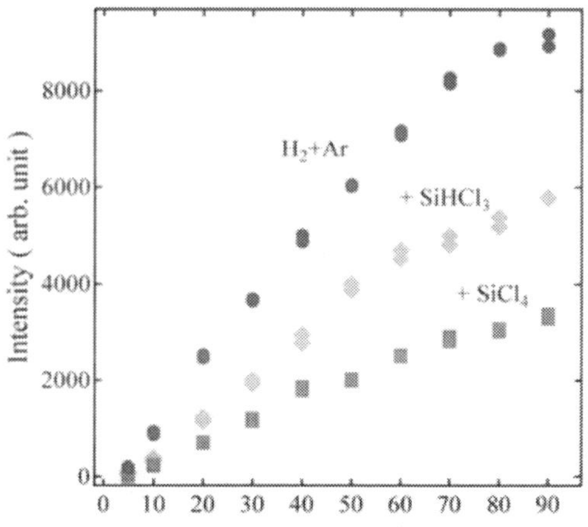

Figure 7. Correlation between H_a intensity and RF power under the conditions of 20mTorr, H_2 and Ar of 30 sccm, respectively. Source materials of $SiHCl_3$ and $SiCl_4$ were carried by Ar of 30sccm.

spectroscopy was also observed.

When $SiHCl_3$ or $SiCl_4$ source was introduced, the intensity of mass number at 36.5 amu corresponding to HCl was increased. We used this variation of HCl intensity as the index for the analysis of reaction. Figure 6(a) shows the variation of HCl and $SiHCl_3$ with and without H_2 ambient as a function of rf power. Intensity of $SiHCl_3$ is constant over 20W without H_2 gas, while it occurs continuously decreased with increase of rf power under H_2 ambient. The effect of H_2 ambient was remarkable in $SiCl_4$ as shown in Fig. 4(b). While no decomposition of $SiCl_4$ is confirmed under no H_2 gas, the reaction is enhanced due to H_2 ambient. Figure 7 shows the variation of H_α intensity under the gas conditions. With increase of RF power, the intensity gradually enhanced. However, the intensity of H_α emission drops at the same rf power when $SiHCl_3$ or $SiCl_4$ source was put into the plasma. Assuming that hydrogen radical must decompose $SiHCl_3$ or $SiCl_4$ before emitting H_α, it is considered that the depression of H_α intensity indicates the effectiveness of hydrogen radical for the decomposition. The H_α intensity was much decreased in case of $SiCl_4$ source. The effect of hydrogen racial seems to be remarkable for $SiCl_4$. It is expected that hydrogen radical can be used to produce Si effectively from $SiCl_4$ which is by-product in Siemens process.

4. Conclusions

We have proposed the use of hydrogen radical to improve the yield of Si production in Siemens process. When hydrogen racial generated from pulsed thermal plasma was applied, it made the temperature of Si production lower for $SiHCl_3$ source. It was found that hydrogen radical worked well for decomposition of $SiCl_4$ by-product in Siemens process. This was consistent with the thermodynamic calculation predicting that the use of hydrogen radical instead of H_2 gas could drastically enhance the yield of Si production.

Acknowledgements

This study was partially supported by JST-JICA project (SATREPS). The authors would like to thank prof. T. Hashimoto, Nihon University, for valuable discussion of the thermodynamic calculations.

References

[1] H. Habuka, M. Katayama, M. Shimada, K, Kikuo, 1994 *Jpn. J. Appl. Phys Part 1*, **33**, 1977
[2] M. P. Tajero-Ezpeleta, S. Buchholz, and L. Mleczko, 2004 *Can. J. of Chem. Eng* **82**,520
[3] S. Wakamatsu and H. Oda, 2005 *Abstract of Jpn. Chem. Conf.* 85, 1, 510
[4] T. Ishigaki, X. Fan, T. Sakuta, T. Banjo, and Y. Shibuya, 1997 *Appl. Phys. Lett.* **71**, 3787
[5] T. Ishigaki, Y. Moriyoshi, T. Watanabe, and A. Kanzawa, 1996 *J. Mater. Res.* **11**, 2811
[6] R. Ye, T. Ishigaki and T. Sakuta, 2005 *Plasma Sources Sci. Technol.* **14,** 387.
[7] R. Ye, T. Ishigaki, H. Taguchi, S. Ito, A.B. Murphy and H. Lange, 2006 *J. App. Phys.*, **100**, 103303
[8] N. Ohashi, T. Ishigaki,N. Okada, H. Taguchi, I. Sakaguchi, S. Hishita, T. Sekiguchi, and H. Haneda, 2003 *J. Appl. Phys.* **93**, 6386
[9] T. Ishigaki, N. Ohashi., H. Taguchi., R. Ye, H. Haneda, and S. Ito, 2006 *Thin Solid Film*, **506-507**, 303
[10] H. Yokokawa, S. Yamauchi and S. Matsumoto "Thermodynamic database MALT for windows with gem and CHD", 2002, *Calphad.* **26**, 155-166.

Removal of H_2S in air by corona discharge under continuous flow condition

Hyun-Woo Park, Chung Hyun Lee and Dong-Wha Park*

Department of Chemical Engineering and RIC-ETTP(Regional Innovation Center for Environmental Technology of Thermal Plasma), Inha University, 253 Yonghyun-dong, Nam-gu, Incheon, 402-751, Republic of Korea.

E-mail: dwpark@inha.ac.kr

Abstract. Hydrogen sulfide (H_2S) gas diluted in air was removed on a large scale using a newly developed wet electrostatic precipitator (Wet-EP) system under the continuous flow condition at atmospheric pressure and room temperature. Acid H_2S is spontaneously neutralized by NaOH and KOH used as basic materials at room temperature. And then, the solid salts generated from neutralization reaction were removed by the electrostatic precipitation method using corona discharge. The experimental variables are reaction ratio between H_2S and basic materials, reaction time and plasma input power for the removal of generated solid salts. The purpose of this work is to determine optimal operating conditions and space velocity. The removal efficiencies of H_2S and dust were over 99% at H_2S initial concentration of 600 ppm and the total flow rate of 2 m^3/min. The optimal space velocities for NaOH and KOH are almost from 12000 to 15000 hr^{-1} at reaction time of 0.06 and 0.048 seconds, respectively. In conclusion, KOH is more efficient than NaOH for removal of H_2S.

1. Introduction

Hydrogen sulfide (H_2S) is one of the odor substances that cause air pollution. The major emission sources of H_2S gas are biogas production process, naphtha process, landfills, sewage treatment process, food waste treatment process and coal gasification process. Human beings have a variety of symptoms such as mental stress, psychological insecurity, petulance, hysteria, hyposmia, headache, vomit and insomnia because of odor substances [1]. Conventional methods for odor control including wet scrubbing, thermal oxidation, catalyst, adsorption, active carbon, ozone oxidation and biofiltration, are limitative technically and economically for the abatment of low concentration odor substances [2-8]. In recent years, studies of H_2S treatment using various plasma systems, such as dielectric barrier discharge, microwave, radio frequency, pulsed corona discharge, gliding arc, have been progressed [9-12]. These methods, however, have problems in large-capacity treatment. In this experiment, a novel wet electrostatic precipitator system (Wet-EP) was used to treat H_2S on a large scale. The Wet-EP system is the combination of two methods which are wet scrubbing for the neutralization reaction of H_2S with basic materials and electrostatic precipitation by corona discharge for the removal of generated solid salts. An ultrasonic atomizer was used for the injection of basic material solutions in order to increase the reaction surface area. The purpose of this work is the confirmation of optimal conditions to remove large amount of H_2S under continuous flow conditions. As variables, the reaction ratio of H_2S with basic materials, reaction time in reaction region and the plasma input power of electrostatic precipitator were controlled.

2. Experimental setup

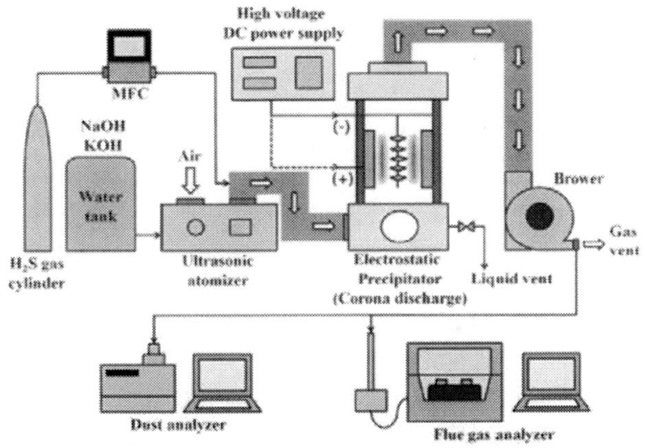

Figure 1. The schematic diagram of the Wet-EP system to remove diluted H_2S gas in air.

Figure 1 shows the schematic diagram of the Wet-ESP system for the removal of dilute H_2S gas in air. The flow rate of H_2S was controlled by a mass flow controller (MFC, Sierra INC., USA) and it was mixed with air at the exit of ultrasonic atomizer at 600 ppm. The total gas flow rate of H_2S-air mixture was controlled by a blower. A DC power supply (Plasmatech Co. Ltd., Korea) which has high voltage and low current was employed for corona discharge. This system consists of two step process. The first step is the spontaneous neutralization reaction of H_2S with basic material solutions. The second step is the electrostatic precipitation process of generated salts and water mists by corona discharge. Gas phase H_2S and liquid phase basic materials spontaneously reacted in chemical reaction region at room temperature and atmospheric pressure. Therefore, solid phase salts and non-reacted basic material solutions were existed in air. These are charged to negative by corona discharge so that negatively charged liquid and solid salts are precipitated at anode. The chemical reaction region was designated in a plastic pipe line with the inner diameter of 0.047 m and the length of 5.2 m before electrostatic precipitator. The type of electrostatic precipitator electrodes are pin-to-plate with the electrodes gap distance of 0.024 m. A cathode is composed of 256 stainless steel pins. An anode is cylindrical plate which is 0.35 m in length and 0.308 m^2 in area, respectively. The inlet and outlet concentrations of H_2S were measured by a flue gas analyzer (MK 9000, RBR Co. Ltd., Germany). Amount of dust with and without the operation of the electrostatic precipitator was analyzed using a dust analyzer (3442, KANOMAX Co. Ltd., Japan) to evaluate the removal efficiencies of H_2S and dusts. Removal efficiencies of H_2S and dusts were calculated as follow equations:

$$\eta_{H_2S}(\%) = \frac{C_{in} - C_{out}}{C_{in}} \times 100 \tag{1}$$

$$\eta_{dust}(\%) = \frac{(m_{off} - m_{atm}) - m_{on}}{m_{off} - m_{atm}} \times 100 \tag{2}$$

where, η_{H2S} and η_{dust} (%) are the removal efficiencies of H_2S and dust, respectively; C_{in} and C_{out} (ppm) are the inlet and outlet concentration of H_2S; m_{on} and m_{off} (mg/m^3) are dust amount before and after the operation of the electrostatic precipitator; m_{atm} (mg/m^3) is dust amount at atmosphere. Amount of dust at atmosphere was 0.155 mg/m^3. The reaction ratio (R) of H_2S and basic materials was calculated as follow equation:

$$R = \frac{\dot{m}_b}{\dot{m}_w} \tag{3}$$

where, $\dot{m}_b$ (mol/min) and $\dot{m}_w$ (mol/min) are input molar flow rates of basic materials and H_2S gas, respectively. In order to evaluate the economic feasibility in each operating conditions, the space velocity (SV) was calculated as follow equation:

$$SV(\text{hr}^{-1}) = \frac{36 \cdot \eta_{H_2S}}{R \cdot t} \tag{4}$$

where, SV (hr^{-1}) is space velocity value in chemical reaction region; t (hr) is reaction time of gas at reaction region. Table 1 indicates the experimental conditions. The range of reaction ratio was from 0.24 to 4.98 and reaction time was varied from 0.01 to 0.06 second. The input power of electrostatic precipitator was controlled from 1 to 180 W. Total gas flow rate and initial concentration of H_2S were fixed at 2 m^3/min and 600 ppm, respectively. As basic materials, NaOH and KOH were used for the neutralization reaction of H_2S.

Table 1 Experiment conditions for the removal of H_2S by using Wet-EP system.

Total flow rate (m^3/min)	2
Initial H_2S concentration (ppm)	600
Type of basic materials	NaOH, KOH
Reaction ratio	0.24 – 4.98
Reaction time (sec)	0.01 - 0.06
Plasma input power (W)	1 - 180
Input voltage (kV)	12 - 22
Input current (mA)	0.1 - 8.5

3. Results and discussion

Figure 2 shows the Gibb's free energy for the neutralization reaction of H_2S with basic materials calculated by thermodynamic equilibrium calculation software (FactSage, CRCT & GTT-technologies, Canada & Germany). Gibbs free energies for all neutralization reactions have negative value at whole temperature range as shown in figure 2. Therefore, it can be occurred to spontaneous neutralization reaction at room temperature and atmospheric pressure. Since ΔG value for neutralization reaction by KOH solution is lower than the reaction by NaOH solution, the reaction between H_2S and KOH is expected to proceed at easier. The major products in the reaction by KOH and NaOH solution are expected to be solid salts such as Na_2S, Na_2S_2 and K_2S.

Figure 3 presents correlation between the removal efficiency of H_2S and the reaction ratio. When H_2S is reacted with KOH solution, removal efficiency is higher compared with NaOH solution. These results are in a good agreement with calculation results on Gibbs free energies in figure 2 consequently, neutralization of H_2S is more favorable to use KOH than NaOH. In addition, the removal efficiencies are increased rapidly with increasing the reaction ratio. When the reaction ratio is 4.98, removal efficiency with KOH basic material is over 99 %.

Figure 4 is calculation results on the space velocity. The space velocity has the opposite tendency with removal efficiency according to the reaction ratio. At the reaction ratio of 4.98, the space velocities for NaOH and KOH were approximately 11700 and 12000 hr^{-1}, respectively. H_2S removal efficiency by KOH was higher than by NaOH at all range of reaction ratio in figure 3, because ΔG

value of reaction by KOH was lower than by NaOH at room temperature in figure 2 Due to these results, reaction by KOH was more progressed easily than reaction by NaOH.

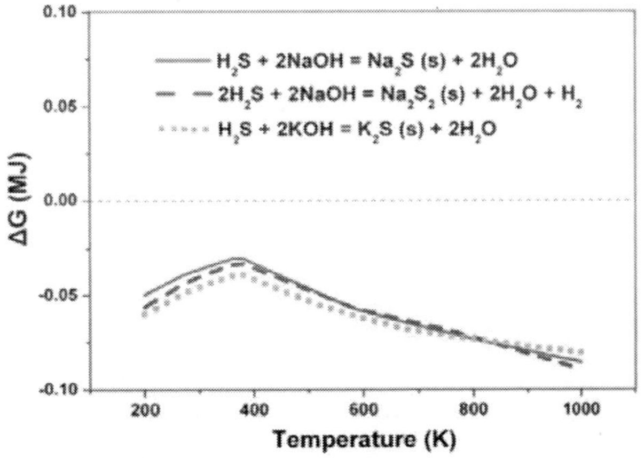

Figure 2. Gibb's free energy calculation result of neutralization reaction of H_2S and basic materials.

H_2S removal efficiency was confirmed at fixed reaction ratio and controlled the reaction time in the reaction region. Figure 5 shows the removal efficiency of H_2S according to the reaction time. Length of plastic pipe line was adjusted in order to control the reaction time. The reaction time was varied from 0.01 to 0.06 second at the fixed reaction ratio of NaOH and KOH with 4.98. Although the removal efficiency with NaOH is sharply decreased as decreasing the reaction time, it with KOH is gradually decreased as decreasing the reaction time. Therefore, it is revealed that KOH has faster neutralization reaction rate and higher chemical activity than NaOH.

Figure 6 shows the space velocity according to the reaction time. The difference of space velocity between NaOH and KOH cases is reduced with increasing the reaction time. On the other hands, the difference of space velocity by NaOH and KOH is reduced increasing the reaction time. The reason is that the difference of H_2S removal efficiency by NaOH and KOH is decreased due to sufficient reaction time. The space velocities for NaOH and KOH were about 11700 and 14900 hr^{-1} at the reaction time of 0.06 and 0.048 seconds, respectively. Therefore, KOH is more effective than NaOH for neutralization of H_2S.

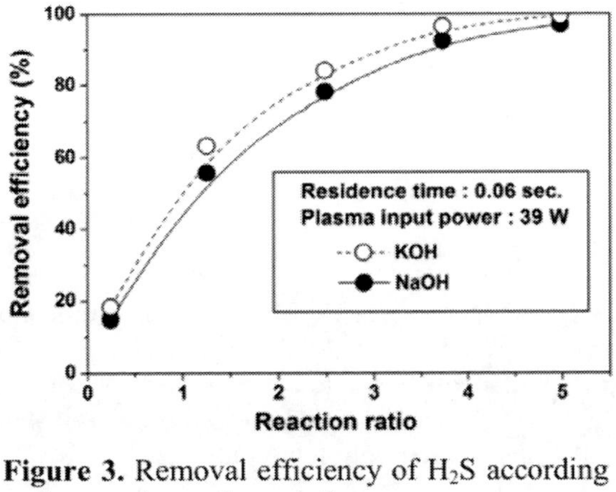

Figure 3. Removal efficiency of H_2S according to the reaction ratio variation.

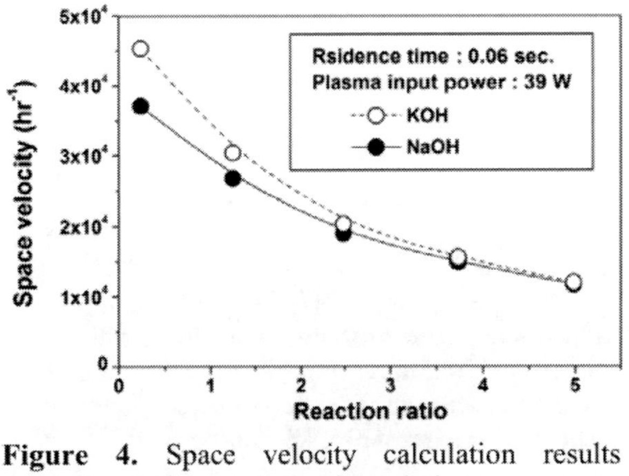

Figure 4. Space velocity calculation results according to the reaction ratio variation.

Journal of Physics: Conference Series **441** (2013) 012004 doi:10.1088/1742-6596/441/1/012004

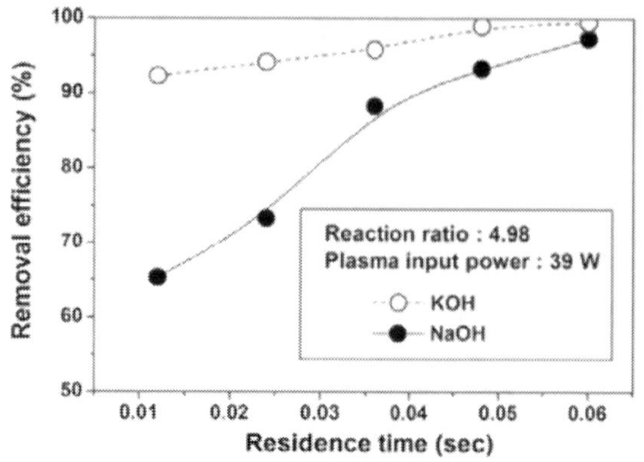

Figure 5. Removal efficiency of H_2S according to the reaction time.

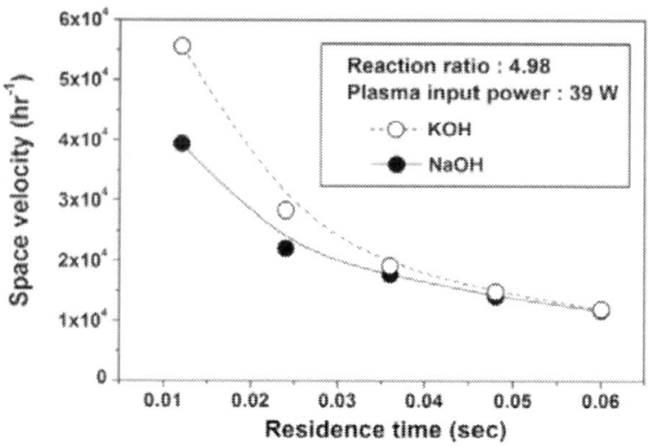

Figure 6. Space velocity calculation results according to the reaction time.

Since large amount of solid salts, un-reacted solution and water mists were generated by neutralization reaction of H_2S with basic materials, the electrostatic precipitator was used to remove them. Figure 7 indicates the removal efficiency of dust according to the plasma input power which was varied from 1 to 180 W. Dust was mostly removed above 30 W. The removal efficiencies are fairly similar each other in figures 7 (a) and (b) for NaOH and KOH, respectively. Amount of exhausted dust was 0.180 mg/m^3 at the input power of 180 W after electrostatic precipitation by corona discharge in both cases. It is very similar with the dust content in atmosphere of 0.155 mg/m^3. Because the amount of exhausted dust without the electrostatic precipitation was over than 140 mg/m^3, the removal efficiencies of dust were over 99% when the input power exceeded 40 W. Therefore, large amount of H_2S gas and dust can be removed simultaneously at room temperature and atmospheric pressure under continuous flow condition using the Wet-EP system.

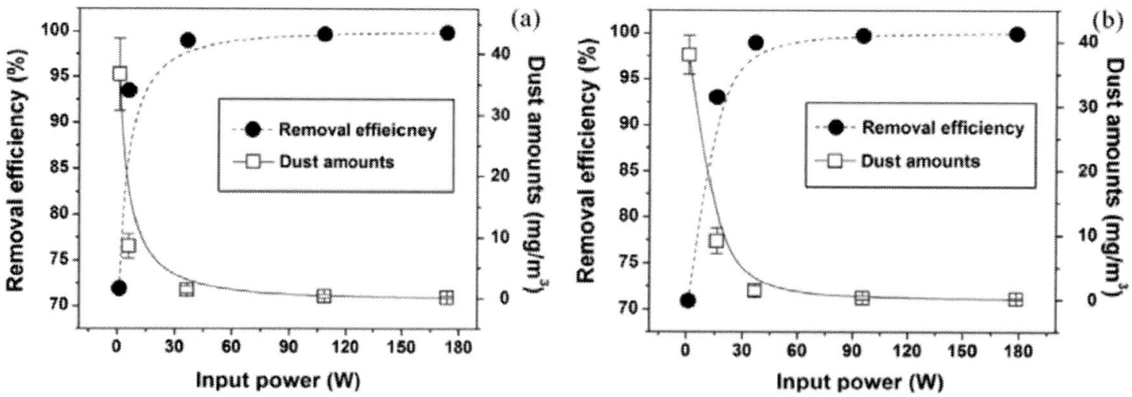

Figure 7. Removal efficiencies of dust according to the plasma input power with (a) NaOH and (b) KOH solutions.

4. Conclusion

A large amount of H_2S gas diluted in air was removed in the Wet-EP system by neutralization reaction and electrostatic precipitation under continuous flow condition. NaOH and KOH were used for neutralization reaction with H_2S gas, and they were spontaneously reacted at room temperature and atmospheric pressure. The removal efficiency of H_2S was 99% at the highest reaction ratio of 4.98. The KOH reacts with H_2S more effectively than NaOH, due to relatively low Gibbs free energy. For high removal efficiency, optimal conditions of H_2S removal reaction by NaOH and KOH were reaction ratio of 4.98 in both cases, reaction time of 0.06 and 0.048 seconds, respectively. Under these conditions, the values of space velocity by NaOH and KOH were 11700 and 14900 hr^{-1}. The input power of plasma was required about 40 W in order to remove dusts completely. In conclusion, H_2S contained contaminant of 2 m^3/min can be efficiently removed by using the low power of 40 W in Wet-EP system.

Acknowledgments

This work was supported by the Regional Innovation Center for Environmental Technology of Thermal Plasma (ETTP) at Inha University designated by MKE (2013).

References

[1] S.S. Shiffman, E.A. Sattely Miller, M.S. Suggs and B.G. Graham 1995 Brain Res. Bulletin **37** 369-375
[2] W. H. Cheng, M. S. Chou, W. S. Lee and B. J. Huang 2002 J. Env. Eng. **128** 313-320
[3] Y. I. Matatov-Meytal and M. Sheintuch 1997 Ind. Eng. Chem. Res. **36** 4374-4380
[4] Z. H. Huang, F. Kang, Y. P. Zheng, J. B. Yang and K. M. Liang 2002 Adsorp. Sci. & Tech. **20** 495-500
[5] S. W. Baek, J. R. Kim, and S. K. Ihm 2004 Catal.Today **93** 575-571
[6] A. H. Wani, R. M. R. Branion, and A. K. Lau 1997 J. Env. Sci. Health **A32** 2027-2055
[7] J. Ruan, W. Li, Y. Shi, Y. Nie, X. Wang and T. Tan 2005 Chemosphere **59** 327-333
[8] I. Traus and H. Suhr 1992 Plasma Chemistry and Plasma Processing **12** 275-285
[9] Zhao Gui-Bing 2007 Chem. Eng. Sci. **72** 2216-2227
[10] E. Krasheninnkov, V. D. Rusanov, S. V. Sanyuk and A. A. Fridman 1986 Zh. Tekh. Fiz. **56** 1104-1109
[11] T. Nunnally, K. Gutsol, A. Rabinovich, A. Fridman, A. Starikovsky, A. Gutsol and R.W. Potter 2009 Inter. J. Hydro. Ener. **34** 7618-7625
[12] S. John, J.C. Hamann, S.S. Muknahallipatna, S. Legowski, J.F. Ackerman and M.D. Argyle 2009 Chem. Eng. Sci. **64** 4826-4834

Water Purification by Using Microplasma Treatment

K Shimizu, N Masamura and M Blajan

Organization for Innovation and Social Collaboration, Shizuoka University, Japan

shimizu@cjr.shizuoka.ac.jp

Abstract. Dielectric barrier discharge microplasma generated at the surface of water is proposed as a solution for water treatment. It is an economical and an ecological technology for water treatment due to its generation at atmospheric pressure and low discharge voltage. Microplasma electrodes were placed at small distance above the water thus active species and radicals were flown by the gas towards the water surface and furthermore reacted with the target to be decomposed. Indigo carmine was chosen as the target to be decomposed by the effect of active species and radicals generated between the electrodes. Air, oxygen, nitrogen and argon were used as discharge gases. Measurement of absorbance showed the decomposition of indigo carmine by microplasma treatment. Active species and radicals of oxygen origin so called ROS (reactive oxidative species) were considered to be the main factor in indigo carmine decomposition. The decomposition rate increased with the increase of the treatment time as shown by the spectrophotometer analysis. Discharge voltage also influenced the decomposition process.

1. Introduction

Water pollution which caused by the industrial development is a problem of modern society. Conventional water treatment technologies are using chemicals, which makes the process expensive [1]. Nonthermal plasma found applicability in various field such as sterilization [2], cancer cell control [3], surface treatment [4], air treatment [5] and biomedical applications [6] due to high performance and low cost [7-9]. There are many plasma types for water treatment such as regional treatment by plasma jet [10-11], pulsed discharge in bubbling water [12-14] and pulse corona discharge using high voltage [15-17]. Needle-plate structure is often used due to the electric field concentration [18] however these methods need expensive and large device to supply the high voltage. The required voltage is in kilovolts order thus also safety is a concern. When considering the implementation of the technology, a small power supply will be required. [19-20].

Plasma discharge could generate various active species, radicals and ions [21]. Ozone, which is a one of the strong oxidizing agent, is used to destroy pollutant that are resistant to other chemical treatment [22]. However persistent substances such as DMSO (dimethyl sulfoxide) cannot be decomposed by ozone [23]. These substances could be decomposed by various active species. Thus plasma is an efficient method to be applied as the water treatment technology.

Our purpose is to apply the water treatment technology by microplasma. Microplasma which is a dielectric barrier discharge at atmospheric pressure could generate active species and radicals at relatively low discharge voltage. In this research, as the microplasma electrodes, a pair of perforated metallic plates covered with a dielectric layer, were used to generate microplasma. Electrodes were placed at small distance above the water and gas passed through the holes of electrode thus flowing towards the surface the water. As the treatment target, indigo carmine diluted in water was used as the organic compound to be decomposed by microplasma treatment. Indigo carmine used in food industry was used as the target to be decomposed and furthermore to evaluate the effect of microplasma treatment.

2. Microplasma electrode and mechanism of water treatment

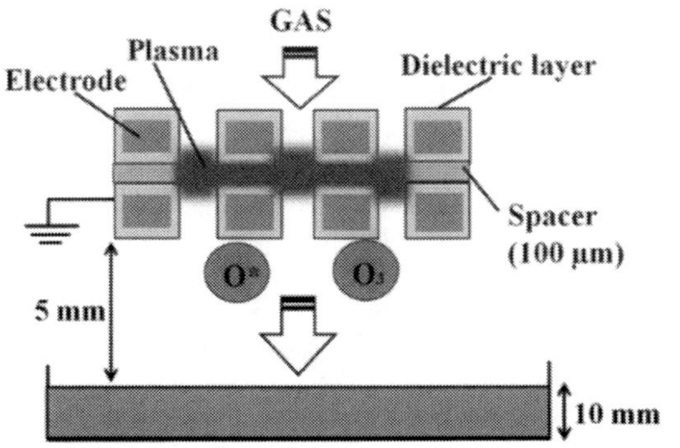

Figure 1. Mechanism of water treatment by microplama. Whole size of the electrode was 60 mm, diameter of the hole was 2.0 mm and flow rate was 10 L/min. $BaTiO_3$ was used as the dielectric barrier and its thickness was 100 μm. Electrodes were placed above the water at a distance of about 5 mm. Active species could be generated such as O_{3s} and O_{3p} between electrodes that affected and decomposed the organic compounds[24].

Figure 1 shows the schematic of microplasma electrode and mechanism of water treatment. It is difficult to generate microplasma in water due to the high electric conductivity at low discharge voltage. Electrodes were placed above the water to generate microplasma in gas phase.

Microplasma electrodes were perforated metallic plates covered with a dielectric layer and faced together at small discharge gap of about 100 μm using a spacer. Only a low discharge voltage which is the one feature of the microplasma, was required due to the narrow discharge gap. The spacer also blocked the gas flow to establish the discharge area. When the voltage was applied between electrodes, microplasma was generated. Discharge voltage depends on the discharge gap and carrier gas. The various active species and radicals generated by microplasma between the electrodes were supplied to the water surface by the carrier gas and reacted to decompose the indigo carmine solution. The distance between the electrode and water surface was about 5 mm and the transit-time for the active species to reach the water surface was about 0.25 ms. Thus allowing that active species and radicals such as O_3 to reach the target to be decomposed. The total amount of the indigo carmine was 200 mL.

3. Experimental setup

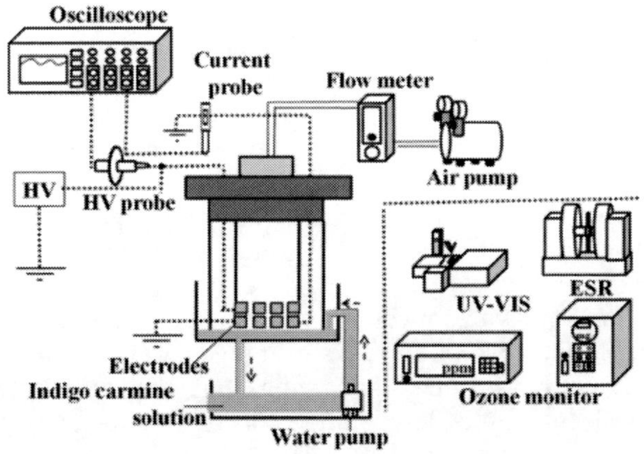

Figure 2. Experimental setup used for the water treatment by microplama. AC power supply, four types of carrier gas and indigo carmine were used this experiment. A Spectrum photometer, ozone monitor and ESR (electron spin resonance analyzer) were used as the evaluation methods.

Figure 2 shows the experimental setup for the water treatment by using microplasma. The carrier gases were supplied in case of the air from an air pump and in case of N_2, Ar and O_2 from gas cylinders. Experiments were carried out at atmospheric pressure at a gas flow rate of 10 L/min. The

Journal of Physics: Conference Series **441** (2013) 012005 doi:10.1088/1742-6596/441/1/012005

indigo carmine solution which had concentration of 10 mg/L was used as the treatment sample. Active species generated by microplasma could react with the target liquid. The water pump placed under the treatment container had the role to assure a continuous flow of water under the electrodes for a uniform treatment of indigo carmine. The water depth beneath the electrodes was about 10 mm as shown in figure 1, small enough to react with the active species generated by the microplasma treatment. A UV-VIS(UV-3100PC, SHIMADZU) was used to measure the absorbance of the liquid and furthermore estimate its organic composition. The ozone concentration in gas phase and liquid phase were analyzed by the gas phase ozone monitor (SOZ-3300, SEKI ELECTRONICS) and the liquid phase ozone monitor (PL-620A, EBARA JITSUGYO), respectively. Ozone concentration is an important factor to consider the water treatment. The ESR(electron spin resonance analyzer)(EMX plus, BRUKER) was used to analyze the radicals dissolved in water by microplasma discharge.

A neon transformer was used as an AC power supply to generate microplasma [25]. Figure 3(a), (b) shows the waveform of the discharge voltage and corresponding discharge current during discharge.

These values contained the voltage drop with the dielectric material.

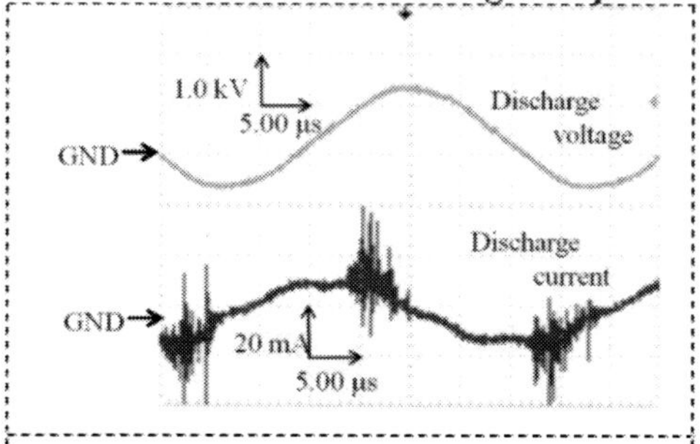

Figure 3(a). Discharge voltage and corresponding discharge current waveforms. Discharge voltage was about 1.00 kV.

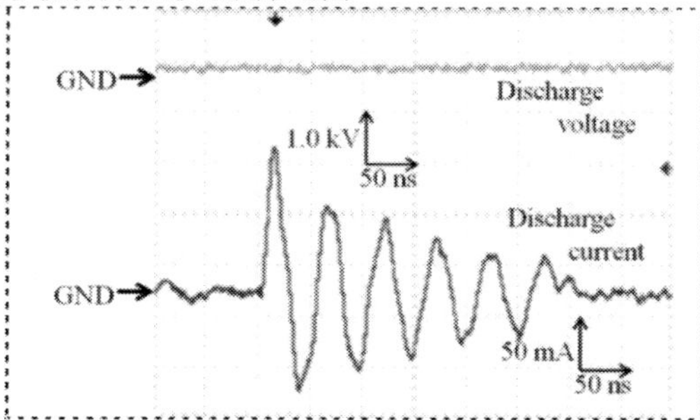

Figure 3(b). This is the short width waveform during discharge. The current peak value was about 150 mA at 25 kHz.

4. Result and Discussion

4.1 Observation of indigo carmine absorbance

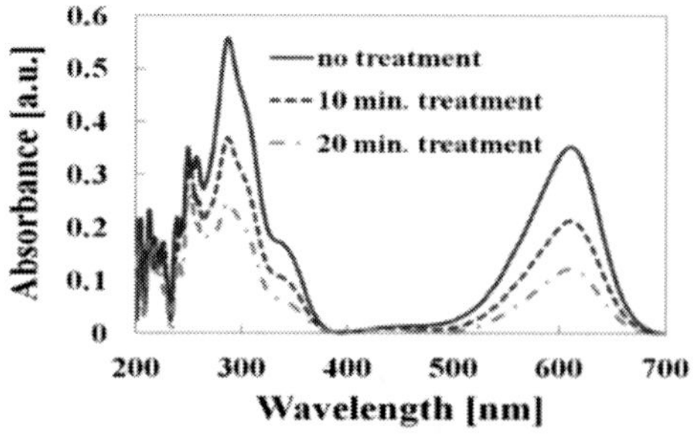

Figure 4. The analysis of absorbance using spectrum photometer. Continuous line corresponds to the non-treated liquid and two dashed lines correspond to the treated liquid. Two absorption peaks of indigo carmine were decreased by microplasma treatment. The decomposition rate increased with the treatment time. Discharge voltage was 1.0 kV, the carrier gas was air.

The absorbance values of indigo carmine solution versus treatment time are shown in figure 4. The peaks at 610 nm and 290 nm were corresponded to the absorption peaks of indigo carmine's organic compound [26]. Peaks' height decreased after microplasma treatment suggesting that the organic compound of the indigo carmine solution was decomposed during microplasma treatment by the various active species.

In gas phase, this reaction was occurred during discharge [27].

$$O + O_2 + M \rightarrow O_3 + M \quad (1)$$

In water these active species were generated during discharge via reactions (2), (3) and (4) [28].Because air was use as carrier gas O and e could reach the water and promote chemical reactions.

$$O\cdot + H_2O \rightarrow \cdot OH + \cdot OH \quad (2)$$
$$e\cdot^- + H_2O \rightarrow \cdot OH + \cdot H + e\cdot^- \quad (3)$$
$$\cdot OH + \cdot OH \rightarrow H_2O_2 \quad (4)$$

Among this active species and radicals, OH and O_3 could play an important role in the decomposition process of indigo carmine [29].

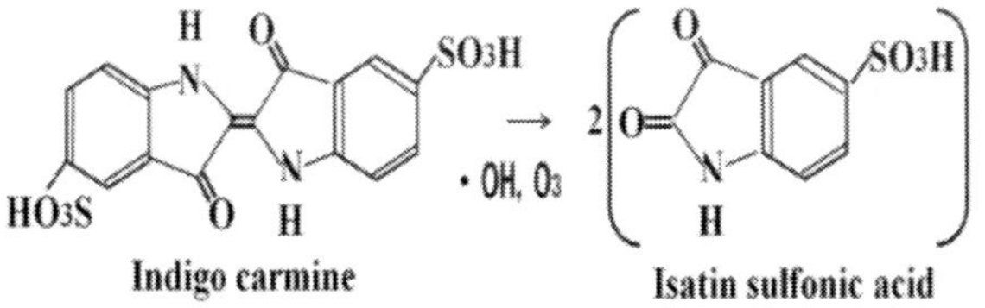

Figure 5. Chemical composition of the indigo carmine and its decomposition process. Left structure corresponding to indigo carmine, right to isatin sulfonic acid. The double bond could be broken by the active species and radicals generated by microplasma [30].

The chemical reaction and structural formula of indigo carmine is shown in figure 5. Indigo carmine has H type chromophoric group, therefore absorbs the visible light of a specific wavelength and gives its colour blue [31]. Indigo carmine has the H type chromophoric group to take on the blue color.The indigo carmine was changed to the isatin sulfonic acid after breaking the H type chromophoric group by the active species and radicals generated microplasma [32]. The double bond of indigo carmine could be broken by microplasma.

4.2Decomposition rates of indigo carmine for various carrier gases.

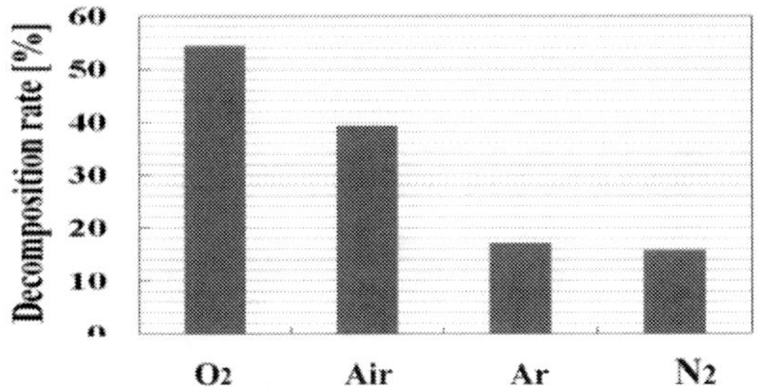

Figure 6. The decomposition rate of indigo carmine for various carrier gases. Oxygen was the most efficient for decomposing indigo carmine. Opposed to the oxygen and air, when using argon and nitrogen the decomposition rates were lower because of the absence of oxygen. Discharge voltage was 1.0 kV

Figure 6. shows the decomposition rates for various carrier gases. Decomposition rate was calculated under the initial and after the treatment value of absorbance as shown in the formula (5). Decomposition rates of the indigo carmine solution by the argon and nitrogen gases were lower because of the absence of the oxygen. Air and oxygen were higher due to the active species of oxygen origin.

$$Decomposition\ rate = \frac{(Initial\ absorbance) - (After\ treated\ absorbance)}{(Initial\ absorbance)} \times 100 \quad [\%] \qquad (5)$$

4.3 ozone concentration in gas phase

Ozone concentration generated from microplasma discharge is shown in figure 7. Air was used as the carrier gas and flow rate was set at 10 L/min. For measuring the dissolved ozone concentration in liquid, treatment time was 10 min. and treatment volume was 200 mL of distilled water. The ozone concentration in gas phase and liquid phase increased with the increase of the discharge voltage. However total ozone generated in gas phase could not be dissolved in liquid phase. About 0.5-2.2% of the total ozone generated in gas phase was dissolved in water. This ozone also affected to decomposition process of the organic compound contained in indigo carmine.

IOP Publishing
doi:10.1088/1742-6596/441/1/012005

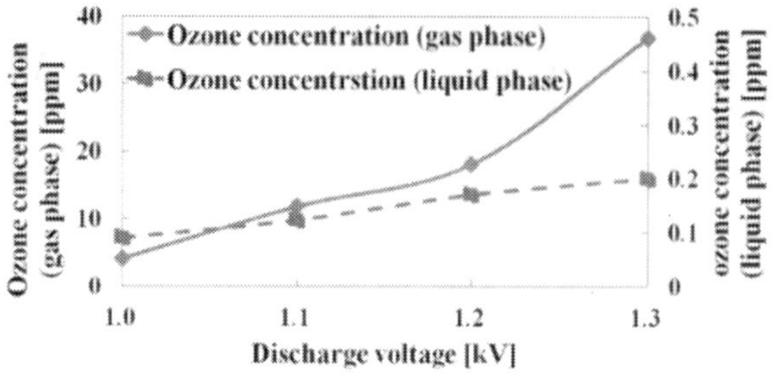

Figure 7. Ozone concentration versus discharge voltage measured in gas phase and liquid phase by microplasma. The ozone concentration increased with the increase of the discharge voltage. Treatment time was set at 10 min. and water temperature was 26℃

4.4 Radical analysis by the ESR method

Figure 8 shows the result of the ESR analysis which detected the dissolved OH radical generated by microplasma in water. In this analysis, spin trap agent 5,5-dymethyl-1-pyrroline N-oxide(DMPO) was used due to the short-lived radical of OH radical in liquid phase [33]. The four peaks shown in figure 8 correspond to the OH radical peaks generated by microplasma and detected by the spin trap agent [34].

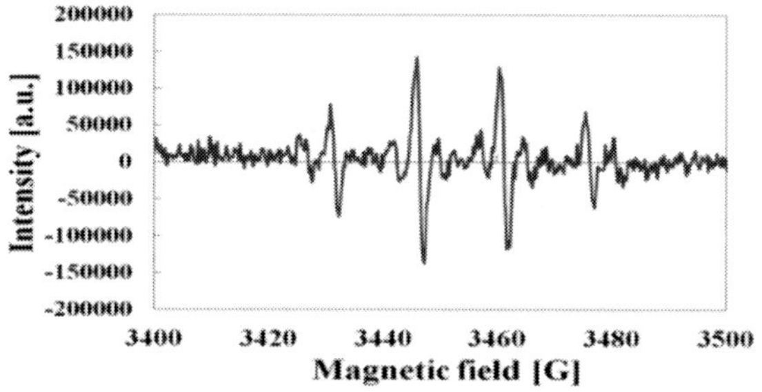

Figure 8. OH signal detected by an ESR measurement. Four peaks corresponding to OH radical were detected by the by the spin trap agent dissolved in water. Discharge voltage was 1.0 kV, treatment time was 5 min. and pure water was used as the treatment target in this analysis.

OH radical could be effective factor to decompose the indigo carmine due to its oxidation process[35].

5. Conclusions

Microplasma could be used as an economical and ecological technology. Water treatment using microplasma was carried out and the following conclusions were obtained:

1. Indigo carmine was decomposed after microplasma treatment and decomposition rate increased with the treatment time.
2. The presence of OH radical could be one of the main factor to decompose indigo carmine. It was dissolved in the water as confirmed by using the ESR.
3. About 0.5-2.2% of the ozone generated in gas phase was dissolved in water at the room temperature.
4. Active species of oxygen origin could be considered as the main factors in indigo carmine decomposition as demonstrated by higher decomposition rates obtained with microplasma generated in oxygen and air.

6. Reference

[1] Zheng X, Blais J, Mercier G, Bergeron M and Drogui P 2007 PAH removal from spiked municipal wastewater sewage sludge using biological, chemical and electrochemical treatment *Chemosphere* **68** 1143-1152.

[2] Tanino M, Wang X, Tahashima K, Katsura S and Mizuno A 2005 Sterilization using dielectric barrier discharge at atmospheric pressure *IEEE Trans. on IAS* **2** 784-788.

[3] Iseki S, Nakamura K, Hayashi M, Tanaka H, Kondo H, Kajiyama H, Kano H, Kikkawa F and Hori M 2012 Selective killing of ovarian cancer cells through induction of apoptosis by nonequilibrium atmospheric pressure plasma *Appl. Phys. Lett.* **100** 113702.

[4] Ceyter N D, Morent R, Leys C, Gengembre L and Payen E 2007 Treatemnt of polymer films with a dielectric barrier discharge in air, herium and argon at medium pressure *Surf. Coat. Tech.* **201** 7066-7075.

[5] Shimizu K, Blajan M and Kuwabara T 2011 Removal of indoor air contaminat by atmospheric microplasma *IEEE Trans. on Ind. Appl.* **47** 2351-2358.

[6] Kong M G, Kroesen G, Morfill G, Nosenko T, Shimizu T, Dijik J V and Zimmermann J L 2009 Plasma medicine: an introductory review *New J Phys.* **11** 115012.

[7] Handa T, Minamitani Y 2009 The effect of a water-droplet spray and gas discharge in water treatment by pulsed power *IEEE Trans. on Plasma Sci.* **37** 179-183

[8] Shih K Y and Locke B R 2011 Optical and electrical diagnostics of the e.ffects of conductivity on liquid phase electrical discharge *IEEE Trans. on Plasma Sci.* **39** 883-892.

[9] Zhao J L, Zhou J C, Su J, Guo H C, Wang X S and Gong W M 2007 Propene epoxidation with in-site H_2O_2 produces by H_2/O_2 non-equilibrium plasma *AIChE J* **53** 3204-3209.

[10] Kuwahata H, Kimura K and Ohyama R I 2010 Decolorization of methylene blue aqueous solution by atmospheric-pressure plasma jet *e-J of Surf. Sci. and nanotech.* **8** 381-383.

[11] Shmelev V M, Saveliev A V and Kennedy L A 2009 Plasma chemical reactor with exploding water jet *Plasma Chem. Plasma Process* **29** 275-290.

[12] Ishijima T, Hotta H, Sugai H and Sato M 2007 Multibubble plasma production and solvent decomposition in water by slot-excited microwave discharge *Appl. Phys. Lett.* **91** 121501.

[13] Shih K Y and Locke B R 2010 Chemical and physical characteristics of pulsed electrical discharge within gas bubbles in aqueous solution *Plasma Chem. Plasma Process* **30** 1-20.

[14] Wang H, Li J, Quan X, Wu Y, Li G and Wang F 2007 Formation of hydrogen peroxide and degradation of phenol in synergistic system of pulsed corona discharge combined with TiO_2 photocatalysis *J Hazardous Materials* **141** 336-343.

[15] Ikoma S, Satoh K and Itoh H 2012 Decomposition of methylene blue in aqueous solution using a pulsed-discharge plasma at atmospheric pressure *Electrical Eng. In Japan* **179** 1-9.

[16] Grymonpre D R, Finney W C and Locke B R 1999 Aqueous-phase pulsed streamer corona reactor using suspended activated carbon particles for phenol oxidation: model-data comparison *Chem. Eng. Sci.* **54** 3095-3105.

[17] Sugiarto A T, Ohshima T and Sato M 2002 Advanced oxidation processed using pulsed streamer corona discharge in water *Thin Solid Films* **407** 174-178.

[18] Sun B, Sato M, Harano A and Clements J S 1998 Non-uniform pulse discharge-induces radical production in distilled water *J electrostatics* **43** 115-126.

[19] Sato M 2009 Degradation of organic contaminants in water by plasma *International Journal of Plasma Env. Sci. and Tech.* **3** 8-14.

[20] Burlica R and Locke B R 2008 Pulsed plasma gliding-arc discharges with water spray *IEEE Trans. on Ind. Appl.* **44** 482-489.

[21] Sato M, Ohgiyama T and Clements J S 1996 Formation of chemical species and their effects on microorganisms using a pulsed high-voltage discharge in water *IEEE Trans. on Ind. Appl.* **32** 106-112.

[22] Minamitani Y, Shoji S, Ohba Y and Higashiyama Y 2008 Decomposition of dye in water solution by pulsed power discharge in a water droplet spray *IEEE Trans. on Plasma Sci.* **36** 2586-2591.

[23] Wu J J, Muruganandham M and Chen S H 2007 Degradation of DMSO by ozone-based advanced oxidation processes *J Hazaradous Materials* **149** 218-225.

[24] Cvelbar U, Krstulovic N, Milosevic S, Mozatic M 2008 Inductively coupled RF oxygen plasma characterization by optical emission spectroscopy *Vacuum* **82** 224-227.

[25] Shimizu K, Blajan M and Tatematsu S 2012 Basic study of remote disinfection and sterilization effect by using atmospheric microplasma *IEEE Trans. on Ind. Appl.* **48** 1182-1188.

[26] Bentouami A, Ouali M S and Menorval L D 2010 Photocatalytic decolourization of indigo carmine on 1, 10-phenanthrolinium intercalated bentonite under UV-B and solar irradiation *J photochem. and photobiol. A: Chem.* **212** 101-106.

[27] Laroussi M and Leipold F 2004 Evaluation of the roles of reactive species, heat, and UV radiation in the inactivation of bacterial cells by air plasmas at atmospheric pressure

International J Spectrometry **233** 81-86.

[28] K Takahashi, Y Sasaki, S Mukaigawa and K Takaki 2010 Purification of high-conductivity water using gas-liquid phase discharge reactor *IEEE Trans. on Plasma Sci.* **38** 2694-2700.

[29] T Kobayashi, T Sugai, T Handa, Y Minamitani and T Nose 2010 The effect of spraying of water droplets and location of water droplets on the water treatment by pulsed discharge in air *IEEE Trans. on Plasma Sci.* **38** 2675-2680.

[30] Kettle A J, Clerk B M and Winterbourn C C 2004 Superoxide convert indigo carmine to isatin sulfonic acid *The J Biological Chem.* **279** 18521-18525.

[31] Vautier M, Guillard C, and Herrman J M 2001 Photocatalytic degradation of dyes in water: case study of indigo and of indigo carmine *J. Catalysis* **201** 46-59.

[32] Flox C, Ammar S, Arias C, Brillas E, Vargas-Zavala A V and Abdelhedi R 2006 Electro-fenton and photoelectron-fenton degradation of indigo carmine in acidic aqueous medium *Appl. Catalysis B Env.* **67** 93-104.

[33] Remya N and Lin J G 2011 Current status of microwave application in wastewater treatment-a review *Chem. Eng. J.* **166** 797-813.

[34] Faure A M, Andersen M L and Nystrom L 2012 Ascrbicacid induced degradation of beta-glucan: hydroxyl radicals as intermediates studied by spin trapping and electron spin resonance spectroscopy *Carbohydr. Polymers* **87** 2160-2168.

[35] Zaied M, Chutet E, Peulon S, Bellakhal N, Desmazieres B, Dachraoui M, Chausse A 2011 Spontaneous oxidative degradation of indigo carmine by thin films of birnessite electrodeposited onto SnO_2 *Appl. Catalysis B Env.* **107** 42-51.

Plasma application for detoxification of Jatropha phorbol esters

S. Kongmany[1], H. Matsuura[2], M. Furuta[2], S. Okuda[2], K. Imamura[2], Y. Maeda[2]

[1] Graduate School of Engineering, Osaka Prefecture University, Gakuen-cho 1-2, Nakaku, Sakai, Osaka 599-8570, Japan
[2] Osaka Prefecture University, Gakuen-cho 1-2, Nakaku, Sakai, Osaka 599-8570, Japan

Email: matsu@me.osakafu-u.ac.jp

Abstract. Atmospheric pressure non-thermal dielectric barrier discharge (DBD) plasma generated by helium gas at high voltage and input power of about 50 W was first applied to detoxification of Jatropha curcas phorbol esters (*J.* PEs) as well as standard phorbol ester (4β-12-O-tetradecanoyl phorbol-13-acetate, TPA) in water and methanol. Plasma irradiation on the solution sample was conducted for 15 min. In aqueous solution, only 16% of TPA was degraded and complete degradation of *J.* PEs was observed. On the contrary, complete degradation of both TPA and *J.* PEs in methanol was achieved by the same plasma irradiation condition. Hydroxyl radical (•OH) generated by plasma irradiation of the solution is expected as the main radical inducing the degradation of PEs.

1. Introduction

Plasma is a promising technology which is useful for various applications such as a tool for green chemistry, environmental analysis and waste management [1]. Beside hot (thermal) plasma, cold (non-thermal) plasma is currently an interesting approach in which the reactive chemical species can be produced at low gas temperature [2]. Non-thermal plasma is known as the mixture of electrons, highly excited atoms, molecules, ions, radicals, photons and so on [3]. Even though many kinds of non-thermal plasma generation devices have been known, dielectric barrier discharge (DBD) get much interest nowadays due to its simple configuration [4]. The type of highly excited atoms, molecules, ions, radicals in the plasma depends on applied gas species [5]. When high-energy electron, highly excited atoms entering into water, hydroxyl radical (•OH), hydrogen atom (•H) and electron (*e*) are formed [6]. The degradation of organic pollutants in aqueous solution by non-thermal plasma is mostly based on oxidation reaction between •OH and the pollutants [7]. Therefore, non-thermal plasma technology is currently classified as one of promising methods used in Advance Oxidation Process (AOP) [8].

Jatropha phorbol esters (*J.* PEs) were toxic compounds found in toxic-genotype jatropha such as *Jatropha curcas* (*J. curcas*) [9]. Phorbol esters are known as tumor promoters [10]. For eliminating the risk to the workers contacting to such toxic oil, methanol is normally used as solvent for extracting *J.* PEs from *J. curcas* oil before biodiesel production [11]. *J.* PEs in the oil was possibly degraded to undetectable level by natural sunlight exposure at least 6 days [12]. Stripping or deodorization at 260 °C under low pressure of about 3 mbar and 1% steam injection can completely degraded *J.* PEs [13]. Under optimized solid state fermentation conditions, P. aeruginosa PseA can completely degrade *J.*

PEs in *J. curcas* seed cake in nine days [14]. Approximately 71.35% of *J.* PEs in *J. curcas* seed cake were successfully removed by γ-irradiation at absorbed dose of 50 kGy [15]. Such methods need long treatment time, high temperature as well as large facility and high dose from γ-irradiation.

The objective of this present work is to apply the dielectric barrier discharge (DBD) non-thermal plasma to degrade/detoxify *J.* PEs in aqueous or methanolic solution. It is expected that the radicals entering into the solution or generated by interaction between plasma species and solvent molecule would be able to degrade the phorbol esters.

2. Materials and methods

2.1. Plasma system

The schematic drawing of plasma system used in this experiment is shown in Figure 1. The plasma source used in the present work is a small and simple one with a low-frequency power supply (Loggy Electronics Ltd, LHV-13AC) and a quartz tube as dielectric barrier [16-17]. In this study, plasma production condition is set as follows: helium gas with a flow-rate of 6 L/min supplied through the quartz tube (inner diameter of 5 mm) from upper inlet to the lower outlet, and the high voltage of 10 kV applied to the electrodes that create barrier discharge between electrodes, and the input power of about 50 W. Under the condition, a plasma plume, 5 cm in length, is ejected from the nozzle into the surrounding air before reaching the surface of solution.

THE GRAPHIC (FIGURE 1) WAS PRINTED WITH THE BEST AVAILABLE COPY.

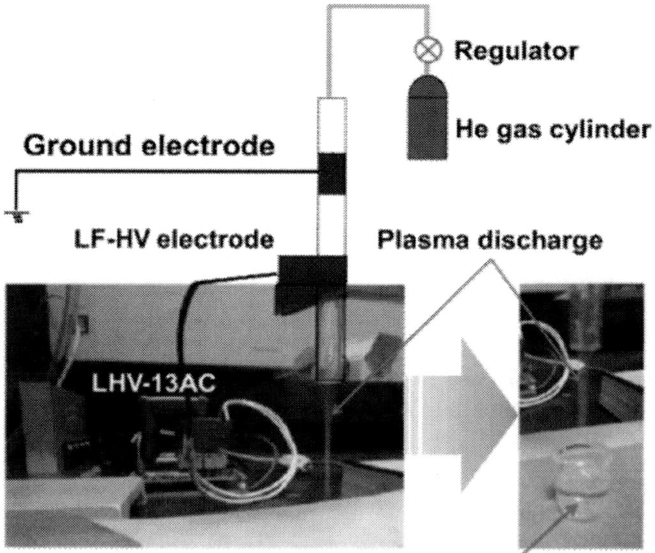

Figure 1 Schematic drawing of DBD plasma configuration used in present study

2.2. Sample preparation, plasma irradiation and analysis

TPA in solid form in ampoule (Wako Pure Chemicals, Tokyo, Japan) was used to prepare stock solution of 1,000 mg/L of TPA by dissolving with methanol (MeOH). This TPA stock solution was further diluted with MeOH to be 100 mg/L of standard solution. *J.* PEs was obtained in our laboratory by methanolic extraction and the extract was then purified by silica-gel column chromatography [18]. The fraction containing *J.* PEs was then analyzed by High Pressure Liquid Chromatography (HPLC) to determine their concentrations. The concentration of each component is evaluated to be as equivalent as that of TPA. The molecular structures of TPA and *J.* PEs are shown in Figure 2.

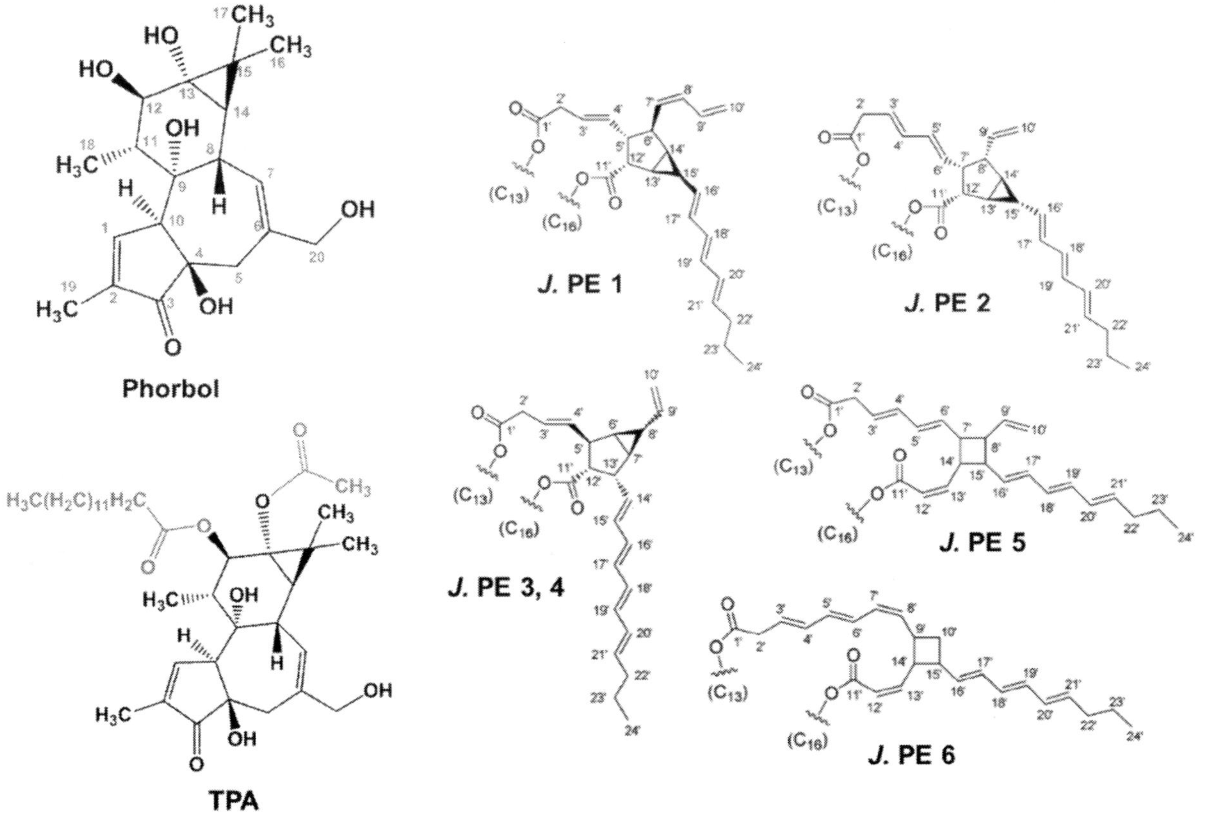

Figure 2 Molecular structures of Phorbol, 4*β*-12-O-tetradecanoyl phorbol-13-acetate (TPA) and *Jatropha* phorbol esters (*J.* PEs) [19].

A hundred micro liter of TPA stock solution (1,000 mg/L) was diluted with water (Milli-Q water) to obtain a 10 μg/mL sample solution in 10-mL-volumetric flask. The methanolic sample solution of TPA was prepared in the same way.

After purification, the 5,758 mg/L of *J.* PEs stock solution was prepared in methanol. The concentration was the total amount of phorbol esters contained in the solution. The sample solutions of both water and methanol were prepared by the same manner as in TPA.

For plasma irradiation, one milliliter of the solution was loaded onto a 1.5-mL-beaker, and then irradiated by the plasma for 15 min. The concentrations of phorbol esters were measured by HPLC equipped with a photodiode array detector (PDA) and an automatic sample injector (SHIMADZU Corporation, Japan). An analytical column was Shim-pack XR-ODSII (2.0 mm i.d × 75 mm). Analytical conditions was as follows: mobile phase was a mixture of water (A) and acetonitrile (B) (Wako Pure Chemicals, Tokyo, Japan) and the run was started at initial mixing ratio of 50:50 (A:B) for 5 min, and in gradient mode to 75:25 for 10 min, and finally to 100:0 for 5 min and hold for 2 min. The column temperature was 35 °C and the injection volume was 5 μL. The phorbol esters were detected in a UV region at the wavelength of 280 nm [13].

3. Results and discussion

Figure 3(a) and 3(b) showed HPLC/UV chromatograms of TPA and *J.* PEs in the aqueous solution respectively. After plasma irradiation, approximately 16% reduction of TPA was observed as shown in Figure 3(a). On the other hand, before plasma treatment four peaks of phorbol esters in aqueous solution are observed but disappeared after plasma irradiation as shown in Figure 3(b).

During plasma irradiation, the oxidative degradation of phorbol esters in the solution by hydroxyl radicals (•OH) may occur through two way. Interaction between high-energy electron (*e**) in the plasma plume and the O_2 and moisture (H_2O) would generated Ozone (O_3) and •OH then entering into the solution via air-solution interface [20].

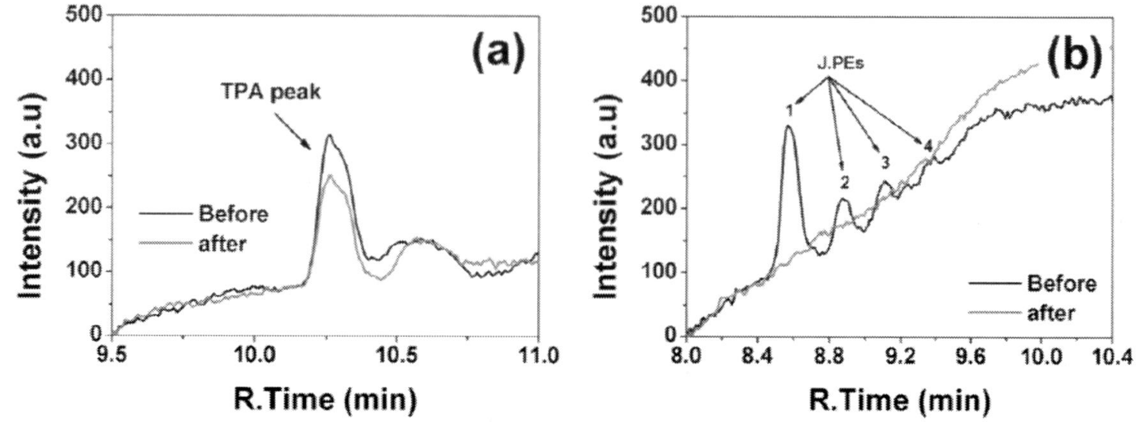

Figure 3 HPLC/UV chromatograms of (a) TPA and (b) 4 peaks of *J.* PEs in aqueous solution before and after plasma irradiation for 15 min.

$$O_2 + e^* \rightarrow 2O\bullet + H_2O \rightarrow 2\,\bullet OH + e \quad (1)$$

$$O_2 + O\bullet \rightarrow O_3 \quad (2)$$

Entering into aqueous solution (Eq. 4), ozone (O_3) dissociates to O_2 and O• (Eq. 5) and the O• react with H^+ in the presence of *e* to form •OH radical [21]. Enhanced by ozone (Eq. 5-6), UV light from plasma discharge would also play a key role on producing •OH radicals [22].

$$O_3 \rightarrow O_2 + O\bullet \quad (3)$$

$$O\bullet + H^+ + e \rightarrow \bullet OH \quad (4)$$

$$O_3 + h\upsilon + H_2O \rightarrow H_2O_2 + O_2 \quad (5)$$

$$H_2O_2 + h\upsilon \rightarrow 2\bullet OH \quad (6)$$

When the plasma electron reaches the surface of aqueous solution (Eq.7), •OH radical is also generated [23].

$$H_2O + e \rightarrow \bullet H, \bullet OH, e, etc. \quad (7)$$

The generated •OH radicals ($6.35 \pm 0.77 \times 10^{-7}$ moles) in water after 15 min plasma treatment was detected by KI-colorimetric method. This observation is not contradicted with the result found in literature [24].

On the contrary, It has been reported that •OH radical is formed under plasma irradiation of methanol [24]. Large amount of •OH radical may be generated that completely oxidized all phorbol esters molecules in the methanolic solution. So we tried similar experiment as Figure 3 with methanol as solvent. In Figure 4, the disappearance of the peaks of TPA as well as *J.* PEs after plasma irradiation was confirmed.

$$CH_3OH + e \rightarrow \bullet CH_3, \bullet OH, e, etc. \quad (8)$$

The decomposition process of PEs depends not only on the kind of PE but also on solvent. It is expected that *J.*PEs was oxidized by •OH radicals through electrophilic reaction. For simplicity, a typical decomposition of TPA is shown in Figure 5.

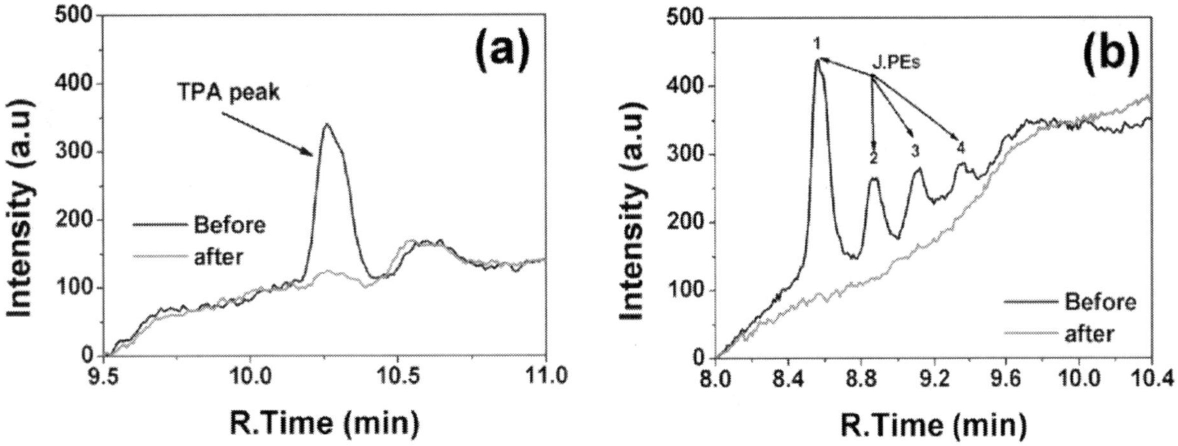

Figure 4 HPLC/UV chromatograms of TPA (a) and 4 peaks of *J.* PEs (b) in methanolic solution before and after plasma irradiation for 15 min.

The •OH radicals attack to the double bonds of ring A and/or B (indicated in Figure 5 as the symbol "A" and "B") in the structure of TPA leading to bond cleavage and ring destruction [26]. These intermediates would follows further oxidation with more •OH radicals producing carboxylic acids, carbon dioxide and water [27].

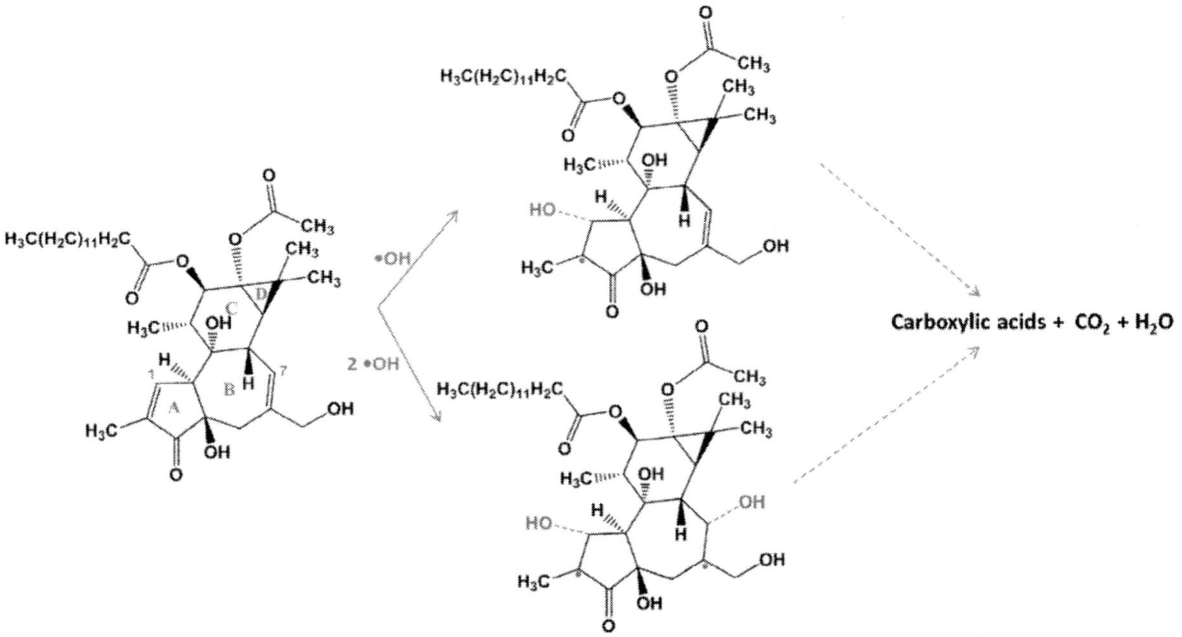

Figure 5 TPA degradation pathways by •OH radical attack.

4. Conclusion

We have firstly demonstrated that atmospheric non-thermal plasma would be usefully applied for degrading phorbol ester (TPA or *J.* PEs) in aqueous or methanolic solution, even though the exact mechanism on the degradation of phorbol ester should be further investigated in detail. After 15-minutes-plasma treatment, 16% and complete degradation of TPA were found in aqueous and methanolic solutions, respectively. Complete degradation of *J.* PEs was achieved in both aqueous and methanolic solutions. The generation of •OH radials ($6.35 \pm 0.77 \times 10^{-7}$ moles) in water after the same plasma treatment condition were confirmed. The degradation of phorbol esters in methanolic solution is more efficient than in aqueous solution. Since •OH radical was not measured directly in our experiment, so more comprehensive study would be necessary. Parameter optimization on the plasma treatment and discharge condition is needed in future study.

Acknowledgements
This study is fully supported by SATREPS (Science and Technology Research Partnership for Sustainable Development) project.

References

[1] Mollah M. *et al* 2000 *Journal of Hazardous Materials vol. B* ***79*** *301–320*
[2] Luigi C. *et al* 2002 *Journal of Materials Processing Technology vol.* ***121*** *373–382*
[3] Liu C.-J. *et al* 1999 *Fuel Processing Technology vol.* ***58*** *119–134*
[4] Gui-Min X. Y. M. *et al* 2008 *IEEE Trans. Plasma Sci. vol.* ***36*** *no.* ***4*** *1352-1353*
[5] Kang W. S. *et al* 2010 *IEEE Trans. Plasma Sci. vol.* ***38****, no.* ***8*** *1982-1990*
[6] Hiroshi K. *et al* 2010 *e-J. Surf. Sci. Nanotech. vol.* ***8*** *381-383*
[7] Yongjun S. *et al* 2008 *Journal of Hazardous Materials vol.* ***150*** *713–722*
[8] Moreau M. *et al* 2008 *Biotechnology Advances, vol.* ***26*** *610–617*
[9] Li C.-Y. *et al* 2010 *Food and Chemical Toxicology vol.* ***48*** *620–625*
[10] Goel G. *et al* 2007 *International Journal of Toxicology vol.* ***26*** *279–288*
[11] Devappa R. K. *et al* 2010 *biomass and bioenergy, vol.* ***34*** *1125-1133*
[12] Ping B. Y. *et al* 2010 *Ecotoxicology and Environmental Safety vol.* ***84*** *268–273*
[13] Harinder M.*et al* 2009 *J. Am. Oil. Chem. Soc. no.* ***86*** *173–181*
[14] Joshi C. *et al* 2011 *Bioresource Technology vol.* ***102*** *4815–4819*
[15] Diwani G. I. E. et al 2011 *Adv.in Applied Sci.Research vol.* ***2*** *no.* ***4*** *221-232*
[16] Teschke M. et al 2005 *IEEE Trans. Plasma Sci. vol.* ***33*** *no.* ***2*** *310-311*
[17] Kang H. R. *et al* 2012 *Proc.Conf. on 11th APCPST/ 25th SPSM* (Kyoto: Japan) ***2-P92***
[18] Imamura K. *et al* 2011 *Proc. Conf. on Annual Meeting of Environmental Chemistry* (Fukuoka: Japan*)* ***p-186***
[19] Hass W. *et al* 2002 *J. Nat. Prod. vol.* ***65*** *no.* ***10*** *586-592*
[20] Mark P.-Cal *et al* 2001 *Environtmental Progress vol.* ***20*** *no.* ***3*** *151–156*
[21] Jibiao Z. *et al* 2008 *Journal of Hazardous Materials vol.* ***154*** *506–512*
[22] Cheng H.-H. *et al* 2007 *J. Environ. Eng. Manage.vol.* ***17*** *no.* ***6*** *427-433*
[23] Tang Q. *et al* 2009 *Chemical Engineering Journal vol.* ***153*** *94–100*
[24] Kanazawa S. *et al.*, 2012 *International Journal of Plasma Environmental Science & Technology, Vol.****6****, No.****2*** *166-71*
[25] Yan Z.-C *et al* 2009 *International Journal of hydrogen energy vol.* ***34*** *48–55*
[26] Almeida L.C. et *al.* 2011 *Applied Catalysis B: Environmental Vol.* ***103*** *21–30*
[27] Wang L. *et al.* 2008 *Journal of Hazardous Materials Vol.* ***154*** *1106–1114*

Journal of Physics: Conference Series **441** (2013) 012007 doi:10.1088/1742-6596/441/1/012007

Influence of pulse width on decolorization efficiency of organic dye by discharge inside bubble in water

S Kawano[1], K Wada[1], T Kakuta[1], K Takaki[1,2], N Satta[3] and K Takahashi[4]

[1] Faculty of Engineering, Iwate University, 4-3-5, Ueda, Morioka, Iwate 020-8551, Japan
[2] Soft-Path Engineering Research Center (SPERC), Iwate University, 4-3-5, Ueda, Morioka, Iwate 020-8551, Japan
[3] Faculty of Agriculture, Iwate University, 3-18-8, Ueda, Morioka, Iwate 020-8550, Japan
[4] Shishido Electrostatic, LTD., 4-7-21, Chigasaki-higashi, Tsuzuki-ku, Yokohama, Kanagawa 224-0033, Japan

Email: takaki@iwate-u.ac.jp

Abstract. Decolorization of an organic dye by discharge in high conductive water using a pulsed power generator and a discharge reactor was investigated. The discharge reactor consisted of a glass tube and a tungsten wire inserted into the glass tube, which was immersed in the water. Room air was injected into the glass tube to generate bubbles in the water. High voltage pulses were generated by an inductive-energy storage system using semiconductor opening switch (SOS) and by a magnetic pulse compression circuit. Fast recovery diodes were used as SOS diode in the inductive-energy storage system. The pulse width was changed in range from 10 to 1200 ns. The high voltage was applied to the tungsten wire. Indigo carmine was employed as a specimen to evaluate decolorization efficiency. Potassium nitrate was used to adjust the solution conductivity. The dye solution was successfully decolorized at 7 mS/cm conductivity. Energy efficiency for decolorization increased from 0.680 to 55.6 mg/Wh with decreasing the pulse width from 1200 to 10 ns owing to the reduction of ohmic loss.

1. Introduction

Water treatment using methods such as filtration, chemical oxidation and bioremediation has been used at a treatment of organic compounds in water [1]. Filtration techniques are widely used because of the high performance, while concentrated waste disposal remains. Chemical oxidation using chlorine and ozone has been shown to produce hazardous by-products such as trihalomethanes, aldehydes and bromate [2, 3]. Ozone is a weaker oxidizing agent than hydroxyl radical and rather selective oxidant. Ozone oxidation and Bioremediation are limited by slow kinetics. A pulsed discharge plasma for treating pollutants in water has been attracted much attention [4]. The production of discharge plasmas with large volume discharge requires pulsed power technologies because the discharges in water demands high electric field. The pulsed discharge makes it possible to instantaneously produce a non-thermal plasma in which various chemical species such as hydroxyl radical exist. These species play important role in degradation of chemical organic compounds [5-7].

In general, an energy efficiency for the degradation of chemical organic compounds in water decreases with increasing conductivity of the water [6-10], because of an increase of energy dissipation by the ohmic loss which doesn't contribute to production of the chemical species [10]. As the result, the energy efficiency for water treatment by the discharge decreases to be low at high conductivity. To solve this problem, gas-liquid separated reactors and short pulse generators have been proposed [7, 11]. The short pulse can produce ozone with high energy efficiency because of a decrease the conductive current [12].

Objective of the present work is to clarify an influence of the pulse width on energy efficiency to decolorize high-conductive solution using pulsed discharge inside bubble. Indigo carmine is employed as specimen to evaluate the energy efficiency. The high voltage pulses are generated by an inductive-energy storage system using semiconductor opening switch (SOS-IES) and by a magnetic pulse compression circuit (MPC). The pulse width is adjusted in range from 10 to 1200 ns. The solution conductivity is adjusted in range from 6 to 7000 μS/cm by mixing potassium nitrate.

2. Experimental setup and procedure

Figure 1 shows a schematic diagram of the pulsed power generator with IES using SOS. The pulsed power generator is consisted of a capacitor C_0, gap switch, a pulsed transformer P.T. (Hitachi Metals, Ltd., FT-3L), primary energy storage capacitor C_1, secondary energy storage inductor L_2, resister R_C and fast recovery diode (Voltage multiplier, K100UF). Fast recovery diodes are used as semiconductor opening switch (SOS). When the gap switch is closed, the energy stored in C_0 is transferred to C_1 through the pulse transformer. The pulsed voltage is generated by LC oscillation in the secondary circuit. The SOS diode is pumped up with forward current in half cycle determined by the inductance L_2 by capacitance C_1. After the half cycle, the current direction changes to reversed direction of the SOS diode. The reversed current is interrupted rapidly by the SOS diode, as the result, the pulse voltage appears between the inductor. The pulse width is controlled to be 10 or 65 ns by adjusting L_2. Figure 2 shows typical waveforms of the output voltage (v_O) without connection to the reactor. The pulse width of the output voltage is 10 ns without external inductor L_2 and increases to 65 ns with external inductor L_2 at 16.2 μH.

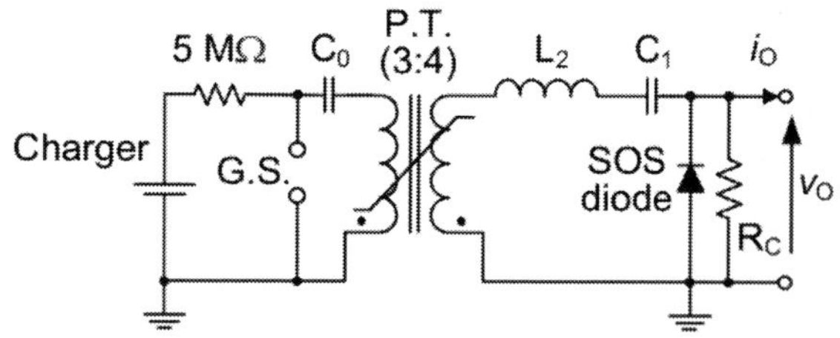

Figure 1. Schematic diagram of the IES pulsed power generator using SOS; C_0 = 2.3 nF, C_1 = 0.5 nF, L_2 = 0 or 16.2 μH, R_C = 20 kΩ.

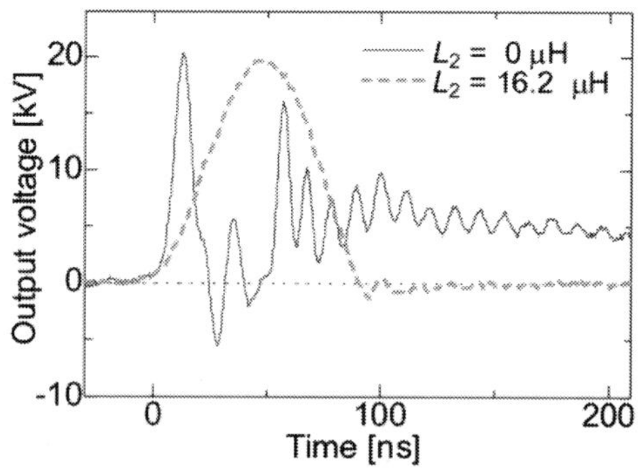

Figure 2. Waveforms of the output voltage of the IES pulsed power generator using SOS without the reactor; L_2 = 0 or 16.2 μH.

Figure 3 shows a schematic diagram of the magnetic pulse compression circuit (Suematsu Electronics CO., LTD., MPC-3000S SP). The capacitor C_0 is charged up to a charging voltage by the charger. The energy stored in C_0 is transferred to C_1 through the pulse transformer P.T. The pulsed voltage is produced at secondary side of the pulse transformer. The pulsed voltage is compressed by saturable inductors (SI_1, SI_2 and SI_4) and capacitors (C_2 and C_3). SI_3 is connected in parallel with C_3 to shorten the pulse width of the output voltage (v_O). The output voltage is rectified by the diode connected to the output terminal. Figure 4 shows typical waveforms of the output voltage (v_O) without connection to the reactor. The pulse width of the output voltage is 400 ns with SI_3 and is 1200 ns without SI_3 in condition of no load condition.

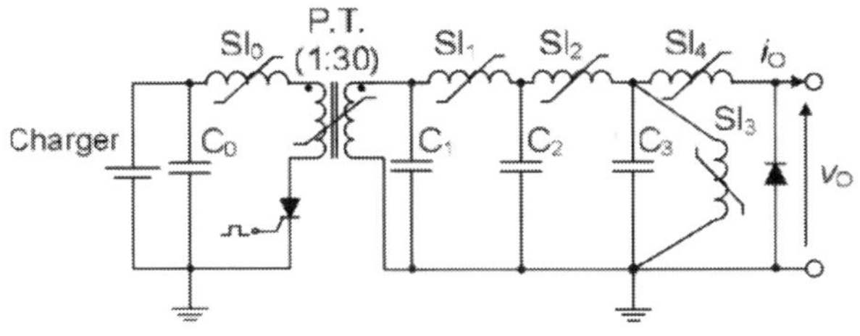

Figure 3. Schematic diagram of a magnetic pulse compression circuit; $C_0 = 2.24$ μF, $C_1 = C_2 = 2$ n F, $C_3 = 0.7$ nF.

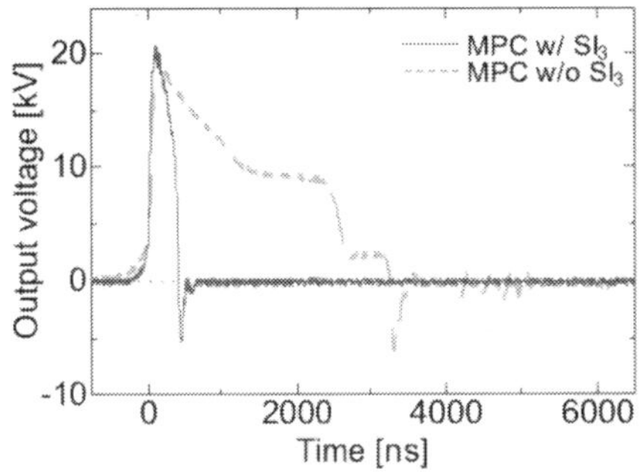

Figure 4. Waveforms of output voltage of the magnetic pulse compression with SI_3 and without SI_3 in condition of no load.

Figure 5 shows a schematic diagram of the discharge reactor. A glass tube containing a stainless steel wire and four glass tubes containing a tungsten wire are immersed in glass beaker (100 mm height, 50 mm diameter) as shown in the figure 5. The inner diameter of the glass tubes is 0.8 mm. The stainless steel wire (0.2 mm diameter) is a ground electrode and is immersed water. The gap length between tip of the ground electrode and tip of the glass tube is 15 mm. Room air is injected into the reactor through another four glass tubes with the gas flow rate in 2 L/min. The gap length between the tip of the tungsten wire and tip of the glass tube is 20 mm. The pulse voltage v_O generated by pulsed power generators is applied to the tungsten wires to generate streamer discharge in the tubes and air bubbles. The output voltage v_O of the pulse generator has peak value of 20 kV. A pulse repetition rates is controlled to be 80 pulses per second.

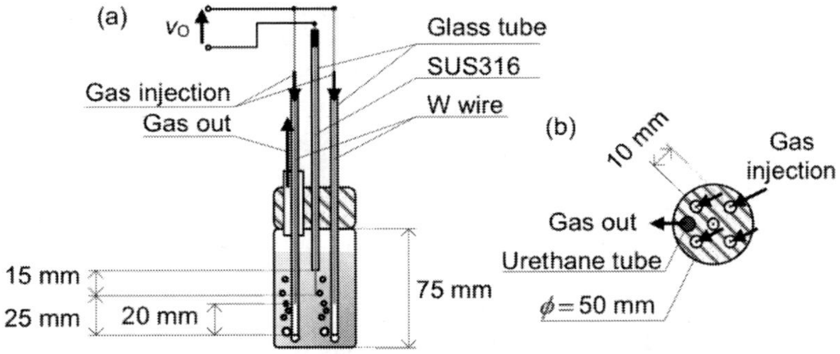

Figure 5. Schematic diagram of the reactor; (a) side view, (b) top view.

A dye solution dissolving indigo carmine (CAS No. 860-22-0) into 100 ml of purified water at concentration of 10 mg/L is used to evaluate by a decolorization efficiency. The initial pH of the solution is 5.5. The electrical conductivity (σ) is adjusted in range from 6 to 7000 μS/cm by mixing potassium nitrate. The absorbance of the dye solution is measured by a spectrometer (Milton Roy Ltd., Spectronic 301) and the decolorization rate (DR) is obtained by following equation,

$$DR = \frac{\text{Absorbance (initial)} - \text{Absorbance (treated)}}{\text{Absorbance (initial)}} \times 100[\%], \tag{1}$$

where the absorbance is obtained at 610 nm in wavelength. The energy efficiency for decolorization of the indigo carmine (ED) is obtained by following equation,

$$ED = \frac{DR \times 36}{J_T} [\text{mg/Wh}], \tag{2}$$

where J_T is total input energy into the reactor calculated by output voltage and current.

3. Experimental Results

Figure 6 shows the decolorization rate as a function of the treatment time for various pulse widths at 6 μS/cm in initial conductivity. The energies per a pulse are 3.45, 2.61, 2.49 and 3.68 mJ at 10, 65, 400 and 1200 ns in pulse width, respectively. The decolorization rate increases with increasing the treatment time. The treatment time at almost 100% decolorization increases with increasing pulse width. The treatment times of 100% decolorization are 5.0, 6.5, 7.0 and 8.5 min at 10, 65, 400 and 1200 ns in pulse width, respectively. The energy efficiencies of 50% decolorization are 100, 88, 83 and 55 mg/Wh at 10, 65, 400 and 1200 ns in pulse width, respectively.

Figure 7 shows the decolorization rate as a function of treatment time for various pulse widths at initial conductivity of 7000 μS/cm. The energies per a pulse are 6.14, 22.9, 56.0 and 160 mJ at 10, 65, 400 and 1200 ns in pulse width, respectively. The treatment times of 100% decolorization are 4.5 and 4 min at 10 and 65 ns in pulse width, respectively. These values are lower than that at 6 μS/cm. On the other hand, the treatment time of 100% decolorization at 7000 μS/cm is much larger than that at 6 μS/cm in the case of 400 and 1200 ns pulse widths. The energy efficiencies of 50% decolorization are 55.6, 38.1, 4.32 and 0.680 mg/Wh at 10, 65, 400 and 1200 ns in pulse width, respectively.

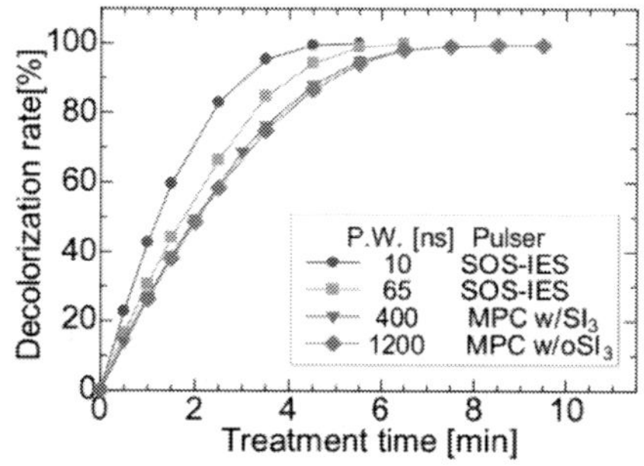

Figure 6. Decolorization rate as a function of treatment time for various pulse widths at 6 μS/cm.

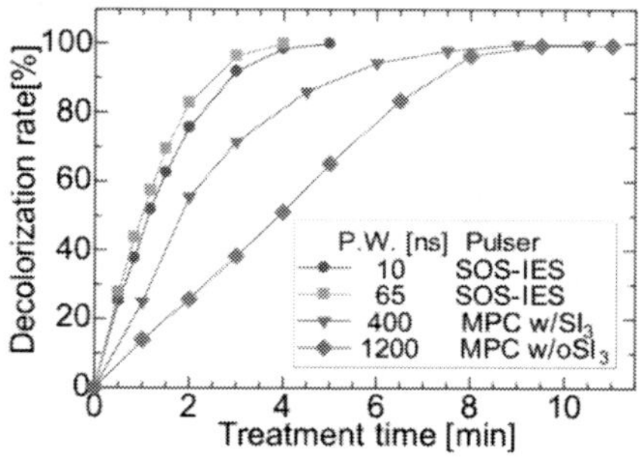

Figure 7. Decolorization rate as a function of treatment time for various pulse widths at 7000 μS/cm.

Figure 8 shows the decolorization rate as a function of initial conductivity for various pulse widths. The decolorization rate at 6 μS/cm is controlled to be 22% by changing the treatment time. The treatment time t_t are 30, 40, 60 and 60 s at 10, 65, 400 and 1200 ns in pulse width, respectively. The decolorization rate increases with increasing conductivity at 10 ns in pulse width. On the other hand, the decolorization rate decreases rapidly at 2000 and 1000 μS/cm conductivity in the case of 400 and 1200 ns in pulse width, respectively. Figure 9 show the energy efficiency for decolorization as a function of conductivity in solution for various pulse widths. The energy efficiency for decolorization increases with decreasing pulse width.

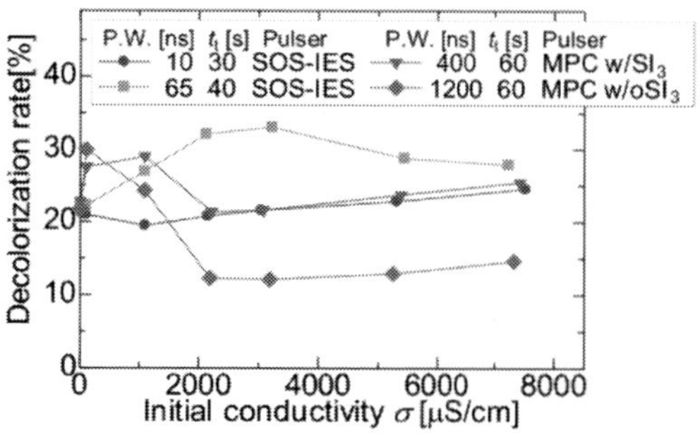

Figure 8. Decolorization rate as a function of initial conductivity for various pulse widths.

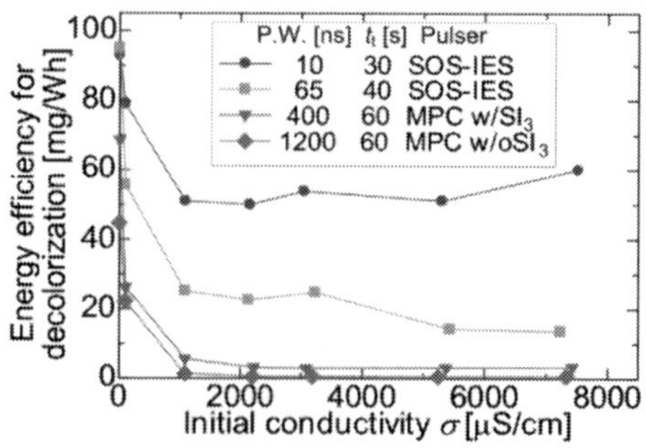

Figure 9. Energy efficiency as a function of initial conductivity for various pulse widths.

4. Discussion

Indigo carmine has high reactivity with ozone [13]. Active species such as ozone (O_3) and oxygen atom (O) are generated by discharge inside air bubbles through the reactions and reaction rates in following equations (3) and (4) [14-16],

$$O_2 + e \rightarrow O + O + e;\ k_3 = 2.1 \times 10^{-9}\ \mathrm{cm^3/s}, \tag{3}$$

$$O + O_2 + M \rightarrow O_3 + M;\ k_4 = 2.5 \times 10^{-35} \exp(970 / T)\ \mathrm{cm^6/s}, \tag{4}$$

where M is the third collision partner and T is the temperature of gas; M takes part in energy absorption, but does not react chemically. Reaction of equation (3) increases with increasing discharge energy [12]. Reaction of equation (4) decreases by increasing T. In figure 8, decreasing the decolorization rate in the condition of the pulse width of 65, 400 and 1200 ns is caused by decreasing amount of ozone generation due to increasing temperature of gas. In addition, ozone is consumed by various reactions because of its high reactivity, as shown in following equations (5) and (6) [13, 5-17],

$$O_3 + O(^1D) \rightarrow O_2 + O_2;\ k_5 = 1.5 \times 10^{-11} \exp(-2218 / T)\ \mathrm{cm^3/s}, \tag{5}$$

$$O_3 + O \rightarrow O_2 + O_2;\ k_6 = 2.23 \times 10^{-12} \exp(-1580 / T)\ \mathrm{cm^3/s}, \tag{6}$$

$$O_3 + {\cdot}OH \rightarrow HO_2 + O_2^-;\ k_7 = 3.76 \times 10^{-13} (T / 298)^{1.98} \exp(-603 / T)\ \mathrm{cm^3/s}. \tag{7}$$

Reactions of equations (5-7) increase with increasing temperature of gas and discharge. Amount of ozone generation is decreased by increasing temperature of gas and discharge.

Figure 10 shows concentration of ozone exhausted from the reactor as a function of conductivity in the solution. The purified water is used instead of the indigo carmine solution. The ozone concentration increases by early increase conductivity, in the case of the pulse width of 10 and 65 ns. On the other hands, it drastically decreases at 1000 μS/cm conductivity in the case of 400 and 1200 ns in pulse width. This decrease is similar to the decrease of the decolorization rate at 400 and 1200 ns in pulse width as shown in figure 8. The ozone concentration at 65 ns in pulse width is higher than 10 ns, because energy per a pulse in 65 ns is higher than 10 ns, for example energy per a pulse in 65 and 10 ns at 7000 μS/cm are 6.14 and 22.9 mJ, respectively. And temperature of gas and discharge increase to 1000 K from 450 K at 10 and 65 ns in pulse width [18, 19]. Therefore, reactions of equations (5-7) hardly increase. In figure 8, the decolorization rate at 65 ns in pulse width is higher than 10 ns. It is same reason in the case of the ozone concentration.

Figure 11 shows the normalized energy per a pulse as a function of initial conductivity for various pulse widths. The normalized energy is obtained as energies divided by the energy of 6 μS/cm in conductivity. The normalized energy increases with increasing pulse width. The streamer discharge propagates from the tip of wire electrode to the surface of bubble [20]. After the streamer discharge reach to surface of the bubble, the output current increases rapidly because of thermalization of the discharge plasma. The amplitude of the discharge current after the plasma thermalization increases

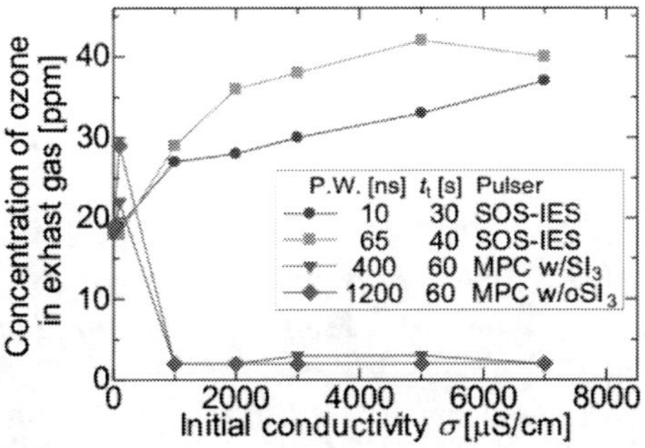

Figure 10. Concentration of ozone in exhaust gas as a function of initial conductivity for various pulse widths.

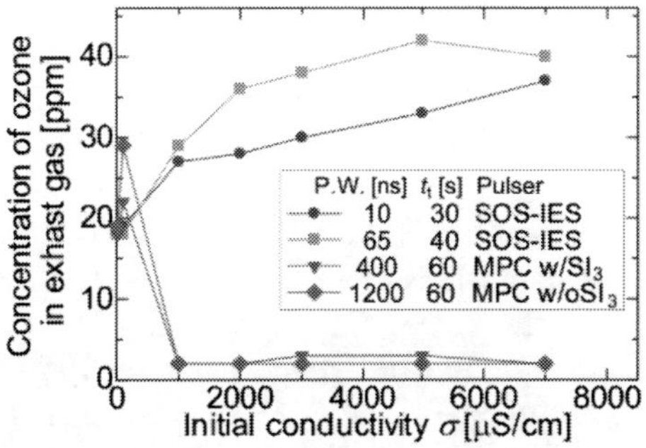

Figure 11. Normalized energy per a pulse as a function of initial conductivity for various pulse widths.

with conductivity because the discharge current is limited by resistance of the water [21, 22]. The consumed energy can also be reduced by shorten the pulse width. As the results, the energy efficiency for decolorization is strongly affected by not only water conductivity but also the pulse width.

5. Conclusion

The water decolorization was carried out at four different pulse widths using two types of the pulse generator. The indigo carmine solutions were decolorized completely by discharge. The time demand for complete decolorization of the dye decreased with decreasing pulse width. The energy efficiency was improved by decreasing pulse width. The short pulse width was effective for high conductive water treatment.

Acknowledgments

The authors are grateful to Dr. Shoji Koide, Faculty of Agriculture, Iwate University and Mr. Ippei Yagi, The University of Tokyo. This work was supported by Grants-in-Aid for Scientific Research (JSPS fellowship No. 20380130).

References

[1] Vandevivere P C, Bianchi R and Verstraete W 1998 *J. Chem. Thechnol. Biotechnol.* **72** pp 289-302
[2] Richardson S D, Thruston A D, JR, Caughran T V, Chen P H, Collette T W, Schenck K M, Lykins B W, JR, Rav-Acha C, Glezer V 2000 *Water Air Soil Poll.* **123** pp 95-102
[3] Gunten U V, Hoigné J 1994 *Environ. Sci. Technol.* **28** pp 1234-1242
[4] Akiyama H, Sakugawa T, Namihira T, Takaki K, Minamitani Y and Shimomura N 2007 *IEEE Trans. Diel. Elect. Insul.* **14** pp 1051-1063
[5] Sato M, *Plasma Sources Sci. Technol.* 2008, **17**, 024021 pp 1-7
[6] Akiyama H. *IEEE Trans. Diel. Elect. Insul.* 2000 **7** pp 646-653
[7] Takahashi K, Takaki K and Satta N 2012 *J. Adv.Oxidat.* **15** 2 pp 365-373
[8] Clements J S, Sato M and Davis R H 1987 *IEEE Trans. Ind. Appls.* **IA-23** 2 pp 224-235
[9] Lukes P, Clupek M, Babicky V and Sunka P 2008 *Plasma Sources Sci. Technol.* **17** 024012 pp 1-11
[10] Dang T H, Denat A, Lesaint O and Teissedre 2008 *Plasma Sources Sci. Technol.* **17** 024013 pp 1-8
[11] Malik M A 2010 *Plasma Chem. Plasma Proc.* **30** pp 21-31
[12] Matsumoto T, Wang D, Namihira T and Akiyama H 2010 *IEEE Trans. Plasma Sci.* **38** 1 pp 2639-2643
[13] Criegee R 1975 *Angew. Chem. Internat. Edit.* **17** 11 pp 745-752
[14] Chang J S, Lawless A Phill and Yamamoto T 1991 *IEEE Trans. Plasma Sci.* **19** 6 pp 1152-1166
[15] Yanallah K , Ziane S H, belasri A and Meslem Y 2006 *J. Mol. Struct.* **777** pp 125-129
[16] National Institute of Standards and Technology last modified February 2012 http://kinetics.nist.gov/kinetics/index.jsp
[17] Locke B R and Thagard S M 2009 *IEEE Trans. Plasma Sci.* **37** 4 pp 494-501
[18] Namihira T, Wang D, Matsumoto T, Okada S and Akiyama H 2009 *IEEE Trans. FM* **129** 1 pp 7-14
[19] Kawano S, Takahashi K, Takaki K, Satta N 2012 *J. Inst. Electrostat. Jpn.* **36** 1 pp 43-49
[20] Takahashi K, Yagi I, Takaki K and Satta N 2011 *IEEE Trans. Plasma Sci.* **39** 11 pp 2654-2655
[21] Yoshinaga K, Okada S, Wang D, Namihira T and Akiyama H 2009 *the 2nd Euro-Asian Pulsed Power Conference* **115** 6 pp 1050-1052
[22] Bruggeman P, Ribežl E, Maslani A, Degroote J, Malesevic A, Rego R and Leys C 2008 *Plasma Sources Sci. Technol.* **17** 025012 pp 1-10

Pulsed Discharge Plasma over a Water Surface Induces Decoloration of Dyes

Wahyudiono[1], **Siti Machmudah**[1,2], **Motonobu Goto**[1]
[1]Department of Chemical Engineering, Nagoya University
Furo-cho, Chikusa-ku, Nagoya 464-8603, Japan
[2]Department of Chemical Engineering, Sepuluh November Institute of Technology
Kampus ITS Sukolilo, Surabaya 60111, Indonesia

Tel: +81-52-789-3392 Fax: +81-52-789-3392
E-mail: mgoto@nuce.nagoya-u.ac.jp

Abstract. It was well known that plasma can be defined as a partially ionized gas composed of ions, electrons and neutral species. Recently, plasma-water interaction has attracted growing interest as it may provide experimental chemists with a quite unique reaction medium. In this work, decoloration of dyes in water as a reaction media and the effects of various parameters with pulsed high-voltage discharge plasma are studied. Such as plasma applied under hydrothermal conditions generates high-energy electrons, ions, and radicals, which in turn may generate new reaction fields, leading to effective organic compounds oxidation for both homogeneous and heterogeneous reactions. Here, we utilize pulse discharge plasma over water surface to study the decoloration of Orange G, Orange II, Congo Red, and Naphthol Blue Black. They were directly fed as starting materials without additives. The experiments were conducted under argon atmospheric at 313 K using a batch type reactor. The products were directly analyzed by UV-Vis (ultra violet-visible) spectrophotometer V-550. The decoloration rate increased with increasing peak pulse voltage and pulse numbers, presumably due to the increased electric field energy. Based on these results, the present system may be promising.

1. Introduction

Dyes or synthetic dyes make our world beautiful, but they bring pollution. Although, the greatest environmental polluter is the chemical industry, of which only a relatively small part pertains to the organic colorants industry, their degradation in wastewater have always been of great interest to the field of environmental protection. Therefore, the removal of these dyes from effluents becomes the major environmental problem of the multiproduct industry [1-5]. Generally, dyes containing wastewaters can be treated in two ways [6]: (i) by chemical or physical methods of dye removal, which refers to the process called decoloration and (ii) by a biodegradation process.

Several advanced oxidation technologies have been utilized for the purpose of degrading organic pollutants in wastewater, such as UV photolysis, photocatalysis, sonochemistry, supercritical water oxidation, electrical discharge plasma technology and electron-beam irradiation [7-13]. In this work, the pulsed discharge plasma over a water surface is considered to be an applicable method for removal of organic pollutants from wastewater [14-15]. In general, several kinds of inorganic salts or other chemicals were added on the decolorization of dyes using advanced oxidation processes [1,16-18]. Here, the synthetic wastewater was directly fed as a starting material without chemical reagents. It is simple and versatile technique that has gained much attention because of its capability and feasibility

Journal of Physics: Conference Series **441** (2013) 012008	doi:10.1088/1742-6596/441/1/012008

in the decomposition of large quantities of wastewater from industries. The utilization of pulsed discharge plasma has demonstrated to be effective at degrading aromatic compounds in aqueous solutions as it not only produces a hydroxyl radical but also atomic species with high oxidation potential [19-21], which in turn become important reactants. These active species include hydrogen peroxide, hydrogen, and oxygen, along with other oxidative species such as hydroperoxyl radicals and reductive species such as hydrogen radicals, superoxide radical anions, and aqueous electrons. The OH radical is particularly known to play an important role in the oxidative purification of water and the degrading of organic compounds [22]. Sugiarto et al. [1] and Sun et al. [23] observed also that the streamer produced by pulsed discharge could induce decomposition of H_2O_2 and produce OH radical, which led to the increase of the degradation efficiency of organic compounds.

Such as pulsed discharge plasma applied to degrade the organic compounds, the degradation of dyes during an electrical discharge in water also agree that OH radicals are the major species responsible for decolorization and the extent to which a dye will be degraded depends on the reactivity between OH radicals and the dye functional groups and substitutes [1-3,16,18,23-29]. When the discharge plasma was conducted on the water surface, the active species products will diffuse in the gas phase and dissolve in the water through the surface layer to react with the organic materials in water. Sugiarto et al. [1,2] and Wang et al.[4] explained that OH radicals played a key role in pulsed discharge systems for decoloration processes of Rhodamine B, Methyl orange, and Chicago sky blue, reacts readily with the carbonate ions generated through the breakdown of the organic compounds, but does not react with organic compounds, or their intermediates during the oxidation processes under the basic solution condition. Sunka et al. [29] investigated decolorization of C.I. Reactive Blue 137 (50 mgL^{-1}) by using corona discharge reactor. They reported that decolorization of reactive dye solution is very low, while the addition of Fe ions in the system improved color removal and in 5 minutes of treatment the complete color removal was obtained. Recently, Sun et al. [30] also reported that the active species generated by discharge in the gas phase include ozone (O_3), OH radical, peroxide (O_2^-) and atomic oxygen (O) can dissolve into the water solution and form hydrogen peroxide in the solution. They found that the concentration of hydrogen peroxide in the solution increased when the O_3 concentration in the gas phase increased. Consequently, these active species generated at the gas-liquid interface helped the rapid decoloration of the organic dye. However, as an explained above, the additives was still used in their decoloration process.

In the present study, experiments have been performed for degrading dyes as organic pollutants using pulsed discharge plasma under argon atmospheric pressure. Orange G, Orange II, Congo Red, and Naphthol Blue Black were selected as the model organic contaminants, which were dissolved in distilled water to form a synthetic wastewater. Under argon atmosphere, pulse discharge plasma might serve as an intense source of ultraviolet radiation [31-35]. Argon is one of the most studied and utilized chemical elements, and dense argon is a promising reaction media alternative. Like carbon dioxide, argon possesses easily obtainable critical points and it is also non-polluting (T_c = 150.9 K, P_c = 4.9 MPa) [36-38]. As reported by Bogaerts et al. [32], when a potential difference is applied between two electrodes under argon atmospheric, the argon gas will break down into electrons and positive ions. This ionization will create new electrons and ions, therefore, the self-sustaining plasma of glow discharge occurred. Mishra et al. [35] found that the relative yield of hydrogen gas from methane increases in the presence of argon. It is suggested that argon could act as a promoter and also had certain influences on the density and energy distribution of electrons, and thus assist the production of various active species during the formation of plasmas. In addition, argon has also catalytic effects on the production of ozone [33].

2. Experimental

The dyes Orange G ($C_{16}H_{10}N_2Na_2O_7S_2$; 80%), Orange II ($C_{16}H_{11}N_2NaO_4S$; 92%), Congo Red $C_{13}H_{22}N_6Na_2O_6S_2$; 85%), and Naphthol Blue Black ($C_{22}H_{14}N_6Na_2O_9S_2$; 85%) commercial products purchased from Sigma Chemical Co. USA, and used without further purification. The synthetic wastewater was prepared by dissolving of each dye with distilled water at 1000 mg/L as an initial

Journal of Physics: Conference Series **441** (2013) 012008 doi:10.1088/1742-6596/441/1/012008

concentration. As a reaction media, water conductivity plays an important role in the generation of arch discharge plasma and on the production of chemically active species [23,39,40].

The reactor contained two inspection windows to monitor the performance of plasma production and was made of stainless steel (SUS-316, AKICO, Japan). The detailed experimental setup and procedure have been described extensively elsewhere [14]. Therefore, the experimental procedure is briefly described here. To begin the experiment, feed containing 1000 mg/L of dye (12 mL) was loaded to the reactor manually. The reactor was repeatedly (three times) purged of air using argon gas with a pressure of about 0.2 MPa; it was then heated up to 313 K.

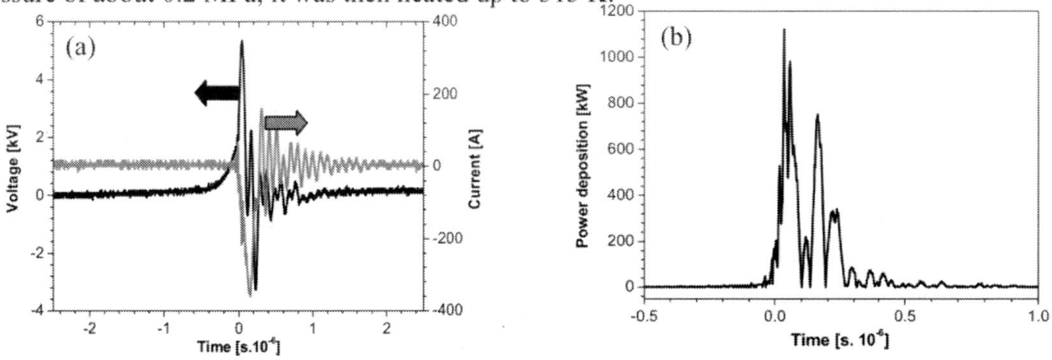

Figure 1. (a). Typical waveform by MPC under argon atmosphere at 0.2 MPa, 313 K with 5 kV peak pulse voltage applied. Black is voltage; grey is current. (b). Electric power deposition at 5 kV peak pulse voltage applied under argon atmosphere at 0.2 MPa, 313 K.

Figure 1(a) shows the waveform profile of pulsed voltage and current under argon atmosphere. To evaluate the energy deposition process of the pulsed discharge plasma, the voltage across the electrode and the discharge current were measured. After the discharge initiation, an electric discharge pulse appears immediately. The energy deposited during one discharge pulse is estimated from the measured voltage and current. Figure 1(b) showed the energy deposition waveform defined by the curve $P = VI$, where P is power (W), V is voltage (V), and I is current (A). After the reaction time had elapsed, the electric current was cut off and the reactor was opened. The products were directly identified by UV-Vis spectrophotometer V-550. The maximum absorbance wavelength (λ max) of Orange G, Orange II, Congo Red, and Naphthol Blue Black are 476, 486, 499, and 620 nm, respectively. The conductivity and pH of the sample liquid was measured using a conductivity meter (ECTestr 11, EUTECH Instruments, Singapore) and pH meter (Compact pH Meter AS-211, Horiba. Ltd, Japan). At 1000 mg/L, the conductivity of Orange G, Orange II, Congo Red, and Naphthol Blue Black were 280, 240, 310, and 290 μS/cm, respectively. On average, the initial pH of them was 9.15, 7.60, 8.18, and 9.40, respectively.

3. Results and Discussion

Generally, the colour of water which polluted with organic colourants could be reduced by the cleavage of the -C = C- bonds, the -N = N- bonds and heterocyclic and aromatic rings occurs. These bonds are intentionally designed to resist biodegradation, i.e. the wastewater originated from reactive dye processes is characterized by poor biodegradability, passing unaffected through conventional treatment systems and discharging into the environment. The absorption of light by the associated molecules shifts from the visible region to the ultraviolet region or infrared region of the electromagnetic spectrum [6]. To estimate the decoloration of dyes after treatment by pulse discharge plasma under argon atmospheric, the liquid products were analyzed by UV-Vis spectrophotometer; this is a useful tool for studying decoloration of dyes, contains, e.g. ethylene group, amine group and its derivatives, since many intermediate products as well as of the resulting products are known to be strongly colored. Figure 2 showed some results for dyes conversion when the pulse discharge plasma was applied. At the same numbers of discharge plasma, the conversion of dyes increased significantly with increasing the peak pulse voltages applied. These figures depicted also that the dyes degradation efficiency increases with increasing peak pulse voltage applied and discharge numbers of plasma. Pulse discharge plasma can be conveniently grouped into localized and extended effects depending on the region. When pulse discharge plasma occurs, the following individual effects occurred simultaneously: over pressure shock wave, strong electrical field, production of various free radicals, and intense ultraviolet radiation [23]. In this case, the arc discharge directly contacted over the water surface where the streamer channel propagates along the surface of the water. They will cleavage active side on the dye structure especially amine group which promoted by the various reactive chemical species such as radicals (OH, H, O, HO_2) and molecular species (H_2O_2, H_2, O_2). With increasing voltage, the electric field strength between the electrode increases resulting in more energetic electrons which dissociate more efficiently water, gas and dye molecules. Therefore, at higher voltage, the discharge produces more active species which enhance the oxidization of dyes yielding lower molecular weight products. It was found also that the electrons wherein the zone have more energy and bombard the water molecules and dye molecules with higher speeds. Wang et al. [4] suggested that the decoloration rate of azo dye increased with increasing peak pulse voltage, presumably due to the increased electric field energy. Based on these results, the dyes degradation increases with increasing peak pulse voltage which leads to the formation of its derived compounds.

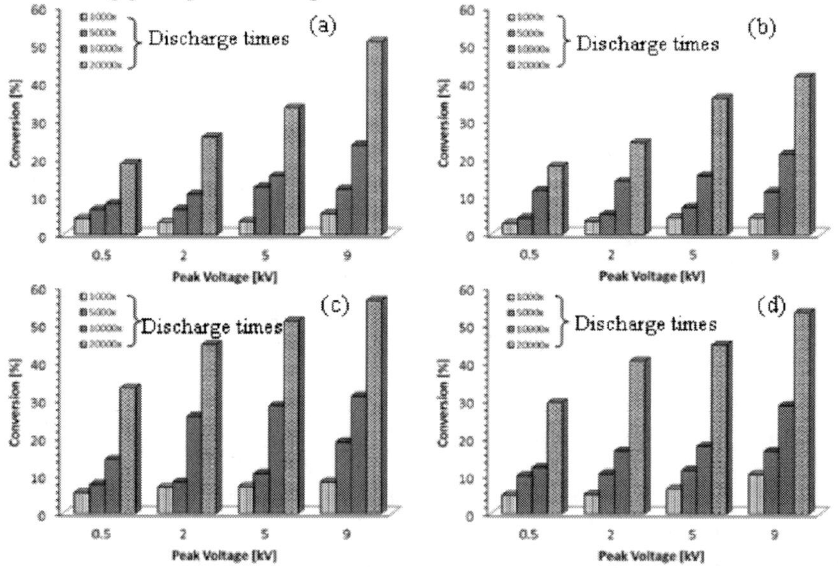

Figure 2. Conversion of dyes after treatment by various peak pulse voltage under argon atmosphere at 313 K (a. Orange G; b. Orange II; c. Congo Red; d. Naphthol Blue Black).

In this work, decoloration of several kinds of dyes has been tested using the pulsed discharge plasma over water surface. The results obtained after treatment with different pulse discharge times are shown in figure 3. In this case, the electric fields which generated by pulse discharge plasma over water surface initiate both chemical and physical processes. The intense electrical discharge disassociates water molecules to yield active radicals, due to collisions of high energy electrons with molecules. These radicals will combine with almost dye compounds and/or dye derived compounds in a very efficient manner. The chemical reactions induced by pulse discharge plasma over water surface are dependent on several factors. For example, the aqueous solution composition can affect the reactions through the presence of electrolytes and radical quenchers. Conductivity affects the electrical discharge over water surface, leading to lower rates of formation of active species, but higher rates of formation of UV light [41].

As shown in the figure 3, the pulse discharge numbers had significant effect on the decoloration rate of dyes. To examine this effect, the experiments were conducted in widely range (1000-20000x discharge numbers). At each reaction conditions, the decoloration rate significantly increased with discharge numbers. For example, the decoloration rate Naphthol Blue Black was around 10% at 1000x discharge times and increased gradually to 53% at 20000x discharge times when the peak pulse voltage applied was 9 kV. In this case, the energy input to the electric field is directly proportional to the numbers of discharge plasma. It could increase the quantity and energy of the high-energy electron. More frequent and energetic electron bombardment with the water molecule can create more active species that would attack dye to its derived compounds. Due to the transfer of the dye compounds from the liquid phase to the gas phase where they can react with the active species of the discharge, especially radicals. Therefore, increasing the number of discharge plasma is useful for dyes decomposition.

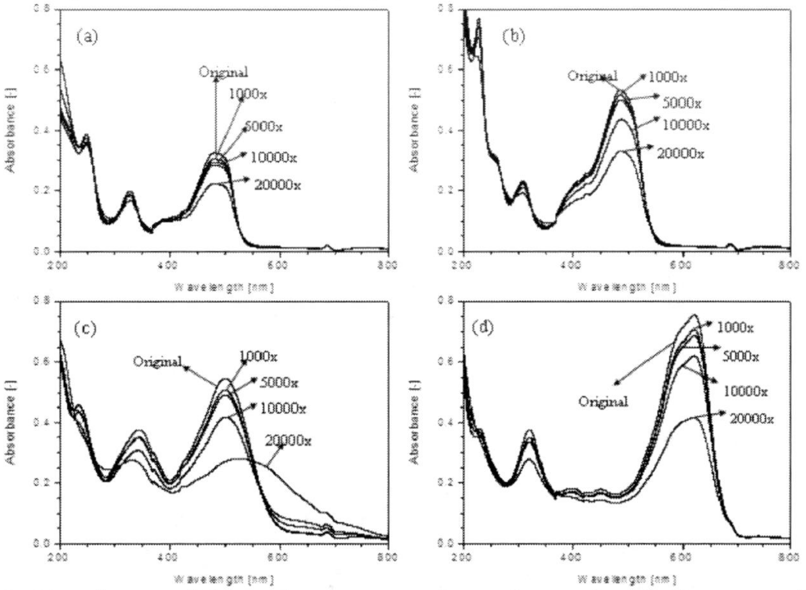

Figure 3. Absorbance of dyes after treatment by various pulse discharge numbers at 313 K with 5 kV peak voltage (a. Orange G; b. Orange II; c. Congo Red; d. Naphthol Blue Black).

Table 1 showed liquid conductivity of feed solutions after treatment by pulse discharge plasma with 5 kV peak voltage. It can be shown that solution conductivity increases greatly after pulse discharge plasma treatment. As reported by Wen et al. [42], many active species, such as •OH, •O, •H, $\cdot HO_2$, and O_2^-, etc., were produced during pulse discharge plasma process in distilled water as reaction media. These active species can react with dye compounds to result in some conductive products which increase the conductivity of solution. The increasing the conductivity of the solution will lead to decrease streamer formation and lower quantities of reactive radicals [23]. As shown in the table 1 that the pH values of solution products decreased significantly due to the pulsed discharge plasma treatment. The pH of the solution is one of important parameters because it can vary the surface charge of the electrode and also shifts the potential of some redox reaction [1-3,43]. As an explained before, the conductivity solution products largely increased with the discharge times of plasma. The observed phenomenon of the dramatic decrease of pH and increase of conductivity corresponds to the several literatures where the active species generated through electrical discharge during treatment process [1-3,23,43,44]. Sugiarto et al. [1] conducted degradation of organic dyes by the pulsed discharge plasma. They reported that the dye molecules decompose into organic acids, aldehydes, resulting in a decrease in pH values and an increase in water conductivity. Mok et al. [3] also reported that the degradation products from Orange II using a gas phase dielectric barrier discharge reactor in water like organic acids could decrease the pH value and increase the solution conductivity.

To evaluate the performance of the pulse discharge plasma systems, the energy yield was estimated [45-48]. Table 1 showed the energy yield values of dye molecules degradation after treatment by pulse discharge plasma. They were changing as a pulse discharge times even under the same reaction condition. The energy yield decreased significantly with increasing the pulse discharge times. These observations are in good agreement with the literature [45-48]. Kim et al. [46] reported that the best results are found in the small specific input energy region where the removal rate is low. They also pointed out that the nature of the energy yield was determined from the relationship between the removed amount of molecules and the specific input energy.

Table 1. Calculated value of energy yield and change of pH and conductivity of the feed solutions after treatment by various pulse discharge numbers with 5 kV peak voltage.

Feed	Discharge times [-]	pH	Conductivity (μS/cm)	Energy yield, G (mg/mJ)	Degradation [wt %]
Orange G	1000	6.74	340	2.81	3.44
	5000	6.04	480	0.51	12.51
	10000	4.07	530	0.25	15.55
	20000	3.42	840	0.10	33.42
Orange II	1000	6.32	280	2.78	4.36
	5000	6.00	310	0.54	7.13
	10000	4.74	540	0.25	15.47
	20000	3.05	750	0.09	36.00
Congo Red	1000	6.52	290	2.70	7.02
	5000	6.12	390	0.52	10.50
	10000	4.10	510	0.21	28.51
	20000	2.02	790	0.07	50.86
Naphthol Blue Black	1000	6.90	290	2.71	6.53
	5000	6.08	320	0.51	11.46
	10000	5.78	690	0.24	17.84
	20000	3.42	900	0.08	44.68

4. Conclusions

The experiment of dyes degradation by pulsed high-voltage discharge plasma over water surface at 313 K, 0.2 MPa argon atmospheres in a batch-type reactor has been performed. Under these conditions, dyes were degraded into its derived compounds with the appearance of light color. UV-Vis spectrophotometer analyzed that the intermediate compounds from the degradation of dyes consist primarily of aromatic compounds which contain nitrogen functional groups. At each reaction conditions, the decoloration rate of Orange G, Orange II, Congo Red, and Naphthol Blue Black significantly increased with peak pulse voltage applied and discharge numbers. The maximum degradation of them was 50.05%, 41.64%, 44.98%, and 53.25%, respectively. With increasing the pulse discharge times, the pH and the conductivity of dye solutions changed clearly. Based on these results, it is proposed that this method is applicable as a medium for dyes degradation despite the stability of the compound and its inability to be easily cleaved under these conditions.

5. Acknowledgement

This work was supported by the Grants-in-Aid for Scientific Research by the Ministry of Education, Culture, Sports, Science and Technology, Japan.

6. References

[1] Sugiarto A T, Ito S, Ohshima T, Sato M and Skalny J D 2003 *J. Electrost.* **58** 135
[2] Sugiarto A T, Ohshima T and Sato M 2002 *Thin Solid Films* **407** 174
[3] Mok Y S, Jo J O and Whitehead J C 2008 *Chem. Eng. J.* **142** 56
[4] Wang H, Li J and Quan X 2006 *J. Electrost.* **64** 416
[5] Yan J H, Liu Y N, Bo Z, Li X D and Cen K F 2008 *J. Hazard. Mater.* **157** 441
[6] Slokar Y M and Marechal A M L 1998 *Dyes and Pigments* **37** 335
[7] Chun H, Wang Y Z and Tang H X 2000 *Chemosphere* **41** 1205
[8] Antonaraki S, Androulaki E, Dimotikali D, Hiskia A and Papaconstantinou E 2002 *J. Photochem. Photobiol.A: Chem.* **148** 191
[9] Doong R A, Chen C H, Maithreepala R A and Chang S M 2001 *Water Res.* **35** 2873

[10] James C W and Thomas J 1995 *Waste Manage.* **15** 303
[11] Giovanni D R and Gabriele D G 2001 *Desalination* **138** 61
[12] Malik M A, Ghaffar A and Malik S A 2001 *Plasma Sources Sci. Technol.* **10** 82
[13] Lin K J, Cooper J W, Nickelsen M G, Kurucz C N and Waite T D 1995 *Appl. Radiat. Isot.* **46** 1307
[14] Wahyudiono, Watanabe H, Machmudah S, Kiyan T, Sasaki M, Akiyama H and Goto M 2012 *Chem. Eng. Proc. Process Intens.* **61** 51
[15] Sharma A K, Locke B R, Arce P and Finney W C 1993 *Hazard. Waste Hazard. Mater.* **10** 209
[16] Li J, Wang T, Lu N, Zhang D, Wu Y, Wang T and Sato M 2011 *Plasma Sources Sci. Technol.* **20** 034019 doi:10.1088/0963-0252/20/3/034019
[17] Dojcinovic B P, Roglic G M, Obradovic B M, Kuraica M M, Kostic M M, Nesic J and Manojlovic D D 2011 *J. Hazard. Mater.* **192** 763
[18] Jiang B, Zheng J, Lu , Liu Q, Wu M, Yan Z, Qiu S, Xue Q, Wei Z, Xiao H and Liu M 2013 *Chem. Eng. J.* **215-216** 969
[19] Chen Y S, Zhang X S, Dai Y C and Yuan W K 2004 *Sep. Pur. Technol.* **34** 5
[20] Sahni M and Locke B R 2006 *Ind. Eng. Chem. Res.* **45** 5819
[21] Wahyudiono, Watanabe H, Kiyan T, Akiyama H, Sasaki M and Goto M 2011 *Trans. Mater. Res. Soc. Japan* **36** 459
[22] Akiya N and Savage P E 2002 *Chem. Rev.* **102** 2725
[23] Sun B, Sato M and Clements J S 1997 *J. Electrost.* **39** 189
[24] Stara Z, Krcma F, Nejezchleb M and Skalny J D 2009 *Desalination* **239** 283
[25] Maehara T, Miyamoto I, Kurokawa K, Hashimoto Y, Iwamae A, Kuramoto M, Yamashita H, Mukasa S, Toyota H, Nomura S and Kawashima A 2008 *Plasma Chem. Plasma Process* **28** 467
[26] Yano T, Shimomura N, Uchiyama I, Fukawa F, Teranishi K and Akiyama H 2009 *IEEE Trans. Dielectr. Electr. Insul.* **16** 1081
[27] Selma M and Takashima K 2008 *IJPEST* **2** 56
[28] Bozic A L, Koprivanac N, Sunka P, Clupek M and Babicky V 2004 *Czech. J. Phys.* **54** C958
[29] Sunka P, Babicky V, Clupek M, Fuciman M, Lukes P, Simek M, Benes J, Locke B R and Majcherova Z 2004 *Acta Physica Slovaca* **54** 135
[30] Sun B, Aye N N, Gao Z, Lv D, Zhu X and Sato M 2012 *J Environ. Sci.* **24** 840
[31] Kosinskaya I V and Polozova L P 1969 *Zhurnal Prikladnoi Spektroskopii* **11** 1151
[32] Bogaerts A, Neyts E, Gijbels R and van der Mullen J 2002 *Spectrochimica Acta Part B* **57** 609
[33] Lukes P, Clupek M, Babicky V, Janda V and Sunka P 2005 *J. Phys. D: Appl. Phys.* **38** 409
[34] Sato M, Soutome T, Mii S, Ohshima T and Yamada Y 2007 *IJPEST* **1** 71
[35] Mishra└ L N, Shibata└ K, Ito H, Yugami N and Nishida Y 2004 *J. Natural Gas Chemist.* **13** 82
[36] Tegeler C, Span R and Wagner W 1999 *J. Phys. Chem. Ref. Data* **28** 779
[37] Liang S and Tilotta D C 1998 *Anal. Chem.* **70** 616
[38] Roder H M, Perkins R A and Nieto de Castro C A 1989 *Int. J. Thermophy.* **10** 1141
[39] Clements J S, Sato M and Davis R H 1987 *IEEE Trans. Ind. Appl.* **1A-23** 224
[40] Sunka P, Babicky V, Clupek M, Lukes P, Simek M, Schmidt J and Cernak M 1999 *Plasma Sources Sci. Technol.* **8** 258
[41] Grymonpre D R, Sharma A K, Finney W C and Locke B R 2001 *Chem. Eng. J.* **82** 189
[42] Wen Y Z, Liu H J, Liu W P and Jiang X Z 2005 *Plasma Chem. Plasma Process.* **5** 137
[43] Zhang R, Zhang C, Cheng X X, Wang L, Wu Y and Guan Z 2007 *J. Hazard. Mater.* **142** 105
[44] Bian W, Zhou M and Lei L 2007 *Plasma Chem. Plasma Process.* **27** 337
[45] Malik M A 2010 *Plasma Chem. Plasma Process.* **30** 21
[46] Kim H H, Prieto G, Takashima K, Katsura S and Mizuno A 2002 *J. Electrost.* **55** 25
[47] Wang H and Chen X 2011 *J. Hazard. Mater.* **186** 1888
[48] Hoeben W F L M, van Veldhuizen E M, Rutgers W R and Kroesen G M W 1999 *J. Phys. D: Appl. Phys.* **32** 133

Journal of Physics: Conference Series **441** (2013) 012009 doi:10.1088/1742-6596/441/1/012009

Transport characteristics of $N_2(A^3\Sigma_u^+)$, $N(^4S^o)$, and $N(^2D^o)$ in a remote nitrogen plasma source

Y Horikawa[1], **T Hayashi**[2], **and K Sasaki**[3]

[1]Department of Quantum Science and Engineering, Graduate School of Engineering, Hokkaido University, Sapporo 060-8628, Japan
[2]Plasma Nanotechnology Research Center, Nagoya University, Nagoya 464-8603, Japan
[3]Division of Quantum Science and Engineering, Faculty of Engineering, Hokkaido University, Sapporo 060-8628, Japan

E-mail: horikawa@athena.qe.eng.hokudai.ac.jp

Abstract. We measured absolute densities of $N_2(A^3\Sigma_u^+)$, $N(^4S^o)$, and $N(^2D^o)$ in the downstream region at various distances from a microwave N_2 plasma source. As a result, it was found that the transport of $N(^4S^o)$ was highly efficient with keeping the densities on the order of 10^{17} m^{-3} at distances up to 12 cm from the discharge region. On the other hand, the densities of $N_2(A^3\Sigma_u^+)$ and $N(^2D^o)$ decayed from 10^{16} to 10^{14} m^{-3} when the distance between the discharge region and the measurement position was lengthened from 6 to 12 cm. The decay curves of the $N_2(A^3\Sigma_u^+)$ and $N(^2D^o)$ densities were different each other. The densities of $N_2(A^3\Sigma_u^+)$ and $N(^2D^o)$ in the downstream region were not owing to the transports but were due to their productions. We discussed the dominant production processes of $N_2(A^3\Sigma_u^+)$ and $N(^2D^o)$ on the basis of the experimental observations.

1. Introduction

Plasma-based reactive ion etching (RIE) [1] is a key technology in the fabrication of ultralarge-scale integrated circuits. However, RIE sometimes causes serious problems in the processing step where plasma-induced damage must be avoided. A mechanism of plasma-induced damage is the bombardment of energetic ions onto the wafer [2]. Using a remote plasma source is a conventional method for avoiding plasma-induced damage caused by ion bombardment. Recently, a new method for damage-less dry etching of silicon using a remote plasma has been reported [3]. In this method, NF_3 is fed into the downstream region of a remote nitrogen plasma source, where the density and the energy of ions are much lower than those in an active plasma. The dry etching of silicon indicates that NF_3 is dissociated into F atoms in the downstream region of the N_2 remote plasma. A question here is the dissociation mechanism of NF_3. A molecular orbital calculation suggests that atomic nitrogen at the metastable $^2D^o$ state and molecular nitrogen at the metastable $A^3\Sigma_u^+$ state can dissociate NF_3 into F atoms, while the collision with atomic nitrogen at the ground state ($^4S^o$) does not result in the production of F atoms [3]. However, the transport characteristics of the metastable states to the downstream region of a remote plasma source have not yet been examined.

In this paper, we measured the densities of $N_2(A^3\Sigma_u^+)$, $N(^4S^o)$, and $N(^2D^o)$ in the downstream region of a remote N_2 plasma source. To compare the absolute densities of the above species, we employed cavity ringdown spectroscopy (CRDS) for measuring the $N_2(A^3\Sigma_u^+)$ density and vacuum ultraviolet absorption spectroscopy (VUVAS) for measuring the $N(^4S^o)$ and $N(^2D^o)$ densities. The measurements were carried out at various distances from the discharge region. The experimental results give us useful

IOP Publishing
doi:10.1088/1742-6596/441/1/012009

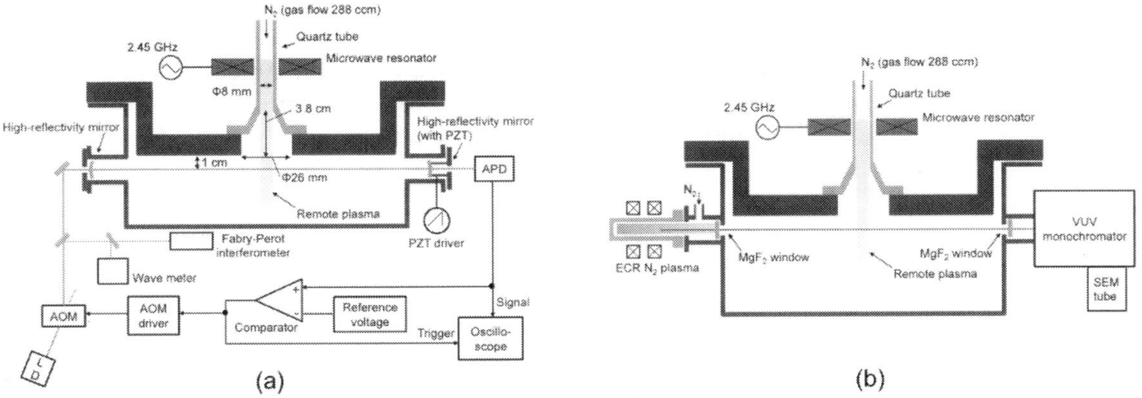

Figure 1. Experimental apparatus for measuring (a) the $N_2(A^3\Sigma_u^+)$ density by cavity ringdown spectroscopy and (b) the $N(^4S^o)$ and $N(^2D^o)$ densities by vacuum ultraviolet absorption spectroscopy in the downstream region of a remote N_2 plasma source.

information for the optimum discharge condition of the remote N_2 plasma to realize high densities of $N_2(A^3\Sigma_u^+)$ and $N(^2D^o)$ in the downstream region.

2. Experimental apparatus

Figure 1 shows the schematic illustrations of the experimental apparatus. Pure N_2 plasmas were produced in a quartz tube by surfatron discharge. The power source was a magnetron generator at a frequency of 2.45 GHz. The magnetron generator was connected to a microwave resonator via an isolator, a power monitor, a stub tuner, and a convertor from rectangular to coaxial waveguides. The microwave power was set at 100 W. The quartz tube with an inner diameter of 8 mm passed through the hole at the center of the microwave resonator. N_2 gas was fed from the top of the quartz tube at a flow rate of 288 ccm. The length of the active discharge region along the quartz tube was dependent on the discharge conditions, and was roughly 1.5 - 2 cm. The quartz tube was connected to a stainless-steel cylindrical chamber with a diameter of 30 cm, where spectroscopic systems for measuring the densities of $N_2(A^3\Sigma_u^+)$, $N(^4S^o)$, and $N(^2D^o)$ were installed. The cylindrical chamber had a hole of 26 mm in diameter on the top plate, and the spatial afterglow of the N_2 plasma was introduced into the chamber via the hole. The microwave resonator was movable, and the following measurements were carried out in the spatial afterglow at various distances from the active discharge region.

The $N_2(A^3\Sigma_u^+)$ density was measured by CRDS as shown in figure 1(a). The details of the CRDS system have previously been reported [4]. Briefly, the spatial afterglow of the N_2 plasma was sandwiched by an optical cavity which was composed of two concave mirrors with high reflectivities. The cavity length was 1 m. We injected a cw diode laser beam into the cavity via an acousto-optic modulator (AOM), and detected the laser intensity transmitted through the cavity. The wavelength of the laser was tuned at around 771.100 nm, which corresponded to the $B^3\Pi_g(v' = 2) - A^3\Sigma_u(v'' = 0)$ absorption band of N_2. The rapid truncation of the laser beam, by triggering AOM at the moment when the cavity length was resonant with the laser wavelength, resulted in an exponential decay of the transmitted laser intensity. And the $N_2(A^3\Sigma_u^+)$ density was deduced from the decay time constant (the ringdown time).

The densities of $N(^4S^o)$ and $N(^2D^o)$ were measured by VUVAS as shown in figure 1(b). The details of the VUVAS system have also been reported elsewhere [5,6]. In brief, the spatial afterglow of the N_2 plasma was placed between an electron cyclotron resonance N_2 plasma (the light source) and a VUV monochromator. Two MgF_2 windows were used for separating the cylindrical chamber, the light source plasma, and the VUV monochromator. The intensities of line emissions yielded from the light source plasma were detected using a secondary-electron multiplier tube via the VUV monochromator. The

Journal of Physics: Conference Series **441** (2013) 012009 doi:10.1088/1742-6596/441/1/012009

$N(^4S^o)$ and $N(^2D^o)$ densities were deduced from the absorption strengths of the $^4S^o - ^4P$ (120.0 nm) and $^2D^o - ^2D$ (124.3 nm) transitions, respectively.

Both the absorption chords of CRDS and VUVAS were located at a distance of 1 cm from the top plate of the chamber. In addition to the absorption measurements, we carried out conventional optical emission spectroscopy using a spectrograph with a focal length of 50 cm and a grating of 600 grooves/mm. The measurement position was 4.8 cm distant from the end of the quartz tube.

3. Results and Discussion

Figures 2 and 3 show the densities of $N_2(A^3\Sigma_u^+)$, $N(^4S^o)$ and $N(^2D^o)$, the gas temperature, and the optical emission intensities of the first positive (N_2), the second positive (N_2), and the first negative (N_2^+) bands. These results were obtained at pressures of 0.5 (figure 2) and 7 Torr (figure 3) at various distances from the discharge region. The origin of the distance from the discharge region (the z axis) is defined as the bottom-side edge of the microwave resonator. The densities of $N_2(A^3\Sigma_u^+)$, $N(^4S^o)$ and $N(^2D^o)$ were estimated from their line-integrated densities on the assumption of the absorption length as 30 cm (the diameter of the cylindrical chamber). The gas temperature was evaluated by fitting the absorption spectrum of the $B^3\Pi_g(v' = 2) - A^3\Sigma_u(v'' = 0)$ transition with the theoretical one which was calculated by assuming the same values for the rotational and translational temperatures (the Doppler broadening). The error bars shown in figure 3(b) indicate the ambiguity in the evaluation of the gas temperature by the spectral fitting. The optical emission intensities plotted in figures 2(c) and 3(c) were obtained by integrating the band spectra along the wavelength.

As shown in figures 2(a) and 3(a), the density of $N(^4S^o)$ was much higher than the densities of $N_2(A^3\Sigma_u^+)$ and $N(^2D^o)$. A higher $N(^4S^o)$ density was observed at a higher gas pressure. The decay of the $N(^4S^o)$ density with distance from the discharge region was negligibly small, keeping the density on the order of 10^{17} m^{-3} at both the pressures of 0.5 and 7 Torr. This result indicates the negligible destruction of $N(^4S^o)$ in the gas phase [7] as well as on the surface of the quartz tube [8]. Accordingly, the lifetime of $N(^4S^o)$ was sufficiently long to be transported from the discharge to downstream regions. Although the production of $N(^4S^o)$ would be expected in the downstream region due to a similar mechanism to those described below, it can be considered that the $N(^4S^o)$ density in the downstream region is dominated by the transport from the active discharge region.

The principal subject of this paper is the transport characteristics of $N_2(A^3\Sigma_u^+)$ and $N(^2D^o)$, since the efficient transport of $N(^4S^o)$ does not contribute to dissociation of NF_3 according to the molecular orbital calculation [3]. The destructions of $N_2(A^3\Sigma_u^+)$ and $N(^2D^o)$ occur both in the gas phase [9–11] and on the surface of the quartz tube. In general, the surface loss probabilities of electronic-excited metastable states are believed to be equal to unity. Although several works reported possibilities that the surface loss probabilities on dielectric surfaces are less than unity ($\alpha = 0.11$ [12] and $\alpha \geq 0.25$ [13] for $He^*(^3S_1)$, $0.5 < \alpha < 0.85$ [13], $\alpha \simeq 0.7$ [14], and $\alpha \simeq 0.72$ [15] for $Ar^*(^3P)$, and $\alpha \geq 0.75$ for $Xe^*(^3P_2)$ [13]), it is unlikely that a surface loss probability is much smaller than unity. When we assume a surface loss probability of unity, the diffusion lifetimes of $N_2(A^3\Sigma_u^+)$ and $N(^2D^o)$ in the quartz tube are estimated to be 10 and 8 μs, respectively, according to their diffusion coefficients reported in literatures [16, 17]. The diffusion lifetimes correspond to the transport lengths of $v_f\tau_D = 9$ and 7 μm in the quartz tube, where τ_D and v_f are the diffusion lifetime and the flow speed of N_2, respectively. Therefore, even though the laminar gas flow avoids the frequent arrivals on the surface of the quartz tube, it is impossible to expect the transport length of several centimeters for $N_2(A^3\Sigma_u^+)$ and $N(^2D^o)$.

As shown in figures 2(a) and 3(a), the densities of $N_2(A^3\Sigma_u^+)$ and $N(^2D^o)$ were much lower than the $N(^4S^o)$ density in the downstream region. Having had considered the experimental result we reported previously [18], it would be expected that the densities of $N_2(A^3\Sigma_u^+)$, $N(^4S^o)$, and $N(^2D^o)$ are on a similar order of magnitude in the active discharge region at a pressure of 0.5 Torr. Hence, the lower $N_2(A^3\Sigma_u^+)$ and $N(^2D^o)$ densities than the $N(^4S^o)$ density suggest losses of $N_2(A^3\Sigma_u^+)$ and $N(^2D^o)$ during their transport. The decays of the $N_2(A^3\Sigma_u^+)$ and $N(^2D^o)$ densities with the distance from the discharge region also suggest their losses during their transport. However, as described in the previous paragraph,

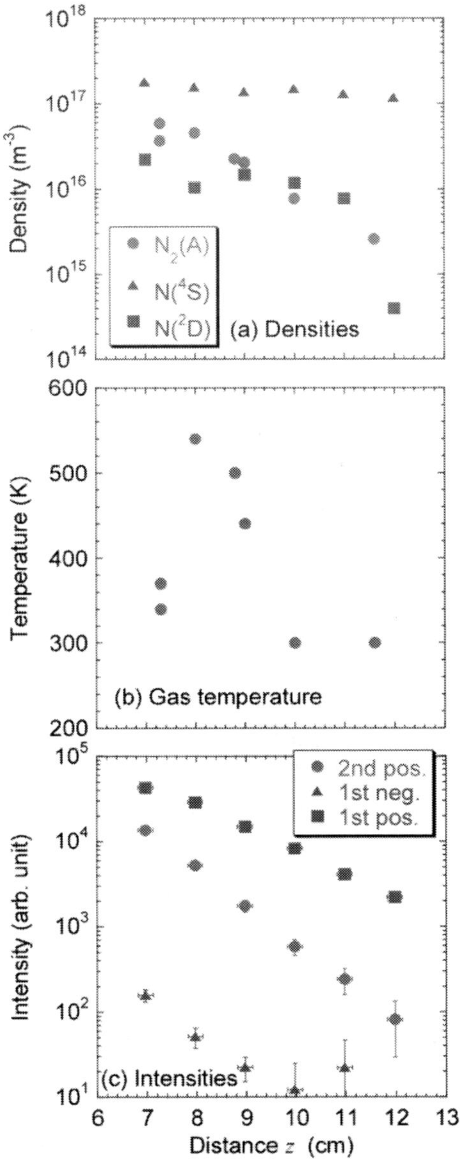

Figure 2. Densities of $N_2(A^3\Sigma_u^+)$, $N(^4S^o)$, and $N(^2D^o)$, the gas temperature, and the optical emission intensities observed in the downstream region of a N_2 remote plasma produced at a pressure of 0.5 Torr.

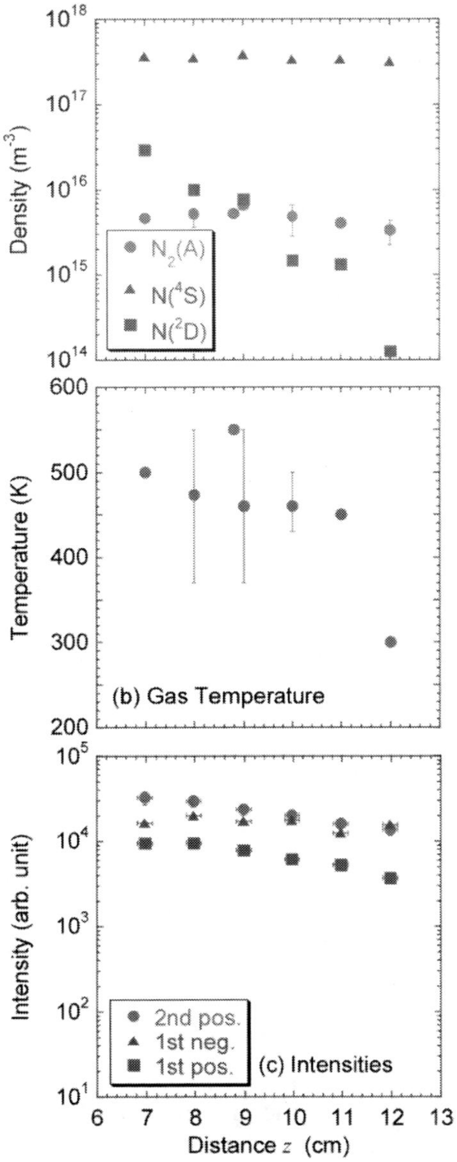

Figure 3. Densities of $N_2(A^3\Sigma_u^+)$, $N(^4S^o)$, and $N(^2D^o)$, the gas temperature, and the optical emission intensities observed in the downstream region of a N_2 remote plasma produced at a pressure of 7 Torr.

the transportation lengths of $N_2(A^3\Sigma_u^+)$ and $N(^2D^o)$ are expected to be short, and the experimentally-observed densities on the order of 10^{15} m^{-3} at a distance of 10 cm from the discharge region cannot be explained by the transport. In other words, the $N_2(A^3\Sigma_u^+)$ and $N(^2D^o)$ densities observed experimentally suggest their productions outside the active discharge region.

The productions of $N_2(A^3\Sigma_u^+)$ and $N(^2D^o)$ are supported by the observations of the optical emission intensities in the downstream region, as shown in figures 2(c) and 3(c). It is noted that no optical emission intensities were observed in the downstream region at a distance of $z \geq 6$ cm from the discharge region, when we employed argon for the discharge. This indicates that the downstream region shown in figures 2 and 3 is the outside of the active discharge region. In contrast, when we employed nitrogen for the discharge, we observed optical emission intensities of the first positive, the second positive, and the first negative bands in the downstream region, which indicate the productions of $N_2(B^3\Pi_g)$, $N_2(C^3\Pi_u)$, and $N_2^+(B^2\Sigma_u^+)$ states, respectively. In addition, we observed gas heating as shown in figures 2(b) ($8 \leq z \leq 9$ cm), which suggests that a kind of power deposition is available in the downstream region.

The production of $N_2(A^3\Sigma_u^+)$ in the flowing afterglow has been investigated intensively by Sadeghi *et al.* (experimentally) [19] and Guerra *et al.* (theoretically) [20]. Guerra and coworkers pointed out the possibility of energy transport to the downstream region by ground-state molecular nitrogen at highly-excited vibrational states. Highly-excited vibrational states are populated by the V-V pumping-up mechanism,

$$N_2(X^1\Sigma_g^+, v) + N_2(X^1\Sigma_g^+, w) \rightarrow N_2(X^1\Sigma_g^+, v+1) + N_2(X^1\Sigma_g^+, w-1). \quad (1)$$

Highly-excited vibrational states thus formed contribute to the production of $N_2(A^3\Sigma_u^+)$ via reactions

$$N_2(X^1\Sigma_g^+, v \geq 10) + N(^2P) \rightarrow N_2(A^3\Sigma_u^+) + N(^4S) \quad (2)$$

and

$$N_2(X^1\Sigma_g^+, v \geq 39) + N(^4S) \rightarrow N_2(A^3\Sigma_u^+) + N(^2D^o). \quad (3)$$

In addition, since the observation of the optical emission intensity at the first negative system of N_2^+ suggests the existence of electrons, electron impact excitation

$$N_2(X^1\Sigma_g^+, v) + e \rightarrow N_2(A^3\Sigma_u^+) + e \quad (4)$$

may partly contribute to the production of $N_2(A^3\Sigma_u^+)$, if electrons are considerably heated in the downstream region. Guerra and coworkers pointed out the importance of $N_2(X^1\Sigma_g^+, v \geq 25)$ as the parent of $N_2(A^3\Sigma_u^+)$ in reaction (4) because of a small amount of electrons with higher energies than the potential energy of $N_2(A^3\Sigma_u^+)$.

Guerra and coworkers proposed similar processes, which originated from highly-excited vibrational states of $N_2(X^1\Sigma_g^+)$, for the productions of $N_2(B^3\Pi_g)$, $N_2(C^3\Pi_u)$, and $N_2^+(B^2\Sigma_u^+)$. Reaction (3) is a production process of $N(^2D^o)$. In addition to reaction (3), because of the low excitation energy of 2.38 eV, electron impact excitation

$$N(^4S^o) + e \rightarrow N(^2D^o) + e \quad (5)$$

may contribute more efficiently than reaction (3) for the production of $N(^2D^o)$. The electron energy distribution function in the afterglow, which is calculated by Guerra *et al.*, contains a large amount of electrons with energies higher than 2.38 eV.

The comparison between the experimental results and the aforementioned reaction kinetics suggests that reaction (3) does not dominate the production process of $N_2(A^3\Sigma_u^+)$ and $N(^2D^o)$. This is because reaction (3) produces $N_2(A^3\Sigma_u^+)$ and $N(^2D^o)$ as a pair, while the $N_2(A^3\Sigma_u^+)$ and $N(^2D^o)$ densities had different decay curves as shown in figures 2(a) and 3(a). In other words, it is considered that the dominant production processes for $N_2(A^3\Sigma_u^+)$ and $N(^2D^o)$ are the reactions (2) and (5), respectively. The decay curve of the $N_2(A^3\Sigma_u^+)$ density at 0.5 Torr was rather steep as shown in figure 2(a), while the decay of the $N_2(A^3\Sigma_u^+)$ density at 7 Torr was negligible as shown in figure 3(a). These decay characteristics of the $N_2(A^3\Sigma_u^+)$ density may be dominated by the density distribution of $N_2(X^1\Sigma_g^+, v \geq 10)$. On the other hand, the decay curve of the $N(^2D^o)$ density observed at 0.5 Torr was relatively flat as shown in figure 2(a), while a rapid decrease in the $N(^2D^o)$ density was observed at 7 Torr as shown in figure 3(a). These decay characteristics of the $N(^2D^o)$ density may be due to the spatial distribution of the electron energy distribution function. The different decay characteristics of the $N_2(A^3\Sigma_u^+)$ and $N(^2D^o)$ density are of importance from the application point of view, since it suggests a possibility of selective "pseudo-transportation" to the downstream region in a remote plasma source.

4. Conclusions

In this paper, we have demonstrated that the densities of $N_2(A^3\Sigma_u^+)$ and $N(^2D^o)$ in the downstream region of a remote N_2 plasma source are on the order of $10^{14} - 10^{16}$ m^{-3}. The $N_2(A^3\Sigma_u^+)$ and $N(^2D^o)$ densities in the downstream region were not obtained by the transport from the active discharge region, but it was due to their productions. The experimental results reveal that the reactions (2) and (5) are the dominant production processes of $N_2(A^3\Sigma_u^+)$ and $N(^2D^o)$, respectively. The different production pathways for $N_2(A^3\Sigma_u^+)$ and $N(^2D^o)$ suggest the possibility of selective "pseudo-transportation" to the downstream region in a remote plasma source. The present experimental results support dissociation of NF_3 and dry etching of silicon in a remote N_2 plasma source. It is noted finally that the productions of $N_2(A^3\Sigma_u^+)$ and $N(^2D^o)$ are expected to be unaffected by the injection of NF_3 into the process chamber, since the backward flow of NF_3 toward the quartz tube is negligible because of the fast gas flow of N_2.

References

[1] De Benedictis S, Dilecce G and Simek M 1993 *Chem. Phys.* **178** 547
[2] Eriguchi K and Ono K 2008 *J. Phys. D Appl. Phys.* **41** 024002
[3] Hayashi T, Ishikawa K, Sekine M, Hori M, Kono A and Suu K 2012 *Jpn. J. Appl. Phys.* **51** 026505
[4] Horikawa Y, Kurihara K and Sasaki K 2010 *J. Phys. Conf. Ser.* **227** 012012
[5] Sasaki K, Kawai Y and Kadota K 1997 *Appl. Phys. Lett.* **70** 1375
[6] Sasaki K, Kawai Y and Kadota K 1999 *Rev. Sci. Instrum.* **70** 76
[7] Yamashita T 1979 *J. Chem. Phys.* **70** 4248
[8] Belmontea T, Lefevrea L, Czerwieca T, Michela H and Ricard A 1999 *Thin Solid Films* **341** 27
[9] Boeuf J P and Kunhardt E E 1986 *J. Appl. Phys.* **60** 915
[10] Piper L G 1989 *J. Chem. Phys.* **90** 7087
[11] Slanger T G and Black G 1976 *J. Chem. Phys.* **64** 4442
[12] Lisitsyn V N, Provorov A S and Chebotaev V P 1970 *Opt. Spectrosc.* **29** 119
[13] Barbet A, Sadeghi N and Pebay-Peyroula J C 1975 *J. Phys. B: At . Mol. Phys.* **8** 1776
[14] Sakurai T, Kubota T, Takahara Y, Inoue Y and Hori H 1999 *Jpn. J. Appl. Phys.* **38** L590
[15] Macko P and Sadeghi N 2004 *Plasma Sources Sci. Technol.* **13** 303
[16] Guerra V and Loureiro J 1997 *Plasma Sources Sci. Technol.* **6** 361
[17] Cernogora G 1980 *Chem. Phys. Lett.* **74** 417
[18] Horikawa Y, Kurihara K and Sasaki K 2010 *Jpn. J. Appl. Phys.* **49** 026101
[19] Sadeghi N, Foissac C and Supiot P 2001 *J. Phys. D: Appl. Phys.* **34** 1779
[20] Guerra V, Sa P A and Loureiro J 2004 *Phys. J. Appl. Phys.* **28** 125

Behaviour of sheath voltage in negative ion plasma

J Y Zhang [1], **R Ichiki** [2] **and Y Kawai** [3]

[1] School of Medicine, Dalian University, Dalian, 116622, P. R. China
[2] Department of Electrical and Electronic Engineering, Oita University, Oita, 870-1192, Japan
[3] Interdisciplinary Graduate School of Engineering Sciences, Kyushu University, Kasuga, Fukuoka, 816-8580, Japan

E-mail: kawai@aees.kyushu-u.ac.jp

Abstract. The sheath voltage in SF_6/Xe plasma was measured with the Langmuir probe and compared with the theoretical value calculated from the sheath equation including negative ions. The negative ion density was estimated from the reduction of the electron saturation current. It was found that the observed sheath voltages agree with the theoretical values qualitatively.

1. Introduction

As well known, there are negative ions in reactive plasmas for CVD [1-2]. Negative ions are confined in plasma without diffusing to discharge electrodes, leading to high negative ion density. Therefore, negative ions may play an important role especially in plasma CVD such as the fabrication of amorphous silicon/microcrystalline silicon for thin film silicon solar cells. However, there are almost no reports on the behaviour of negative ions in plasma CVD because of the difficulty of negative ion measurements in reactive plasma. The negative ion density is widely measured utilizing a laser photodetachment technique [3]. The simplest method to estimate the negative ion density is to use the Langmuir probe, that is, the negative ion density normalized to the ion density is easily estimated from the reduction of the electron saturation current [4]. We have proposed the sheath voltage method to estimate dominate ions [5], that is, when both the sheath voltage and the electron temperature are known, we can estimate dominant ion species. Furthermore, when there are negative ions, the sheath voltage decreases and Bohm sheath is modified [6]. Thus, the sheath voltage provides interesting information about negative ions as well as positive ions. In fact, we observed the decreases in the sheath voltage in SF_6/Ar double plasma and found that the observed sheath voltages agree with those calculated using the sheath equation including negative ions [5]. Here, in order to establish the sheath voltage methods in negative ion plasma, we performed the experimental on the sheath voltage measurements in SF_6/Xe double plasma. The use of Xe gas has an advantage that provides higher electron density plasma, leading high etching rates or high deposition rates.

2. Experimental

The experiments were performed using a double plasma device [7-8], as shown in figure 1. The diameter and length of the stainless vacuum chamber were 500 mm and 1000 mm, respectively. In this experiment, the separation grid (SG) was removed such that the chemical composition in the plasma was uniform in all regions. Xe and SF_6 gases were introduced into the chamber with the mass flow controllers separately. The flow rate of Xe gas was kept at 6 sccm (8×10^{-4} Torr) and that of SF_6 gas was regulated from 0 to 0.1 sccm. The double plasma, produced by dc discharges between the filaments (F) and the chamber wall (A), was confined by the multidipole cusp–mirror magnetic field generated by permanent magnets. The accelerating voltage and discharge current were 50 V and 60 mA, respectively. A 0.6-cm-diam planar Langmuir probe (P) was used to measure the plasma parameters. The Langmuir *I-V* curve was measured with the X-Y recorder (the input impedance: 1 MΩ). The probe surface was cleaned periodically using ion bombardment in pure Xe plasma. Mass spectra of positive and negative ion species were analyzed with the QMS (Hiden EQP-500).

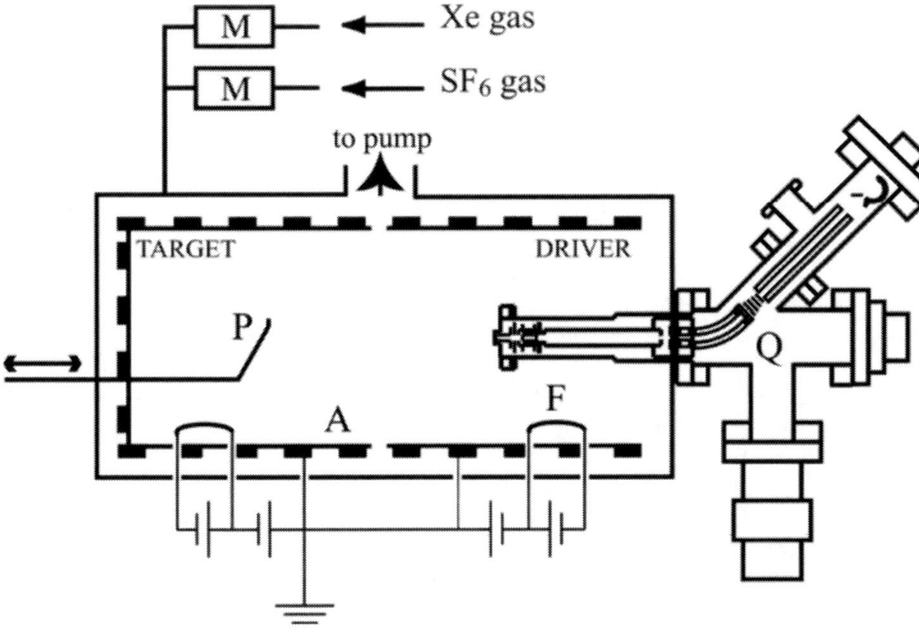

Figure 1 A schematic of DP machine. Here P and Q denote the Langmuir probe and the QMS.

The negative ion density normalized to the ion density, n_-/n_+, was estimated from the following equation [4]:

$$\frac{n_-}{n_+} = 1 - \frac{I_{es}}{I_{eso}} \qquad (1)$$

where I_{es} and I_{es0} is the electron saturation current with negative ions and without negative ions, respectively.

3. Results and discussion

At first a SF_6/Xe double plasma was produced and the plasma parameters were measured as a function of $SF_6/(Xe+SF_6)$ with the Langmuir probe, where the ratio $SF_6/(Xe+SF_6)$ is that of the mass flow rate. The plasma parameters were: Te=0.09-0.2 eV and $n_+\sim3.9\times10^{14}$-3.6×10^{15} m^{-3}. The QMS spectra showed that dominant positive ions and negative ions are Xe^+ and F^-, respectively. When $SF_6/(Xe+SF_6)$ was increased, the electron saturation current rapidly decreased. We estimated the negative ion density normalized to the ion density, n_-/n_+, as a function of

$SF_6/(Xe+SF_6)$ from Eq. (1). Figure 2 shows that when $SF_6/(Xe+SF_6)$ is increased, the negative ion concentration rapidly increases.

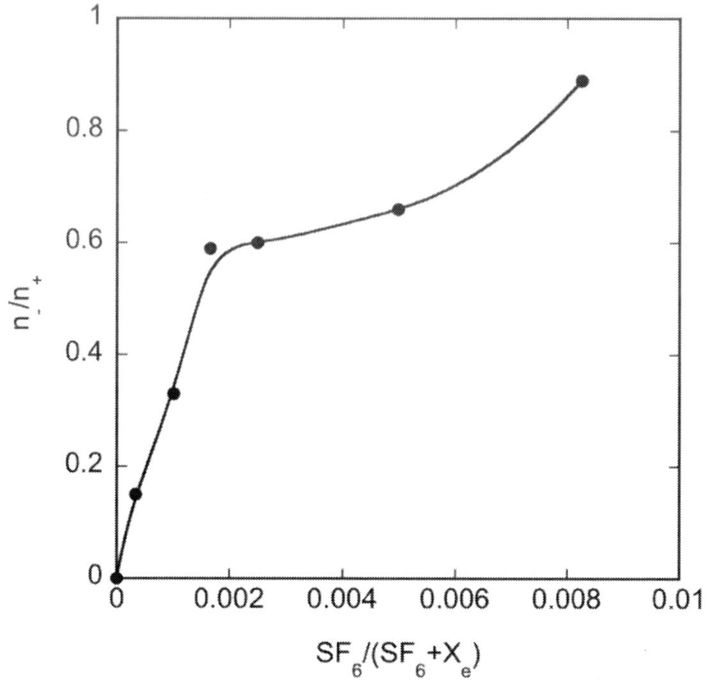

Figure 2 The dependence of the negative ion concentration n_-/n_+ on $SF_6/(Xe+SF_6)$. Here the curve was plotted to fit the experimental results using the least-squares analysis.

Then, we measured the sheath voltage as a function of $SF_6/(Xe+SF_6)$ and plotted the sheath voltage as a function of negative ion concentration n_-/n_+ in figure 3. Here the sheath voltage was estimated from the potential difference between the plasma potential and the floating potential that were measured with the probe. The solid line in figure 3 is the theoretical sheath voltage that was calculated using the sheath equation derived by Shindo and Horiike [9].The sheath equation and the Bohm criterion voltage V_B are as follows:

$$n_e \exp(-\frac{eV}{kT_e})(\frac{kT_e}{2\pi m})^{1/2} + n_- \exp(-\frac{eV}{kT_-})(\frac{kT_-}{2\pi M_-})^{1/2} = \{\frac{2(kT_+ + eV_B)}{M_+}\}^{1/2}\{n_e \exp(-\frac{eV_B}{kT_e}) + n_- \exp(-\frac{eV_B}{kT_-})\} \quad (2)$$

$$n_e \exp(-\frac{eV_B}{kT_e})\{\frac{1}{kT_e} - \frac{1}{2(kT_+ + eV_B)}\} + n_- \exp(-\frac{eV_B}{kT_-})\{\frac{1}{kT_-} - \frac{1}{2(kT_+ + eV_B)}\} = 0 \quad (3)$$

Journal of Physics: Conference Series **441** (2013) 012010 doi:10.1088/1742-6596/441/1/012010

Here M_- is negative ion mass and T_+ and T_- is the temperature of positive ions and negative ions, respectively. As described in Ref. [9], the sheath voltage is very sensitive to T_+ and T_-. In this calculation, $T_+=T_-$ was assumed and the result of figure 4 was used. As seen from figure 3, the observed sheath voltages agrees with the theoretical ones for $T_+=0.03$ eV except for high negative ion concentrations. One of the reasons for this deviation may be due to the assumption of T_+ and T_-. Although we did not measure T_+ and T_-, assumed ion temperature $T_+=0.03$ eV is reasonable since Te is lower than 0.2 eV.

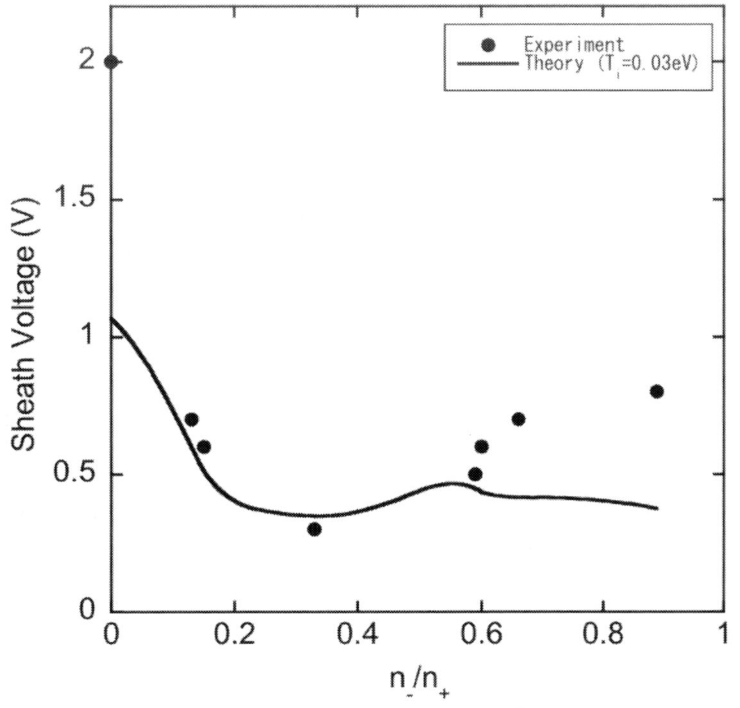

Figure 3 The dependence of the sheath voltage on the negative ion concentration n_-/n_+, where the curves are the theoretical values calculated from the sheath equation, Eqs.(2) and (3) .

When there are negative ions in plasma, the ion saturation current decreases because the sheath width broadens out [10]. We plotted the ion saturation current I_{is} as a function of negative ion concentration. Figure 4 indicates that the ion saturation current decreases proportional to n_-/n_+. This tendency was shown numerically by Shindo and Horiike [10]. However, measured ion saturation currents did not agree with the theoretical values. Moreover, the sheath equation is a nonlinear equation, so that there are many solutions [9]. In fact, three solutions existed in SF_6/Ar plasma, and the experimental values agreed with the calculated ones. However, as seen in fig. 3, there is one solution for the present parameter region in SF_6/Xe plasma. Thus, more detailed discussions about the sheath structure in a negative ion plasma will be necessary for the comparison between the experiment and the theory.

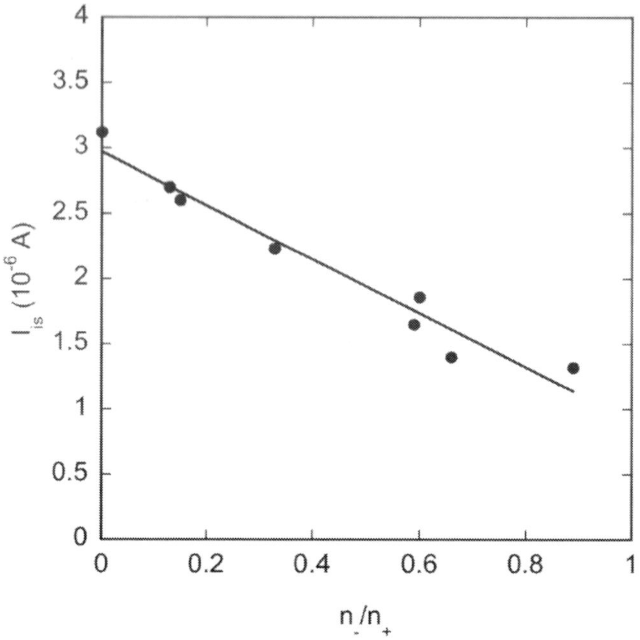

Figure 4 The dependence of the ion saturation current on the negative ion concentration n_-/n_+. Here the curve was plotted to fit the experimental results using the least-squares analysis.

4. Conclusion

We produced a SF_6/Xe double plasma to study the characteristics of a negative ion plasma with the Langmuir probe. Here the negative ion concentration was estimated from the reduction of the electron saturation current. It was found that measured sheath voltages agree with the theoretical values for T_+=0.03 eV qualitatively and the ion saturation current decreases proportional to the negative ion concentration.

References

[1] Matsuda A and Tanaka K 1982 *Thin Solid Films*, **92** 171
[2] Lieberman M A and Lichtenberg A J 1994 *Principle of Plasma Discharges and Materials Processing*, Chap.6 (John Wiley & Sons, Inc.)
[3] Bacal M, Hamilton G W, Bruneteau A M and Doucet H J 1979 *Rev. Sci. Instrum.*, **50** 2288
[4] Ichiki R, Yoshimura S, Watanabe T, Nakamura Y, and Kawai Y 2002 *Phys. Plasmas*, **9** 4481
[5] Zhang J Y, Ichiki R and Kawai Y 2012 *Proc. of XXI Europhysics Conference on the Atomic and Molecular Physics of Ionized Gases (Porto)* P1.5.1
[6] Amemiya H 1990 *J. Phys.D:Appl.Phys.*,**23** 999
[7] Shindo M, Uchino S, Ichiki R and Kawai Y 2001 *Rev. Sci. Instrum.*, **72** 2288
[8] Fukumasa , Hosoda M and Naito H 1992 *Rev. Sci. Instrum.*, 63 2696
[9] Shindo H and Horiike Y 1991 *Jpn. J. Appl. Phys.*, **30** 161
[10] Shindo H and Horiike Y 1993 *Jpn. J. Appl. Phys.*, **32** 5109

Observation of a microwave bandgap in a one-dimensionally periodic low-pressure plasma structure

Kazunori Takahashi[1,2,3] **and Kazuaki Miyamoto**[2]

[1]Department of Electrical Engineering, Tohoku University, Sendai 980-8579, Japan
[2]Department of Electric Engineering and Computer Science, Iwate University, Morioka 020-8551, Japan
[3]Soft-Path Engineering Research Center, Iwate University, Morioka 020-8551, Japan

E-mail: kazunori@ecei.tohoku.ac.jp

Abstract. One-dimensionally periodic structure of plasma density is produced by a 400 kHz capacitively coupled discharge in 15 Pa argon. The maximum density can be increased up to above 10^{10}-10^{11} cm^{-3} by increasing the rf power. The density ratio between the higher and lower density layers is maintained at about 0.7 in the present configuration. The transmittance of the microwave in the GHz range is measured and simply calculated. The results indicate the presence of the bandgap near about 3 GHz, which originates in the periodic structure.

1. Introduction

Photonic crystals consisting of periodic structures of dielectric materials are well known to have unique characteristics, e.g., a band gap for a specific frequency (wavelength) relating to its lattice constant [1, 2]. Although recent experimental investigation has demonstrated a measurement of a near-field of the light in nano-scale solid crystal [3], the direct measurement of the wave field inside the crystals, which is important and useful for clarification of the detailed physics and for finding a novel phenomenon such as nonlinear effects, has not been reported because of its experimental difficulties.

As plasma charged particles interact with electromagnetic waves and the wave dispersion relation is resultantly different from that in vacuum, the dielectric constant of the plasma medium depends on the plasma parameters. In the simplest plasma with no magnetic field, high frequency electromagnetic waves interact mainly with electrons near the electron plasma frequency. Recently, plasma photonic crystals (PPCs) have been reported experimentally, theoretically, and numerically for the purpose of active control of microwaves and millimeter waves [4, 5, 6, 7, 8, 9]. One of advantages of the PPCs is that the dielectric constant can be controlled by external experimental parameters such as discharge power, gas pressure, and so on. In the previous and undergoing experiments, the PPCs consisting of atmospheric pressure discharge and relatively higher pressure discharges are used for the medium, where the plasma is periodically produced in millimeter scale [4, 5]. Although the medium with different dielectric constant can appear in the gas phase for the case of the PPCs, the direct measurement of the wave field inside the crystals still remains an experimental challenge because the insertion of the diagnosis tool into the crystals perturbs the discharge operation.

More recently, production of centimeter-scale periodic plasma structure in low pressure discharge has been attempted for the direct measurement of the wave field inside the PPCs [10]. In such a configuration, it is expected that the band gap due to the periodic plasma structure appears in the range of GHz frequency; then small wave detectors can be inserted inside the crystals with no disturbance of the discharges. However, the maximum plasma density in the previous experiment using a dc argon discharge is only 10^9 cm^{-3}; hence the dielectric constant for GHz waves is unchanged from that for the vacuum. In order to perform the PPC experiments for the GHz microwaves in such a low pressure plasmas, the density should be increased up to 10^{10} - 10^{11} cm^{-3} at least.

In this paper, a one-dimensionally periodic plasma structure is produced by a 400 kHz capacitively coupled discharge and its density is increased up to above 10^{10} cm^{-3} at the rf power of about 400 W. Further, a transmittance of microwaves in the range of GHz frequency is preliminarily measured; the results indicate a presence of a bandgap in a few GHz and compared with a simple model.

2. Experimental Setup

Experiments are performed in a 60-cm-width, 45-cm-height, and 35-cm-depth grounded vacuum chamber shown in Fig. 1(a). The chamber is evacuated to a base pressure of 10^{-3} Pa by a diffusion/rotary pumping system, after that argon gas is introduced from the side port of the chamber and the gas pressure is maintained at 15 Pa. Living and grounded electrodes shown in Fig. 1(b) are installed inside the chamber. The living electrode is an aluminum plate and powered by a 400 kHz rf power supply through an impedance matching circuit. The rf power can be increased up to about 500 W and the net power can be estimated from the forward and reflected rf powers. Rectangular plastic insulators and grounded aluminum shields are mounted on the living electrodes at 3 cm intervals. Another grounded electrode is located at the upper side, 7 cm away from the living electrode. In such a configuration, the discharge can occur through a capacitively-coupled discharge process only in the region with no insulator and grounded shields; then one-dimensionally modulated plasma-density structure can be formed. The x, y, and z axes and the origin point $(x, y, z) = (0, 0, 0)$ of the axes are defined as indicated in Fig. 1.

The axially movable Langmuir probe (LP) is inserted along the z axis at $(x, y) = (0, 3.5$ cm$)$ from the side port of the chamber to diagnose the plasma density. The density is estimated from the ion saturation current of the LP with the measured electron temperature of about 2 eV, assuming a Maxwellian electron energy distribution.

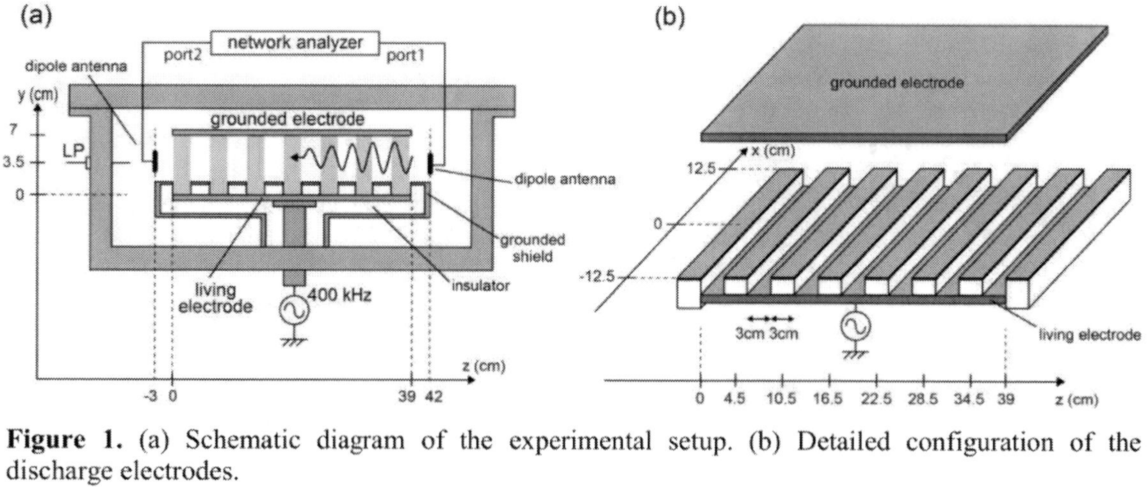

Figure 1. (a) Schematic diagram of the experimental setup. (b) Detailed configuration of the discharge electrodes.

The measurement of a transmittance of microwaves in the range of GHz frequency is performed by two dipole antennas located at $(x, y, z) = (0, 3.5\text{ cm}, -3\text{ cm})$ and $(0, 3.5\text{ cm}, 42\text{ cm})$, where these antennas have a maximum gain at ~4 GHz. The transmittance coefficient S_{21} between the two antennas including the periodic plasma structure is measured by a network analyzer. To minimize effects of the peripheral chamber wall, electrodes, transmission line, and the antenna characteristic, the transmittance T relating to the plasma structure is defined as

$$T = S_{21\,plasma\,on} / S_{21\,plasma\,off}, \tag{1}$$

where $S_{21\,plasma\,on}$ and $S_{21\,plasma\,off}$ are the measured S_{21} with plasma on and off, respectively.

3. Experimental Results

Figure 2 shows the axial profile of the plasma density (open squares) measured by the LP for the net rf power of 180 W, where the solid line is added as a visual guide of the density profile. It is found that the periodic modulation of the plasma density can be formed. The peak positions correspond to the region with no insulators and no grounded shields on the living electrode, while the trough of the density exists in the region with the insulators and grounded shields. As a result, the high density region can be formed at 3 cm intervals, which is dominated by the electrode configuration.

Figure 3(a) shows the plasma densities at the peak ($z = 25.5$ cm) and trough ($z = 22.5$ cm) positions as a function of the net rf power. The densities in both the regions increases with the increase in the rf power; the maximum density at the peak position is found to be ~ 8×10^{10} cm^{-3} at 400 W rf power, which is higher than the previously reported dc discharge [10] by one order of magnitude. Figure 3(b) shows the density ratio of n_{min} at $z = 22.5$ cm (at the trough position) to n_{max} at $z = 25.5$ cm (at the peak position) obtained from Fig. 3(a). The result shows that the density ratio is about 0.7 and is unchanged by the rf power in the present configuration. Hence, it is demonstrated that the one-dimensionally periodic structure of the plasma density of 10^{10}-10^{11} cm^{-3} is produced by the 400 kHz capacitively coupled discharge, with controlling the absolute value of the maximum density and remaining the density ratio between the peak and trough positions of about 0.7.

The measurement of the transmittance given by Eq. (1) is performed with the rf power as a parameter and the results are plotted in Fig. 4. Transmittances depression less than -20 dB are observed below ~2 GHz. The upper frequency of this transmittance depression seems to increase with an increase in the rf power as indicated by dotted lines in Fig. 4. Further, the transmittance depressions

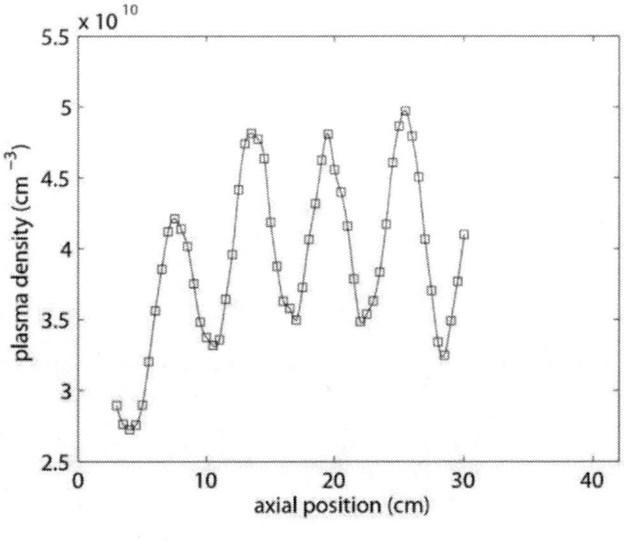

Figure 2. The axial profile of the plasma density measured along z axis and $(x, y) = (0, 3.5\text{ cm})$ for the net rf power of 180 W (open squares). The solid line is added as a visual guide.

at ~2 – 2.2 GHz and ~2.5 – 3 GHz are found to be enhanced by the rf power as indicated by arrows; the frequency of the transmittance depression also seems to be increased by the rf power. Above 3 GHz, it is found that the transmittance is unchanged by the rf power. These changes of the transmittance by the rf power are discussed with a simple model in Sec. 4.

4. Theoretical model

Here an analytical model of the transmittance is described to discuss the experimental results shown in Fig. 4. Now let us consider the simple configuration shown in Fig. 5(a), consisting of the plasma layers with different dielectric constant, which are numbered from 0 to $N+1$. The forward and reflected waves are labelled as E_i and E_r in the layer 0, and as A_j and B_j in the layer j. The transmitted waves propagating in the layer $N+1$ is labelled as E_t. Taking account of the boundary conditions between the layer j and $j+1$, the Maxwell equation gives the relation between these waves as

$$\begin{pmatrix} E_i \\ E_r \end{pmatrix} = M_0 \begin{pmatrix} A_1 \\ B_1 \end{pmatrix}, \qquad \begin{pmatrix} A_j \\ B_j \end{pmatrix} = M_j \begin{pmatrix} A_{j+1} \\ B_{j+1} \end{pmatrix}, \qquad \begin{pmatrix} A_N \\ B_N \end{pmatrix} = M_N \begin{pmatrix} E_t \\ 0 \end{pmatrix}, \tag{2}$$

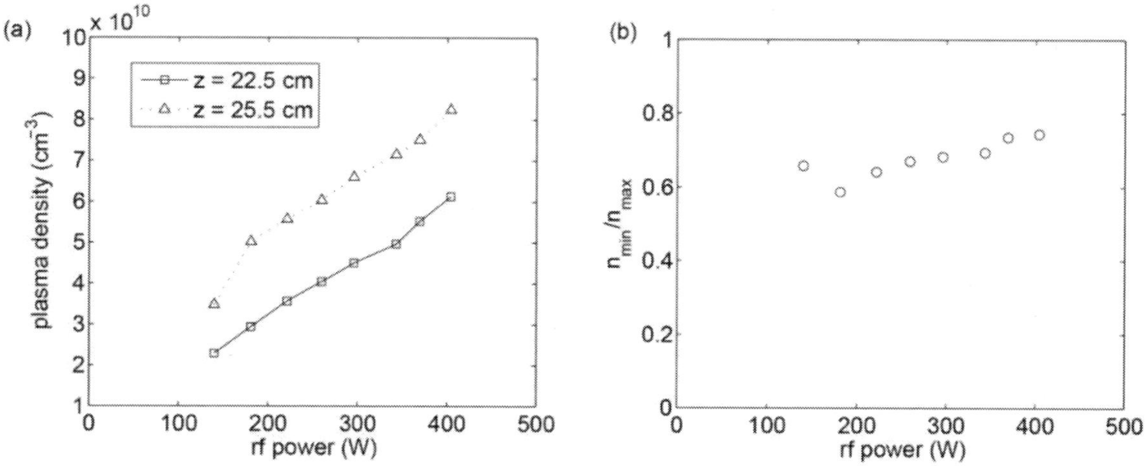

Figure 3. (a) Plasma densities at $z = 22.5$ cm (open squares, n_{min}) and at $z = 25.5$ cm (open triangles, n_{max}) corresponding to the peak and trough positions, respectively, versus net rf power. (b) Density ratio $n_{min} = n_{max}$ obtained from Fig. 3(a).

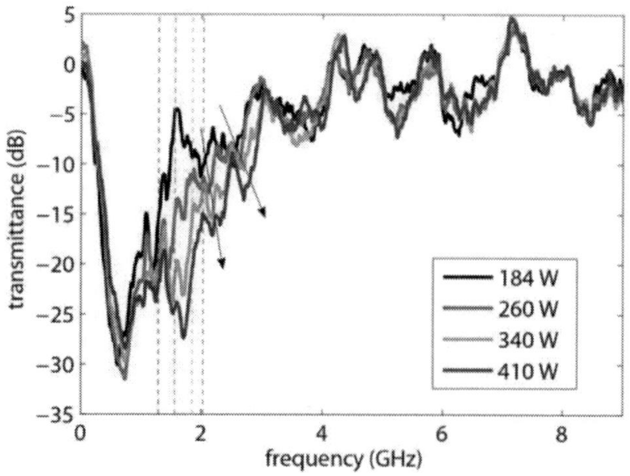

Figure 4. Measured transmittance T for the net rf power of 184 W (black solid line), 260 W (red solid line), 340 W (green solid line), and 410 W (blue solid line), respectively. The dotted lines show the upper frequencies of the transmittance depression below 2 GHz. The arrows indicate that the transmittance depression between 2 GHz and 3 GHz. No change by the rf power is detected above 3 GHz.

position z_j as

$$M_j = \begin{pmatrix} \frac{k_j + k_{j+1}}{2k_j}\exp\{-i(k_j - k_{j+1})z_j\} & \frac{k_j - k_{j+1}}{2k_j}\exp\{-i(k_j + k_{j+1})z_j\} \\ \frac{k_j - k_{j+1}}{2k_j}\exp\{i(k_j + k_{j+1})z_j\} & \frac{k_j + k_{j+1}}{2k_j}\exp\{i(k_j - k_{j+1})z_j\} \end{pmatrix} . \quad (3)$$

Equation (2) and (3) can be rewritten as

$$\begin{pmatrix} E_i \\ E_r \end{pmatrix} = M \begin{pmatrix} E_t \\ 0 \end{pmatrix}, \quad (4)$$

$$\text{where } M \equiv M_0 M_1 M_2 \cdots M_N = \begin{pmatrix} M_{11} & M_{12} \\ M_{21} & M_{22} \end{pmatrix} . \quad (5)$$

The transmittance can be calculated from the above equations as $T = |E_t / E_i|^2 = |1 / M_{11}|^2$. In plasma medium, the dispersion relation of the electromagnetic waves gives the local wave number as $k = (\omega / c)\ \{1 - (\omega_{pe} / \omega)^2\}^{1/2}$, where $\omega_{pe} = (e^2 n_p / \varepsilon_0 m_e)^{1/2}$ is the electron plasma frequency relating to the plasma density.

In the present model, the lower and higher density layers are mutually set at 3 cm intervals to compare with the experiments, where the density is assumed to be uniform in each layer. The density ratio of $n_{min} = n_{max} = 0.7$ is used according to Fig. 3(b), the number of the layers is chosen as $N = 13$, and the layers labeled by even and odd numbers are set as the lower and higher density layers, respectively. The higher density n_{max} is chosen as a parameter to compare with Fig. 4.

Figure 5(b) shows the calculated transmittance with the plasma density n_{max} in the high density layer as a parameter, whose densities correspond to that for the case of each rf power in Fig. 4. Around 2 GHz, a large depression of the transmittance is found to exist, which is close to the electron plasma

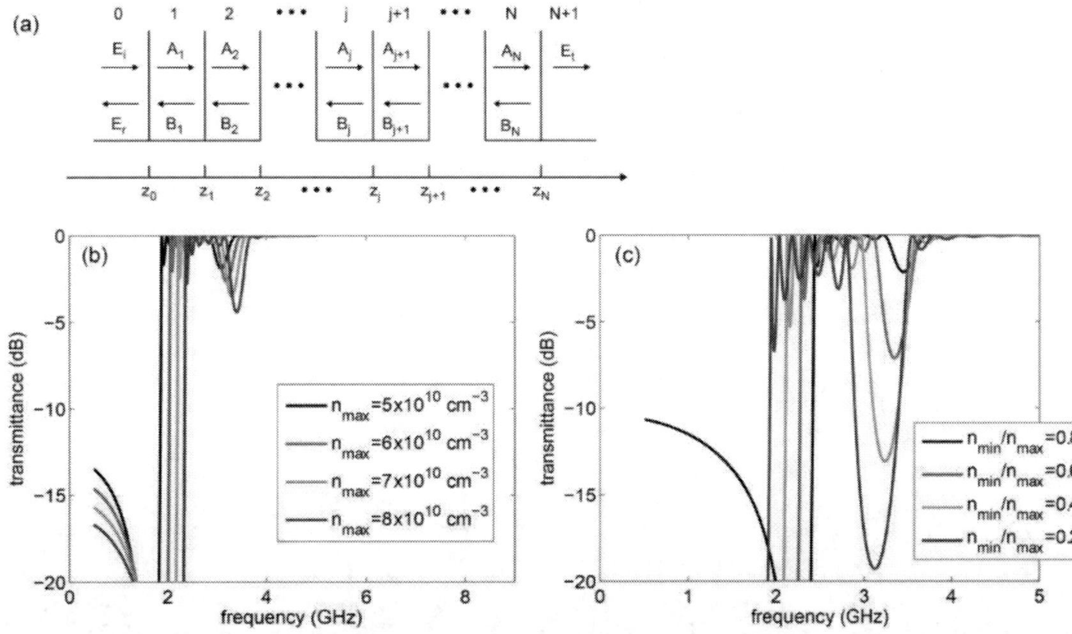

Figure 5. (a) Schematic diagram of the analytical model. (b) Calculated transmittance with the maximum density n_{max} as a parameter, where the density ratio is maintained at $n_{min} / n_{max} = 0.7$. (c) Calculated transmittance with the density ratio n_{min} / n_{max} as a parameter, where the maximum density is maintained at $n_{max} = 8 \times 10^{10}$ cm^{-3}.

frequency for each density and is similar to the depression indicated by the dotted lines below ~2 GHz in Fig. 4. Furthermore, a small depression appears around 3 GHz in Fig. 5(b); the frequency at this depression also shifts to the high frequency side with the increase in the plasma density. It is confirmed that this depression is due to the periodic structure of the plasma density by calculation with no periodic structure (not shown here); the behavior of the transmittance depression qualitatively agrees with the results in Fig. 4, although only the single depression is observed at about 3 GHz in the calculation, respectively. As the wave does not propagate along z axis exactly in the experiment because of the boundary condition of the chamber wall and the electrodes, some propagation modes are considered to be superimposed. In addition, there might be a density gradient along the x axis, which has not been measured yet due to the absence of a vacuum port. In such a situation, the microwaves will diverge and change their propagation direction. These might be a reason why the two depressions at the different frequencies are detected in Fig. 4. The measurements of the detailed density profile and the microwave and integrating the measured density profile into the analytical and/or numerical calculations of the wave propagation are required in near future. Above 3.5 GHz in Fig. 5, no depression is observed as well as the experimental result in Fig. 4. Based on the above discussion, we can deduce that the experimentally observed depression of the transmittance originates in the one-dimensionally periodic plasma-density modulation shown in Fig. 2.

Moreover, the effect of the density ratio $n_{min} = n_{max}$ is calculated and plotted in Fig. 5(c), where the maximum density is chosen as $n_{max} = 8 \times 10^{10}$ cm^{-3}. It is clearly found that the transmittance depression can be enhanced by the large density ratio, i.e., large density modulation. Hence it would be better to have a larger density ratio for further PPCs experiments. The production of such profile, the direct measurement of the wave field inside the PPCs, and finding a novel nonlinear phenomenon still remain further subjects.

5. Conclusion

The plasma density of 10^{10}-10^{11} cm^{-3} is spatially and periodically modulated in centimeter scale by mounting rf shield on a living electrode. The measured and calculated transmittances of the microwave suppose the presence of the bandgap for the microwave range at about 3 GHz. The calculated result also shows that the bandgap can be enhanced by increasing the density ratio between the lower and higher density layers.

Acknowledgement

The authors would like to thank H. Chiba for his technical help for the construction of the chamber. This work is partially supported by a Challenging Exploratory Research (23654197) from MEXT in Japan.

References

[1] John S, 1987 *Phys. Rev. Lett.* **58** 2486.
[2] Yablonovitch E, 1987 *Phys. Rev. Lett.* **58** 2059.
[3] Burresi M, Engelen R J P, Opheij A, Oosten D van, Mori D, Baba T and Kuipers L, 2009 *Phys. Rev. Lett.* **102** 033902.
[4] Sakai O, Sakaguchi T and Tachibana K, 2005 *Appl. Phys. Lett.* **87** 241505.
[5] Sakai O and Tachibana K, 2007 *IEEE Trans. Plasma Sci.* **35** 1267.
[6] Sakai O, Sakaguchi T and Tachibana K, 2007 *J. Appl. Phys.* **101** 073304.
[7] Guo B, 2009 *Phys. Plasmas* **16** 043508.
[8] Fan W, Zhang X and Dong L, 2010 *Phys. Plasmas* **17** 113501.
[9] Sakai O, Naito T and Tachibana K, 2010 *Phys. Plasmas* **17** 057102.
[10] Takahashi K, Terui T, Chiba H and Fujiwara T, 2011 *IEEE Trans. Plasma Sci.* **39** 2504.

Surface-charge measurements in microgap dielectric barrier discharge using bismuth silicon oxide crystals

S Mukaigawa[1,2], **H Matsuda**[1], **H Fue**[1], **R Takahashi**[1], **K Takaki**[1,2], **T Fujiwara**[1,2]

[1]Faculty of Engineering, Iwate University, Morioka 020-8551, Japan

[2]Soft-Path Engineering Research Center (SPERC), Faculty of Engineering, Iwate University, Morioka 020-8551, Japan

e-mail: mukaigaw@iwate-u.ac.jp

Abstract. A surface-charge measurement system based on the Pockels effect in bismuth silicon oxide dielectric crystals was constructed for measuring the surface-charge density of the dielectrics in the microgap dielectric barrier discharge.We re-examined the calculation methods of the surface-charge density from the voltage applied to the BSO crystal, obtained by laser interferometry. The charge calculated using the circuit equation coincided with the that obtained using the discharge current. Under certain experimental conditions, the maximum values of the surface charge density in the discharge cell with and without a protection glass were +2.0 and +2.5 nC/cm^2, respectively.

1. Introduction

Dielectric barrier discharge is used in many fields such as plasma surface treatment, lighting, and plasma display panels, and is often used in the production of microplasma. The charging effect of the inner wall in microplasma is more important than that in macroscale discharge. The charge accumulated on the dielectric surface is an important parameter controlling the minimum value of the sustaining voltage and its uniformity [1-3]. Because the surface dielectric barrier charges accumulated in the half-cycle of the driving voltage favor the reappearance of the discharge, surface charges play a key role in the generation and stabilization of emerging patterns. Thus, it is essential to measure the quantity and distribution of surface charges in self-organized discharge [4-6].

Recently, the measurement of surface charge in bismuth silicon oxide $Bi_{12}SiO_{20}$ (BSO) crystals using the Pockels effect has become one of the most important methods for understanding the generation of self-organized plasma discharge and plasma photonic crystal applications [7-16]. When performing charge measurements by the laser interference method using the Pockels effect in BSO, the directly observable physical quantity is the intensity of laser light, and the physical quantity deduced from this intensity is the voltage applied to a BSO crystal. When deducing the electric charge accumulated in a BSO crystal surface from this voltage, some researchers tacitly assume that $Q = CU_{BSO}$, where Q is the charge accumulated on the BSO crystal, C is the capacitance of the BSO crystal, and U_{BSO} is the voltage applied to the BSO crystal induced by the applied electric field and others employed the inverse calculation from the potential distribution to the charge distribution on the dielectric surface [17, 18].

In this study, we re-examined the simplified calculation method of the surface-charge density from the voltage applied to the BSO crystal obtained by laser interferometry. In the surface-charge

Journal of Physics: Conference Series **441** (2013) 012012 doi:10.1088/1742-6596/441/1/012012

measurements of these electro–optic crystals, the crystal acts as the dielectric barrier. In the present study, we also introduced a glass layer on the surface of the BSO crystal to prevent damage from plasma irradiation and investigated the influence of the glass layer on the measured surface charge.

2. Basic principles and experimental setup

The refractive indices of electro-optic materials such as single-crystal BSO change when an electric field is applied. Birefringence causes the refractive index to differ at orthogonal directions in a crystal. When an electric field is applied to BSO, the phase retardation $\Delta\phi$ between the ordinary and extraordinary rays is given by

$$\Delta\phi = \frac{2\pi}{\lambda} n_0^3 r_{41} U_{BSO} , \quad (1)$$

where λ is the wavelength of light, n_0 is the normal refractive index, and r_{41} is the coefficient of the Pockels effect.

Figure 1 shows the schematic of the optical measurement system. It consists of a He-Ne laser with a wavelength of 632.8 nm, an analyzer, a quarter-wave plate, a polarizer, a band pass filter, a semiconductor detector (Hamamatsu Photonics C6386-01), and a discharge cell. When the incident light passes through the polarizer, the light becomes linearly polarized as only the +45°orientation can pass through it. Subsequently, after the quarter-wave plate, it changes to circularly polarized light. When electric field is applied to the BSO crystal, light phase retardation appears and the light changes from circularly polarized to elliptically polarized. Therefore, light intensity changes after it passes through the analyzer at -45° orientation. The output light intensity I is given by the following equation:

$$I = \frac{I_0}{2}\left\{1 + \sin\left(\frac{2\pi}{\lambda} n_0^3 r_{41} U_{BSO}\right)\right\}, \quad (2)$$

where I_0 is the intensity of the linearly polarized light. According to Eq. (2), the voltage applied to the BSO crystal, U_{BSO}, can be derived from the light intensity. When the phase retardation $\Delta\phi$ is small, the light intensity I is approximated by the following equation:

$$I = \frac{I_0}{2}\left(1 + \frac{2\pi}{\lambda} n_0^3 r_{41} U_{BSO}\right) . \quad (3)$$

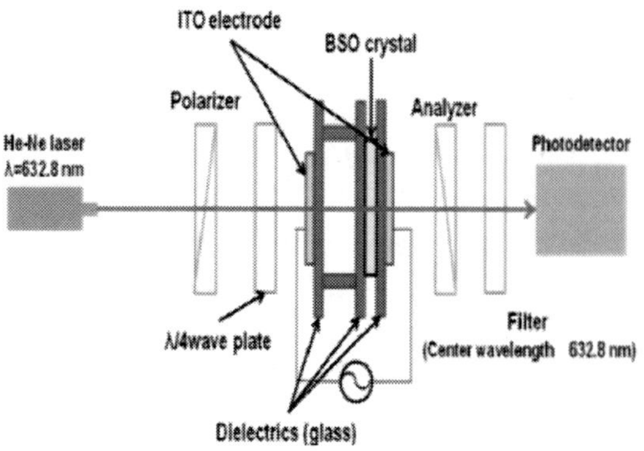

Figure 1. Experimental apparatus.

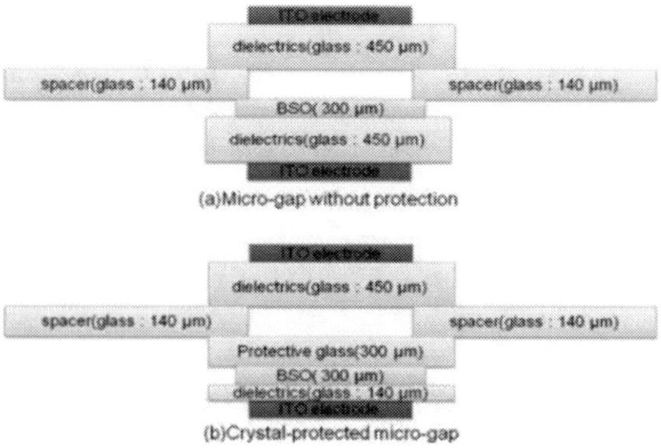

Figure 2. Structure of the microgap discharge cell.

Figure 2 shows the structure of the microgap discharge cell. It consists of a dielectric layer (glass), a pair of spacers (glass), and the BSO crystal. The thickness of the BSO crystal a_{BSO} was 300 μm, the relative permittivity ε_{BSO} was 56, the normal refractive index n_0 was 2.56, and the coefficient of Pockels effect r_{41} was 5 pm/V. Influence of optical rotation in BSO is negligible because of the thickness of the BSO sample [19] .An antireflective coating for light of 632.8 nm wavelength was applied to the BSO crystal surface. In case (a) that the BSO crystal was not protected from the discharge by the glass layer, a 450-μm-thick glass plate was used for the upper and lower dielectric layers. In case (b) that the BSO crystal was protected from the discharge by a glass layer, a 300-μm-thick glass plate was placed between the spacers and the BSO crystal, and 450-μm- and 140-μm-thick glass plates were used for the upper and lower dielectric layers, respectively. This was done to equalize the voltage applied to the BSO crystal by making the resultant capacity of the discharge cell almost the same. The relative permittivity of the glass plates was 7. The discharge gap length was controlled by the thickness of the spacer, which was 140 μm long. Indium Tin Oxide (ITO: resistivity 10-20 Ω/cm^2, area 10×15 mm^2) was deposited as an electrode on the surface of the dielectric. The ITO coated one side of the dielectrics. We applied a sinusoidal voltage with 100 kHz frequency and 1000 V amplitude to the upper ITO electrode. The lower ITO was grounded. The discharge was generated between the gaps. We introduced helium gas into the gap with a flow velocity of 14.0 m/s. The applied voltage and current were measured using a high voltage probe (Tektronix P6015A) and Rogowski coil (Pearson CT2877). The discharge current was calculated from the measured current.

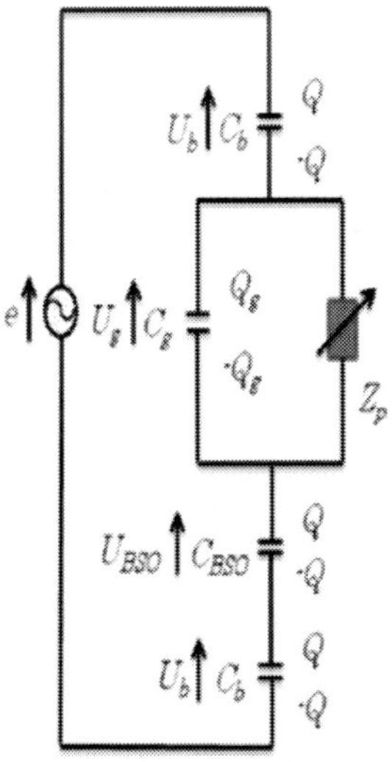

Figure 3. Equivalent circuit of the microgap discharge cell.

3. Calculation of surface charge

When applying the experimental optical system to measure the surface charge, it is necessary to accurately evaluate the relation between the surface charge density and voltage applied to BSO(U_{BSO}). The simplest calculation method assumes that all the electric lines of force of the electric charge accumulated on the BSO crystal surface pass through the BSO crystal. The relation between the surface-charge density σ obtained by this method and voltage U_{BSO} is expressed as follows:

$$U_{BSO} = U_{BSO}^{ex} + \frac{\sigma}{C_{BSO}}, \tag{4}$$

where C_{BSO} is the capacitance of the BSO crystal per unit area, which is expressed as $C_{BSO} = \varepsilon_0 \varepsilon_{BSO} / a_{BSO}$; ε_0 is the vacuum permittivity; and U_{BSO}^{ex} is the voltage of the BSO crystal induced by the external electric field. Since the voltage component due to the surface charge and the voltage component due to the external electric field are contained in U_{BSO}, in order to measure the surface charge by laser interferometry, the effect of the voltage due to the external electric field on the laser light intensity I must be removed. Subsequently, the laser light intensity I_r measured in the absence of surface charge (reference light) is subtracted from the intensity of the laser light I measured when the BSO crystal is charged. I_r is expressed as follows:

$$I_r = \frac{I_0}{2}\left(1 + \frac{2\pi}{\lambda} n_0^3 r_{41} U_{BSO}^r\right), \tag{5}$$

where U_{BSO}^r is the voltage applied to the BSO crystal without surface charge. Assuming $U_{BSO}^{ex} = U_{BSO}^r$ and using Eqs. (3) and (5), the following equation is obtained:

$$\sigma = (I - I_r)\frac{2}{I_0}\frac{C_{BSO}}{2\pi n_0^3 r_{41}/\lambda}. \tag{6}$$

The method for determining the relation between the surface-charge density and voltage U_{BSO} applied to the BSO crystal using a circuit equation is an alternative method. The equivalent circuit of the discharge cell in figure 2 is shown in figure 3. The circuit equation is expressed in terms of the surface-charge density σ_t obtained by this method and the voltage e applied to the electrodes, as follows:

$$\begin{cases} e = U_{BSO} + 2U_b + U_g \\ Q - Q_g = \sigma_t \\ Q = C_b U_b = C_{BSO} U_{BSO} \\ Q_g = C_g U_g \end{cases}, \tag{7}$$

where C_b is the capacitance of the dielectrics per unit area, C_g is the capacitance of the discharge gap per unit area, U_b is the voltage applied to the dielectrics, U_g is the voltage applied to the discharge gap, Q is the charge density in the dielectrics and BSO crystal, and Q_g is the charge density accumulated in the discharge gap (the difference between Q and σ_t). The following equation is obtained from Eq. (7):

$$U_{BSO} = \frac{C_T}{C_{BSO}}\left(e + \frac{\sigma_t}{C_g}\right), \tag{8}$$

where C_T is the series capacitance of C_{BSO}, C_g, and the two C_b, and is expressed as $1/C_T = 1/C_{BSO} + 1/C_g + 2/C_b$. The first term in Eq. (8) is equal to the voltage of the BSO crystal calculated by the capacitive voltage division in series capacitors. Although $U_{BSO}^{ex} = U_{BSO}^{r}$ had to be assumed in the first method, it is automatically filled in the present method. The second term in Eq. (8) differs from that in Eq. (6) C_T/C_g times because the electric lines of force emitted by the surface charge are divided into the BSO crystal and discharge gap side. Using Eqs. (3), (5), and (8), the following equation is obtained:

$$\sigma_t = (I - I_r)\frac{2}{I_0}\frac{C_{BSO}}{2\pi n_0^3 r_{41}/\lambda}\frac{C_g}{C_T}. \tag{9}$$

The charge density σ_t differs C_g/C_T times from the surface charge density σ calculated by the first method. The value of C_g/C_T is 1.85 with the structure of figure 2(a), and is 2.12 with the structure of figure 2(b).

4. Results

Figure 4 shows the applied voltage and discharge current waveform in the discharge cell without the protection glass. The discharge current was calculated following Liu et.al. [20]. We define the starting time (t=0) as the time at which the driving voltage is 0 V. Double peak current occurs per half period of the driving voltage. The first peak appears at t =0.4 μs, and the second appears at t =1.6 μs. The discharge current reaches the maximum value of 20.8 mA at 0.4 μs. The time width at which the current waveform reaches more than 20 % of the maximum value is 0.3 ns. A glow discharge covers the entire discharge area when the current pulse appears.

The time variations in the laser light I and reference light I_r are shown in figure 5 with the applied voltage waveform. When we observed the reference light I_r at the applied voltage of 1000V, air was used to prevent the formation of discharge in the microgap. There is an obvious difference in the waveform I and I_r, resulting from the surface charge on the BSO crystal.

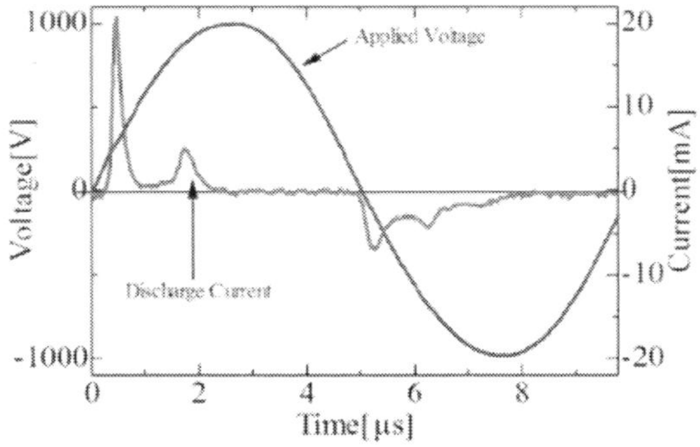

Figure 4. Applied Voltage and discharge current waveform in case of glow DBD in atmospheric pressure helium; amplitude V_p=1000 V, frequency f=100 kHz.

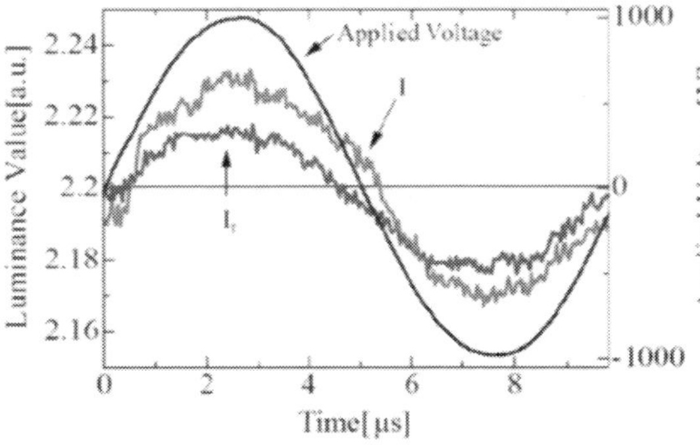

Figure 5. Intensity of the output laser light I and reference light I_r.

The time variation in the difference in the laser light and reference light, $I-I_r$, in the half period of the positive voltage is shown in figure 6 with the discharge current waveform. The value of $I-I_r$ is negative from t =0to 2.1μs. The value of $I-I_r$ begins to increase at t = 0.4 μs when the discharge current starts to flow, and reaches the maximum thereafter. Similarly, the $I-I_r$ synchronized with the discharge current at the half period of the negative voltage (t =5-10 μs). The change in the value of $I-I_r$ and the movement synchronization of the charge due to the discharge current imply that the information on the amount of electric charges on the dielectric surface is included in $I-I_r$.

Figure 7 shows the calculated surface-charge density and the discharge current waveform. Three methods were used to calculate the surface-charge density. That is, the charge was calculated using Eqs. (6), and (9), and the time integration of the discharge current. The BSO crystal surface was negatively charged before the electric discharge starts at t = 0.4s. During this period, the values of the surface-charge density calculated by Eq. (6), Eq. (9), and the integration of the discharge current were -0.8 nC/cm^2, -1.4 nC/cm^2 and -2.4 nC/cm^2, respectively. At the time t = 0.4 μs of the first discharge, the charge density increased simultaneously and the discharge stopped at t = 0.7 μs. The charge density did not change during t = 0.7-1.7 μs. The values of the surface-charge density calculated by Eq. (6), Eq. (9), and the integration of the discharge current were +1.0 nC/cm^2, +1.9 nC/cm^2, and +0.5 nC/cm^2, respectively. When the second discharge started at t = 1.7 μs, the charge density increased again and reached a maximum. The values of the surface-charge densities calculated using Eq. (6), Eq. (9), and the integration of the discharge current were +1.4 nC/cm^2, +2.5 nC/cm^2 and +2.5 nC/cm^2, respectively. Then, the charge density did not change till t = 5 μs, and the BSO crystal surface was positively charged. In the half period of the negative voltage (t = 5-10 μs), the charge density changed by synchronizing with the discharge current. The minimum values of the surface-charge density calculated using Eq. (6), Eq. (9), and the integration of discharge current were -0.8 nC/cm^2, -1.5 nC/cm^2, and -2.6 nC/cm^2, respectively. Then, charge density did not change till t = 10 μs, and the BSO crystal surface was negatively charged. A comparison of the calculation methods of the surface charge density shows that the calculation of the surface charge using the discharge current coincides with that using Eq. (9), whereas the calculation results based on Eq. (6) differ from those of the other two methods. Despite the simple assumption that all the electric lines of force of the electric charge accumulated on the BSO crystal surface pass through the BSO crystal, the calculation based on this assumption is not a good approximation.

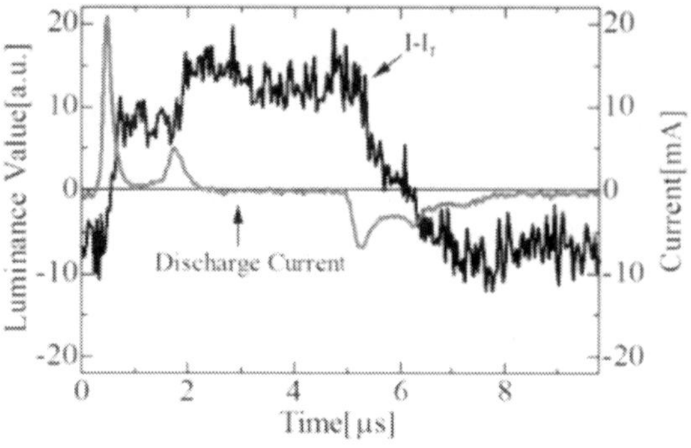

Figure 6. Difference between laser light and reference light $I-I_r$.

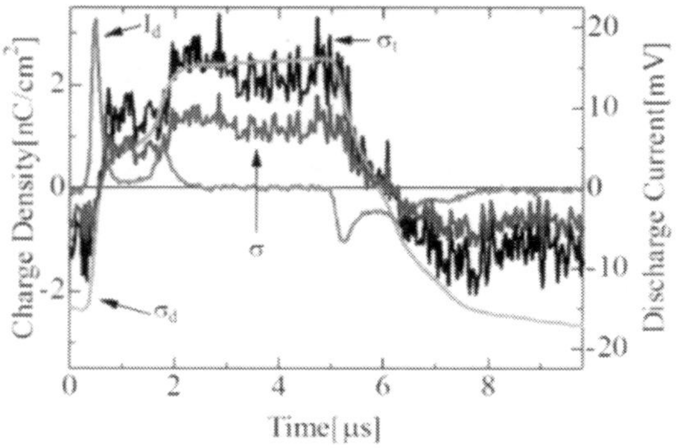

Figure 7. Comparison of the surface-charge density calculated with Eq. (6) (σ), Eq. (9) (σ_l), and by integrating the discharge current I_d (σ_d).

Figure 8 shows the time variation in the surface-charge density in the discharge cell with and without protection of the BSO crystal. The calculation based on Eq. (9) was used to evaluate the surface charge. When the discharge cell with protection glass was used, the large noise components were contained in the output voltage of the semiconductor detector. It is expected that a malfunction of the antireflective coating on the BSO surface due to the direct junction between the BSO crystal and cover glass causes multiple reflections. The maximum values of the surface-charge density in the discharge cell with and without the protection glass were +2.0 nC/cm^2 and +2.5 nC/cm^2, respectively. Despite equalizing the parameters of the equivalent circuit of the electric discharge cell, there was a difference in the surface density of the charge. The reason of this difference is that the state of the plasma changed because of the differences in the structure of the plasma/solid interface (e.g. secondary electron emission coefficient) [21].

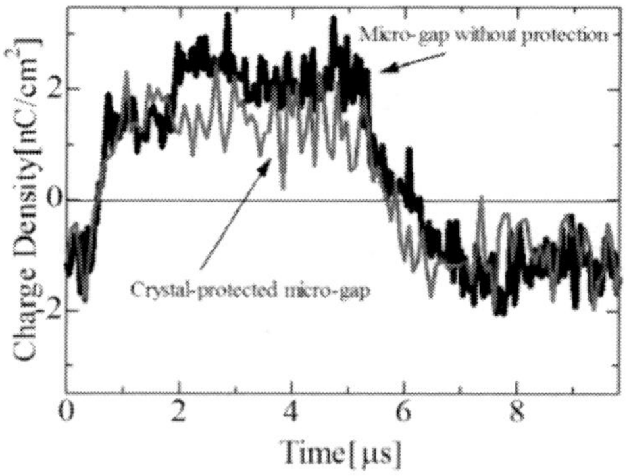

Figure 8. Comparison between the surface-charge density in the discharge cell with and without protection of the BSO crystal.

5. Conclusion

The calculation method of the surface-charge density using laser interferometry and a BSO crystal was re-examined. It was observed that the signal $I-I_r$ synchronizes with the discharge current. Thus, the change in signal $I-I_r$ corresponds to the movement of the surface charge to the opposite side of the dielectric surface in the discharge region. Information regarding the surface-charge density was extracted by subtracting the signal of the reference light I_r from the output light intensity I. Comparing the calculation methods of the surface-charge density, the calculation results of the surface charge using the discharge current coincide with those based on the circuit equation; however, the calculation results based on the assumption that all the electric lines of force of the electric charge accumulated on the dielectrics surface pass through the BSO crystal differ from those based on the other two methods. Furthermore, the surface-charge measurements using the BSO crystal were applied to a BSO crystal that was protected from discharge by a glass layer. When the applied voltage was 1000V and the flow velocity was 14.0 m/s, the maximum values of the surface-charge density in the discharge cell with and without the protection glass were +2.0 nC/cm^2 and +2.5 nC/cm^2, respectively.

Acknowledgement
This work was supported by JSPS KAKENHI Grant Numbers 23540570, 24540530.

References

[1] Kogelshatz U 2003 *Plasma Chem. Plasma Process.* **23** 1.
[2] Gibalov VI and Pietsch GJ 2000 *J. Phys. D* **33** 2618.
[3] Kogelshatz U 2002 *IEEE Trans. Plasma Sci.* **30** 1400.
[4] Duan X, Ouyang J, Zhao X and He F 2009 *Phys. Rev. E* **80** 016202.
[5] Stollenwerk L, Amiranashvili Sh, Boeuf JP and Purwins H-G 2007 *Eur. Phys. J. D* **44** 133.
[6] Chirokov A, Gutsol A, Fridman A, Sieber K, Grace J and Robinson K 2005 *IEEE Trans. Plasma Sci.* **33** 300.
[7] Stollenwerk L,Laven JG, and Purwins H-G 2007 *Phys. Rev. Lett.* **98** 255001.
[8] Abolmasov SN, Abo R, Shirafuji T and Tachibana K 2006 *Japanese J. Appl. Phys.* **45** 8255.
[9] Sugimoto K, Takahashi K, Shimomura O and Sakurai T 2003 *J. Phys. D, Appl. Phys.* **36** 2887.
[10] Takeuch N, Hamasaki T, Yasuoka K and Sakurai T 2011 *J. Electrostat.* **69** 87.
[11] Takeki S, Yoda H, Tetayama T, Ishii K and Murakami Y 2007 *Japanese J.Appl. Phys.* **46** 3596.
[12] Jeong DC, Bae HS and Whang KW 2005 *J. Appl. Phys.* **97** 013304.
[13] Gegot F, Callegari Th, Aillerie M and Boeuf JP 2008 *J. Phys. D: Appl. Phys.* **41** 135204.
[14] Mu HB, Zhang GJ, Komiyama Y, Suzuki S, Miyake H, Tanaka Y and Takada T 2011 *IEEE Trans. Dielect. Insul.* **18** 485.
[15] Zhu Y, Takada T, Sakai K and Tu D1996 *J. Phys. D : Appl. Phys.* **29** 2892.
[16] Kawasaki T, Terashima T, Suzuki S, and Takada T 1994 *J. Appl. Phys.* **76**, 3724.
[17] Kumada A, Fukuda Y, Kamada Y and Hidaka K 2006 *IEEJ Transactions on Electrical and Electronic Engineering*, 2006 **1**, 95
[18] Tanaka D, Matsuoka S, Kumada A and Hidaka K 2009 *J. Phys. D: Appl. Phys.* **42,** 075204
[19] Pellat-Finet P *Optics Communications* 1984 **50** 275.
[20] Liu S and Neiger M 2003 *J. Phys. D: Appl. Phys.* **36** 3144.
[21] Fue H, Hasegawa T, Sato T, Mukaigawa S, Takaki K and Fujiwara T 2011 *IEEE Trans. Plasma Sci.* **39** 2140.

Journal of Physics: Conference Series **441** (2013) 012013 doi:10.1088/1742-6596/441/1/012013

Effects of metastable atoms on breakdown voltage in Argon DBD

T Yoshinaga and H Akashi

National Defense Academy of Japan, 1-10-20 Hashirimizu, Yokosuka, Kanagawa 239-8686, Japan

yoshinag@nda.ac.jp

Abstract. A one-dimensional fluid calculation is performed to find out the effects of metastable atoms in alternating-current (AC) driven dielectric barrier discharge (DBD) of argon. Dependences of the breakdown voltage and the density of the metastable state atoms on the driving frequency and the pressure are investigated to clarify the importance of the secondary electron emission by metastable atoms and the metastable-metastable ionization reaction. The results suggest that such processes can be significant when the driving frequency becomes several hundred Hz under several Torr.

1. Introduction

DBD is applied in a variety of chemical plasma processes and light sources to generate non-equilibrium plasmas [1]. Dielectric layers play an important role by suppressing excessive current which leads to the damaging of electrodes. They are also important since the electrons necessary for self-sustaining discharge are yielded via secondary electron emission. While the dielectric materials (e.g. MgO) are regarded to have higher secondary electron emission coefficient (γ) than metals, those effects have not been evaluated well. Two major experimental methodologies have been attempted to estimate γ. One is a direct measurement in which the accelerated ion beam is bombarded and the emitted electron current is measured [2]. This method can measure γ defined as the ratio of the electron yield to the injected ion flux, while the injected ions tend to have higher energy levels (over tens of eV) than that of interest in DBD operation (at most ~ 10 eV). The other method measures the breakdown voltage ($V_{\rm bd}$) between two parallel planar electrodes and deduce the secondary ionization coefficient (γ') using the Townsend's criterion for the self-sustaining discharge [3-6]. This method is rather close to the practical discharges in terms of the energy level of bombarding particles. While γ and γ' are strongly connected with each other, those values would not be exactly the same since γ' includes the backscattering effect via collisions with the background gas particles. The experimental approaches often take into account the contributions only from the ions. However, there exist other kinds of excited species in actual discharges. For example, it is well known that the effect of metastable atoms cannot be neglected especially in AC-driven DBDs because of their long lifetimes [7].

In this work we examine the effect of the secondary electron emission by metastable atoms and the metastable-metastable ionization reaction by investigating the discharge characteristics such as $V_{\rm bd}$ in relation to the driving frequencies (f) and pressures (p_0) using a one-dimensional numerical model.

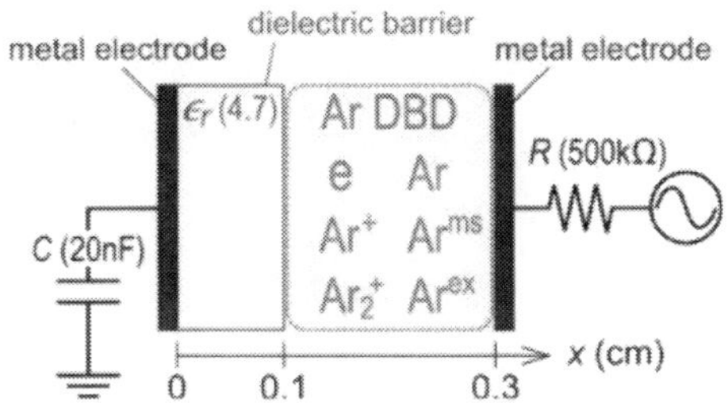

Figure 1. Schematic of one-dimensional fluid model for argon DBD.

Table 1. Reaction processes included in the present model.

No.	Reaction	Rate coeff. (cm^3s^{-1})	Ref.
1	$e + Ar \rightarrow Ar^{ms} + e$	$k_{ms}(\varepsilon_e)$	[8,9]
2	$e + Ar \rightarrow Ar^{ex} + e$	$k_{ex}(\varepsilon_e)$	[8,9]
3	$e + Ar \rightarrow Ar^{+} + 2e$	$k_{i}(\varepsilon_e)$	[8,9]
4	$e + Ar^{ms} \rightarrow Ar^{+} + 2e$	$k_{cui}(\varepsilon_e)$	[8,10]
5	$Ar^{ms} + Ar^{ms} \rightarrow Ar^{+} + Ar + e$	5×10^{-10}	[11]
6	$Ar^{ex} \rightarrow Ar + h\nu$	10^{-9} [a]	[8]
7	$Ar^{+} + 2Ar \rightarrow Ar_2^{+} + Ar$	2.5×10^{-31} [a]	[8,12]
8	$Ar_2^{+} + e \rightarrow 2Ar$	8.8×10^{-8}	[13]

[a] Dimensions of reaction 6 and 7 are (s^{-1}) and (cm^6s^{-1}).

2. Numerical Model

A simple one-dimensional fluid model is adopted in this study. Figure 1 shows the discharge gap configuration with the external circuit. The grounded side of the parallel plane electrodes with cross section of 4 cm^2 is covered with the dielectric material with relative permittivity (ϵ_r) at 4.7 and thickness (d_b) of 0.1 cm. The discharge gap (d) is 0.2 cm. The plasma is assumed to be uniform on the plane parallel to the electrodes, so that the transport parameters vary only with the distance from the electrodes. It is also assumed that the plasma is consisted with electrons (e), ions (Ar^+, Ar_2^+), metastable atoms (Ar^{ms}) and other excited atoms (Ar^{ex}) besides the ground-state argon atoms (Ar). There are an external resistor (R = 500 kΩ) connected on the powered side and a capacitor (C = 20 nF) on the grounded side as experiments [6]. The continuous-wave (CW) sinusoidal voltage with amplitude at 500 V is applied. The one-dimensional equation of continuity in discharge space is denoted by,

$$\frac{\partial n_j}{\partial t} = -\frac{\partial \Gamma_j}{\partial x} + S_j\,, \tag{1}$$

$$\Gamma_j = W_j n_j - D_j \frac{\partial n_j}{\partial x}\,, \tag{2}$$

where the subscript j represents the particle species (e, Ar^+, Ar_2^+ and Ar^{ms}). n_j, Γ_j, S_j, W_j, and D_j represent the particle density, flux, production (or loss) term, drift velocity and diffusion coefficient of the particle j, respectively. Only W_j of charged particles can have finite values. The advection of Ar^{ex} is not calculated in this study since the de-excitation dominates because of its short lifetime (esmitated as 10^{-9} s). The reactions for S_j included in the present calculation are listed in table 1. The above swarm parameters are deduced using the Monte-Carlo simulation with the same cross-sections described in [8]. The conservation of electron energy is considered as follows,

$$\frac{\partial (n_e \varepsilon_e)}{\partial t} = -\frac{5}{3}\left(\Gamma_e \varepsilon_e - D_e n_e \frac{\partial \varepsilon_e}{\partial x}\right) - eE\Gamma_e + S_\varepsilon\,. \tag{3}$$

Here, ε_e is the mean energy of electrons, E the electric field and S_ε the gain and loss term of electron energy via inelastic reactions. It is assumed that the electron velocity distribution is Maxwellian. Time evolutions of the particle densities and the electron energy are computed using equations (1) – (3) coupled with the Poisson's equation,

$$\frac{\partial E}{\partial x} = e\left(n_{Ar^+} + n_{Ar_2^+} - n_e\right)/\epsilon_0\,. \tag{4}$$

At the plasma boundaries, the secondary electrons flow into the discharge space both from the metal electrode and the dielectric surface. The fluxes Φ_{2nd} are given by,

$$\Phi_{2nd} = \gamma_i\left(\Gamma_{Ar^+} + \Gamma_{Ar_2^+}\right) + \gamma_m\Gamma_{Ar^{ms}} + \gamma_p\Gamma_p\,, \qquad (5)$$

where γ_i, γ_m and γ_p represent the secondary electron emission coefficient of ions, metastable atoms and photons, respectively. Γ_p is the photon flux onto the electrodes. The values of γ will be different depending on the influx species and electrode materials.

3. Results and discussion

Figure 2 shows waveforms of a discharge driven at $f = 20$ Hz under $p_0 = 5$ Torr. Positive polarity of the discharge current corresponds to the direction in which the positive voltage is applied on the powered electrode and the positive ions flow towards the grounded dielectric electrode ($-x$ direction in figure 1). Secondary electron emission coefficients (γ_i, γ_m, γ_p) are all set at 0.01 for both of the metal and dielectric electrodes. The breakdown occurs with the inrush discharge current (I) with a peak at ~ 10 μA at t ~ 0.1 s as shown in figure 2 (c). The gap voltage (V_g) is ~ 250 V at this moment. After this breakdown I immediately decreases and V_g stops increasing and is kept constant at ~ 235 V until the source voltage (V_s) starts to decrease as shown in figures 2 (c) and (d). There remains a small current flowing after the breakdown and the positive charge of ions accumulates on the dielectric barrier (q) as shown in figure 2 (c). This charge compensates the excessive voltage impressed between the discharge gap despite the increasing V_s. The spatially averaged electron density ($\langle n_e\rangle$) increases and decreases along with I, while the spatially averaged density of metastable atoms ($\langle n_{ms}\rangle$) shows a quite slow decrease after the breakdown due to the long lifetime and the small diffusion coefficient as shown in figure 2 (a). The spatially averaged electron energy ($\langle \varepsilon_e\rangle$) has risen upto ~ 10 eV before breakdown and remains roughly constant until V_g starts to decrease at t ~ 0.115 s as shown in figure 2 (b). Then, $\langle \varepsilon_e\rangle$ decreases to almost zero when the polarity of V_g is reversed (t ~ 0.12 s). The magnified waveforms when breakdown occurs [figures 2 (e) and (h)] show that I around 1 μA increases exponentially with t as well as $\langle n_e\rangle$ with the time constant at $\tau_{1\mu A}$ [$= I/(dI/dt)$] ~ 5 μs. As shown in

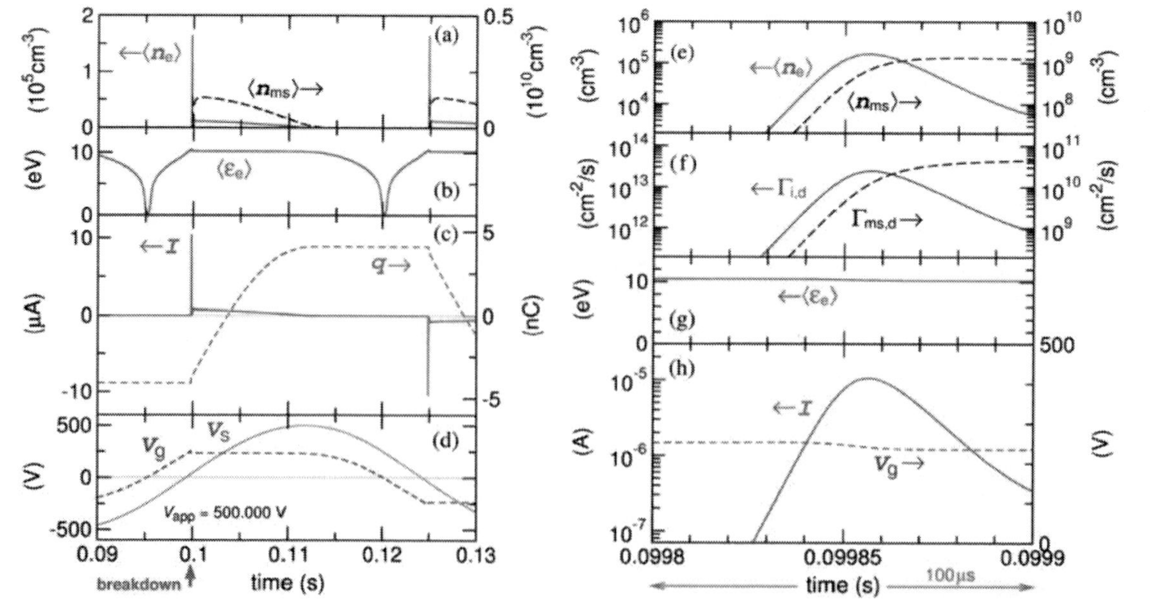

Figure 2. Typical waveforms of a discharge driven at 20 Hz. (a): Spatially-averaged density of electrons $\langle n_e\rangle$ and metastable atoms $\langle n_{ms}\rangle$. (b): Spatially-averaged electron energy $\langle \varepsilon_e\rangle$. (c): Discharge current I and accumulated charge on the dielectric layer q. (d): Applied voltage V_s and gap voltage V_g. (e)-(h): Magnified waveforms around breakdown event. (e): Density ($\langle n_e\rangle$ and $\langle n_{ms}\rangle$). (f): Particle flux of ions $\Gamma_{i,d}$ and metastable atoms $\Gamma_{ms,d}$ onto the dielectric layer. (g): Electron energy ($\langle \varepsilon_e\rangle$). (h): Discharge current I and gap voltage V_g.

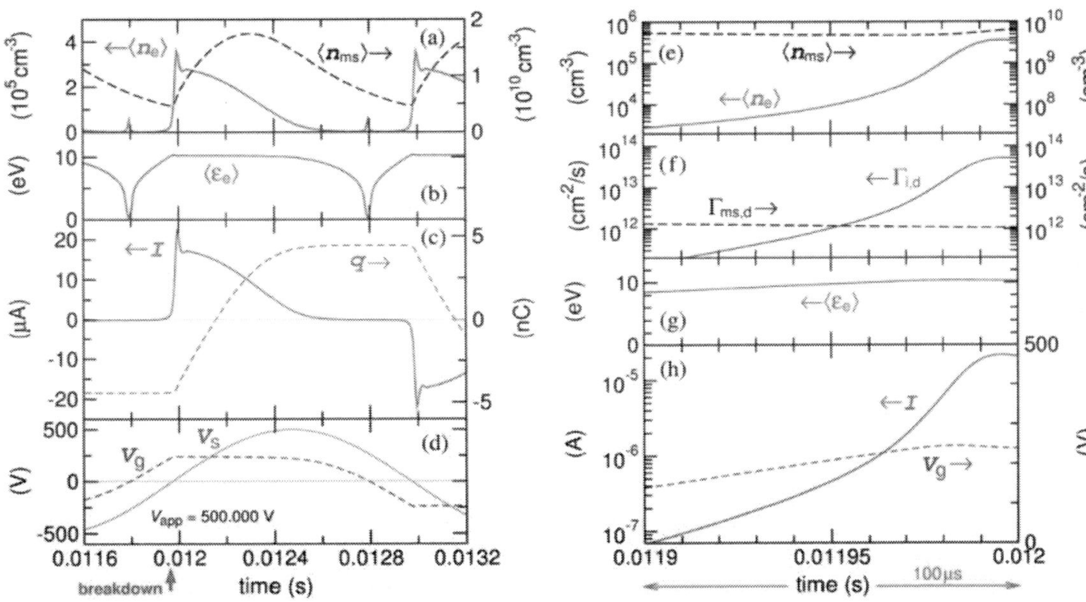

Figure 3. Discharge waveforms driven at 500 Hz. Plotted parameters are the same as shown in figure 2.

figure 2 (f) the influx of ions flowing onto the dielectric layer ($\Gamma_{i,d}$), which shows an exponential increase as I and $\langle n_e \rangle$, is found to be 10^3 - 10^4 times higher than the metastable atoms' flux ($\Gamma_{ms,d}$). Here, $\Gamma_{i,d}$ represents the summed flux of monomer and dimer ions ($\Gamma_{Ar^+} + \Gamma_{Ar_2^+}$), while the monomer ions flux is several hundred times higher than the dimer ions flux. Since the three-body reaction will hardly occur under such low pressure discharge, dimer ions' density is low compared with the monomer ions' density. The secondary electron emission by photons can be negligible since the photon flux ($\Gamma_{p,d}$) is estimated at ~ 1 - 100 cm^{-2}/s, being much less than even $\Gamma_{ms,d}$ in this discharge. Therefore, ions are the most significant species for the secondary electron emission at the breakdown as long as γ_i does not become several orders of magnitudes smaller than γ_m.

Figure 3 shows the waveforms of a discharge driven at f = 500 Hz, while other discharge conditions are the same with the discharge shown in figure 2. Similarly to the discharge at f = 20 Hz, I increases rapidly when V_g reaches some particular level (breakdown occurs) and V_g becomes almost constant (between 230 - 240 V) after the breakdown as shown in figures 3 (c) and (d). After the inrush of the initial current, however, I does not decrease immediately and forms a long tail in the order of 10 µA for ~ 300 µs. A long decay time can be found also in $\langle n_e \rangle$ as shown in figure 3 (a). The time evolution of $\langle n_{ms} \rangle$ is of the largest difference from the case at f = 20 Hz. It appears to increase and decrease independently of the other parameters and remains finite throughout the discharge cycle [see figure 3 (a)]. The long decay time of $\langle n_{ms} \rangle$ would be originated by that the metastable atoms' motion is not governed by the electric field but only by the slow diffusion adding to their long lifetimes. Both the metastable-metastable ionization (reaction 5 in table 1) and the secondary electron emission can occur throughout the discharge due to the remaining metastable atoms, yielding electrons and ions. This suggests that $\langle n_{ms} \rangle$ would be a significant parameter in high frequency operations. Since $\Gamma_{ms,d}$ is found to be similar to $\Gamma_{i,d}$ (~ 10^{12} cm^{-2}/s) around the breakdown (t ~ 0.01195 s) the metastable atoms would have a comparable contribution to the ions for the secondary electron emission as shown in figure 3 (f). I and $\langle n_e \rangle$ do not increase exponentially but in a complex and slower way in contrast to the discharge at 20 Hz as shown in figures 3 (e) and (h). The time constant at 0.1 µA ($\tau_{0.1\mu A}$) is ~ 30 µs, while it decreases to $\tau_{1\mu A}$ ~ 13 µs at 1 µA and $\tau_{10\mu A}$ ~ 8 µs at 10 µA.

Since it is difficult to define the breakdown voltage exactly, we simply focus on the voltage when I reaches 1 µA ($V_{1\mu A}$). Figure 4 (a) shows the dependence of $V_{1\mu A}$ on f with different γ_m. Without the

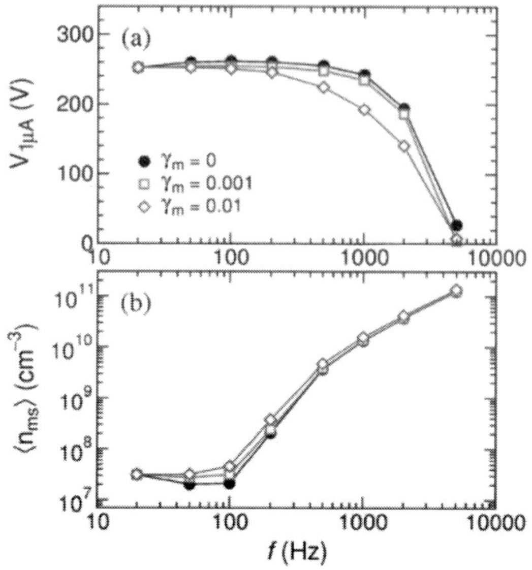

Figure 4. Dependence of (a) breakdown voltage $V_{1\mu A}$ and (b) metastable atoms density $\langle n_{ms}\rangle$ on driving frequency with different γ_m.

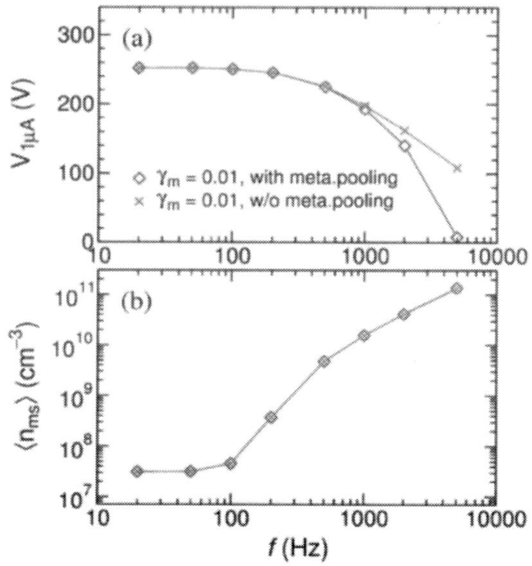

Figure 5. Comparison of frequency characteristics between the cases with and without the metastable-metastable ionization reaction process.

secondary electron emission by metastable atoms ($\gamma_m = 0$), $V_{1\mu A}$ is independent of f when f is lower than ~ 200 Hz, while it decreases with f over 500 Hz. $\langle n_{ms}\rangle$ at the same moment (I = 1 μA) increases with f when f becomes higher than 100 Hz as shown in figure 4 (b). Since the decrease of $V_{1\mu A}$ becomes remarkable as γ_m becomes higher, it is suggested that the secondary electron emission by the metastable atoms strongly affects the breakdown voltage under such higher frequency range. The condition under which $V_{1\mu A}$ appears to decrease at f = 500 Hz with $\gamma_m = 0.01$ corresponds to the condition when the secondary electron flux by the metastable atoms ($\gamma_m \Gamma_{ms,d}$) becomes comparable to that by ions ($\gamma_i \Gamma_{i,d}$) in figure 3. On the other hand, $V_{1\mu A}$ drops when $\langle n_{ms}\rangle$ becomes ~ 10^{10} cm^{-3} even with $\gamma_m = 0$, suggesting that the other effects of metastable atoms (e.g. metastable-metastable ionization or cumulative ionization from metastable atoms) are also taking place. Note that $V_{1\mu A}$ becomes almost zero at f = 5 kHz with any γ_m as shown in figure 4 (a). This indicates that the discharge becomes continuous and the current-carrying species remain in the discharge space throughout one discharge cycle.

Figure 5 compares the case with and without the metastable-metastable ionization process with γ_m = 0.01. The values of $V_{1\mu A}$ are quite similar under $f \lesssim 1$ kHz as shown in figure 5 (a). The difference between the two cases appears only under $f \gtrsim 2$ kHz, which is higher than the frequency level that the effect of γ_m appears. This suggests that higher $\langle n_{ms}\rangle$ is needed for the metastable-metastable ionization process than γ_m effect to affect $V_{1\mu A}$. This process would not be the primary loss mechanism of metastable atoms since $\langle n_{ms}\rangle$ takes the same value in both cases as shown in figure 5 (b). Considering the discharge driven at f = 500 Hz, $\langle n_e\rangle \sim 2\times10^4$ cm^{-3} and $\langle n_{ms}\rangle \sim 4\times10^9$ cm^{-3}. The rate coefficient of cumulative ionization (k_{cui}) is roughly estimated at ~ 10^{-7} cm^{3}s^{-1} from $\langle \varepsilon_e\rangle \sim 10$ eV around breakdown [see figure 3 (b)], besides the rate coefficient of metastable-metastable ionization (k_{mm}) is assumed at 5×10^{-10} cm^{3}s^{-1} (see table 1). The reaction rates of these processes become $\langle k_{cui} n_e n_{ms}\rangle \sim 8\times10^6$ cm^{-3}s^{-1} and $\langle k_{mm} n_{ms}^2\rangle \sim 8\times10^9$ cm^{-3}s^{-1}, respectively, showing that the metastable-metastable ionization process dominates the loss processes of metastable atoms. Therefore, the metastable atoms would be lost mainly via the diffusion process (the diffusion coefficient is assumed as D_{ms} = 50 cm^{2}s^{-1}Torr^{-1} [14]) from the discharge space since the metastable-metastable ionization is not the dominant loss mechanism of metastable atoms.

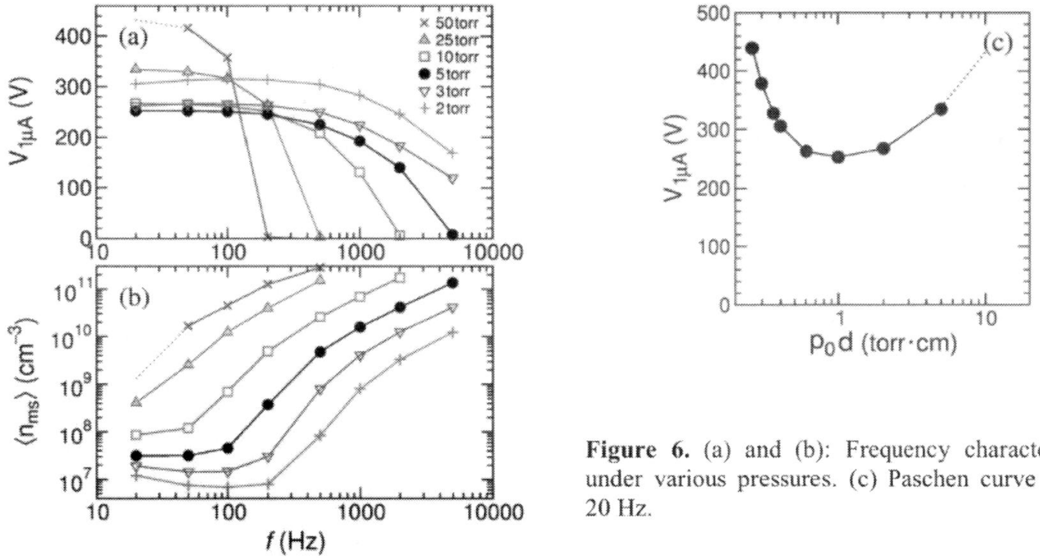

Figure 6. (a) and (b): Frequency characteristics under various pressures. (c) Paschen curve at f = 20 Hz.

Figures 6 (a) and (b) show the frequency dependences under different p_0. The frequency at which $V_{1\mu A}$ decreases becomes lower under higher p_0. The increase of $\langle n_{ms} \rangle$ with p_0 would be caused from the smaller diffusion coefficient of metastable atoms under higher p_0. Although a breakdown occurs at V_g ~ 430 V under f = 20 Hz and p_0 = 50 Torr, I does not reach 1 μA. By plotting $V_{1\mu A}$ at f = 20 Hz against $p_0 d$ the Paschen curve can be obtained as shown in figure 6 (c). This characteristics resembles the experimentally obtained one (figure 10 in [4]) although the values in detail are different. The pressure of 5 Torr corresponds to the pressure at Paschen minimum ($p_0 d$ = 1 Torr·cm).

4. Summary

The effects of metastable atoms in argon DBD between two parallel planar electrodes are investigated using one-dimensional fluid model. The breakdown voltage decreased with f when f exceeds several hundred Hz. Under p_0 = 5 Torr, the effect of γ_m first appears at f ~ 500 Hz. Next the metastable-metastable ionization process takes place at ~ 2 kHz. Under $f \lesssim$ 200 Hz the effects of metastable state atoms do not appear. The metastable atoms' density builds-up higher and the breakdown voltage decreases more largely under higher p_0. These results suggest that γ_m of the dielectric materials in DBD might be evaluated as well as γ_i by examining the breakdown voltage under various p_0 and f.

References

[1] Kogelschatz U 2003 *Plasma Chem. Plasma Process.* **23** 1
[2] Choi E-H et al. 1998 *Jpn. J. Appl. Phys.* **37** 7015
[3] Auday G, Guillot Ph and Galy J 2000 *J. Appl. Phys.* **88** 4871
[4] Suzuki S and Itoh H 2004 *Jpn. J. Appl. Phys.* **43** 7234
[5] Suzuki S and Itoh H 2007 *Jpn. J. Appl. Phys.* **46** 1129
[6] Suzuki S, Sekizawa T, Kashiwagi Y and Itoh H 2011 *Jpn. J. Appl. Phys.* **50** 106002
[7] Lymberopoulos D P and Economou D J 1993 *J. Appl. Phys.* **73** 3668
[8] Akashi H, Y Sakai and Tagashira H 1994 *J. Phys. D: Appl. Phys.* **27** 1097
[9] Sakai Y, Sawada S and Tagashira H 1986 *J. Phys. D: Appl. Phys.* **19** 1741
[10] Ton-That D and Flannery M R 1977 *Phys. Rev. A* **15** 517
[11] Rokni M, Jacobb J H, Mangano J A and Brochu R 1977 *Appl. Phys. Lett.* **31** 79
[12] Tinck S, Boullart W and Bogaerts A 2009 *J. Phys. D: Appl. Phys.* **42** 095204
[13] Otsuka S, Tochikubo F and Uchida S 2006 *Jpn. J. Appl. Phys.* **45** 7881
[14] Phelps A V 1953 *Phys. Rev.* **89** 1202

Journal of Physics: Conference Series **441** (2013) 012014 doi:10.1088/1742-6596/441/1/012014

CFHall Code Validation with 3D3V Weibel Instability Simulation

A.Yu. Perepelkina, I. A. Goryachev, V. D. Levchenko

Keldysh Institute of Applied Mathematics, RAS, Miusskaya sq., 4, Moscow, 125047, Russia

E-mail: mogmi@narod.ru

Abstract. Despite the existence of massively parallel codes and their application to plasma simulation, in the field of computer simulations the unsolved problems still remain. The necessity of various approximations in practical problems, for example, decreasing of the simulation dimension and physical parameter calibrations, arises, but such approximations can make the simulation results questionable.

While aiming for the maximal efficiency of computational algorithms it is possible to manage without the majority of popular approximations and to broaden the applicability limits of the numerical research. Such possibility arises with the use of the Locally Recursive non-Locally Asynchronous (LRnLA) algorithms as the base for a program code. This was used as a basis for CFHall plasma simulation code development. CFHall is presented in this work along with the brief description of the underlying algorithms. The applicability of the CFHall to plasma instability simulation is demonstrated on the example of weibel instability simulation.

1. Introduction

During the years of the existence of the numerical experiment in plasma physics it has shown its benefits in theoretical research for the verification of the arising hypotheses, as well as for the development of plasma technologies. Plasma simulation is used to obtain information which is impossible to get in the laboratory experiments, for simulation of stars and galaxies scale, or for small scale simulations, for which the probing devices are inefficient, or for simulation of the systems, where the presence of the probing devices distorts the plasma properties. While aiming for the advance of plasma technologies numerical experiment allows to observe the system changes caused by calibration of control parameters. Such calculation enables to save on the production of various devices to achieve the best device construction.

Fully kinetic codes are chosen for the simulation in the systems where the plasma distribution function cannot be approximated by its equilibrium state. Since the early 1980s a variety of plasma simulation codes has been developed, and the development of computer systems has always been accompanied by program code enhancements. However, despite the appearance of the massively parallel codes and their use on supercomputers, problems unsolved by numerical experiments still remain.

To overcome one of such problems, namely, to investigate the anomalous electron transport in Hall Effect thruster [1], the fully kinetic 3D3V code for plasma simulation CFHall has been developed.

The essence of the problem is in the fact that electron transport to anode in the Hall effect plasma thrusters is not explained within the theoretical approach. Plasma instabilities are considered the main reason of the process. However due to a large parameter range in the Hall thruster system (the ratio of plasma frequency to ion travel time, Debye radii to channel size) the necessary mesh size and time steps quantity becomes too large for a simulation with the existing simulation codes on the existing systems, including supercomputers. Therefore to obtain the estimates of Hall thruster parameters various approximations are commonly used. For example, lower dimensions, physical parameter calibration, hydrodynamic approach. One important drawback of the use of above approximations is the inability of adequate description of plasma instabilities while the instabilities play a decisive role. So such codes cannot be used for investigation of anomalous transport.

By improving the efficiency of the computational algorithms at the code's base it is possible to overcome the arising limits of the given computer systems. This had become the main idea in the development of the CFHall code. While aiming for the maximal efficiency of computational algorithms it is possible to manage without the majority of popular approximations and to broaden the applicability limits of the numerical research. The use of Locally Recursive non-Locally Asynchronous algorithms (LRnLA) [3] enables such possibility.

At the same time the code in question isn't limited by the above problem. It may be used for plasma research problems in various fields of plasma theory and technology, including the whole range of numerical experiment application in problems where the kinetic plasma description is necessary. Among the problems for which the code has been tested there are simple study problems such as Langmuir oscillations, Larmor circumrotation, skin effect, as well as numerical experiments for specific applications, such as laser interaction with sub-critical plasma layer, wake-field plasma acceleration, plasma transport through the crossed electric and magnetic fields fields.

In this work we demonstrate the CFHall code along with the basic algorithms in its base, and its applicability to plasma instability simulation is demonstrated on the example of the Weibel instability.

2. Code features

We use Vlasov equation for collisionless plasma dynamics and Maxwell equations for fields evolution. The numerical methods in use are PIC (Particle-in-Cell) [4] and FDTD (Finite Difference Time Domain). The main idea of the project is to maximize the efficiency of the computational algorithms.

For the numerical schemes with local dependencies the common way of data storage in multidimensional arrays require excessive time expenses for data access operations. Recursive data storage as in a Morton Z-curve [5] provide a solution to this problem. The Z-curve was implemented as a "cubeLR" data structure. In CFHall code we utilize Z-curve storage along 2 of the 3 spatial dimensions. In each cell of a 3D cubeLR structure a one cell wide column of field values along Z axis is stored.

LRnLA algorithms provide the superior dependency graph traversal rules by taking advantage of memory subsystem hierarchy, from on-chip CPU cash and up to disk and network memory. The benefits of the algorithms may be explained by the locality of the sequentially processed data. While providing a noticeable acceleration for sequential calculations, it is also optimized for distributed computations and is adaptable to parallel computations on any levels. In the CFHall code the LRnLA algorithm family ConeFold is utilized. Similarly to Z-curve storage, it may be considered as a recursively divided blocks, but in operation space (fig. 1).

CFhall also utilizes AVX/SSE vectorization for further acceleration. The component values along z axis are stored in a vectorized array. Array elements are stored in 4-element vectors. Operations on such vectors take the same amount of time as operations on scalars (double or float). Therefore it is possible make the calculation up to 4 times faster. The actual acceleration is slightly less due to costs of vector formation.

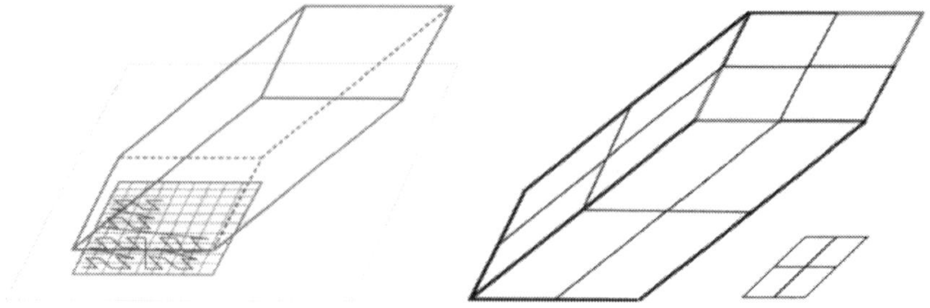

Figure 1. cubeLR visualization in data space and ConeFold visualization in operation space (left). ConeFold is recursively divided similarly to cubeLR (right).

3. Weibel instability

In the current study the Weibel instability is chosen to demonstrate the ability of plasma instability simulation, as well as computing time. The instability is named after E.Weibel who described it in 1959 [6] and arises in plasmas with anisotropic velocity distribution. Its mechanism has been explained on a simple model of two couterstreaming plasmas [7]. According to it one should take into account 2 cold charged particle streams flowing in opposite directions and a small magnetic field perturbation $B_x e^{iky}$ had appeared (see fig. 2). The Larmor rotation of plasma particles causes the periodical ($\sim e^{iky}$) unbalancing of the currents, which in turn adds to magnetic field growth due to Amper law. Thus the further exponential growth of the perturbation is explained. Since the wavelength of the most unstable modes are comparatively small, the filamentary structure is expected in the resulting oscillations.

Such magnetic field formation had become of interest as a possible reason of a stable magnetic field generation in plasmas, containing anisotropic velocity distribution.

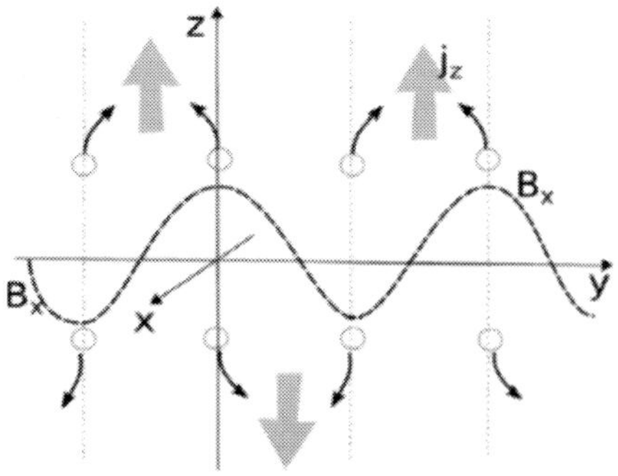

Figure 2. Weibel instability growth illustration. Initially there are two couterstreaming plasmas along z axis and a small perturbation $B_x e^{iky}$.

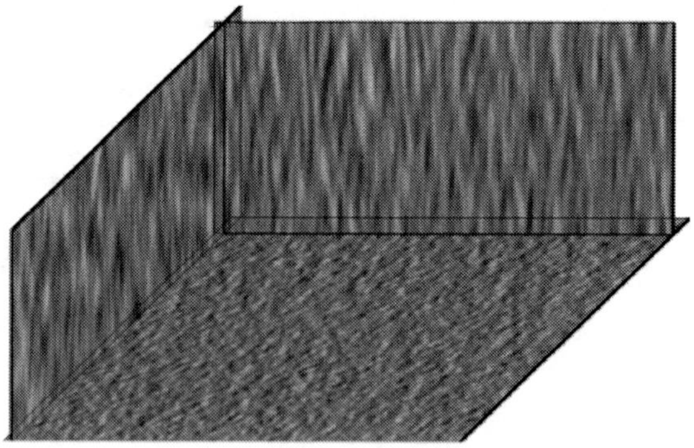

Figure 3. 3D distribution of the $B_x(x, y, z)$ field. $t = 11.52$, $B_x \sim 0.1$. As expected from theory, filaments along z axis are observed.

4. Problem statement and simulation results

In the current study the rise of the instability is observed in a system of the two colliding plasma streams. Plasma distribution function is written as

$$f(\vec{v}) = f_m(v_x)f_m(v_y)f_m(v_z - 0.5) + f_m(v_x)f_m(v_y)f_m(v_z + 0.5), \quad (1)$$

Journal of Physics: Conference Series **441** (2013) 012014 doi:10.1088/1742-6596/441/1/012014

where f_m is a Maxwell distribution function with temperature equal to 0.005. We use the dimensionless units. Electron masses m_e for mass, electron charges e for charge, light speed c for velocities. From these the units for distances c/ω_p and for fields $m_e\omega_p c/e$ follow. All initial field values are zero. For the estimate of the maximal instability increment on the linear stage the formula from [8] was used. For the chosen parameters the most unstable mode has the wavenumber $k \sim 35$ ($\vec{k} \perp \vec{B}$), and its increment is $\gamma \sim 0.592$. The mesh size and time and space steps were varied. Following variants are of note to the current paper.

The first one included $128 \times 128 \times 120$ mesh size, 10 electrons from each stream in each cell, time steps equal to $dx = dy = dz = 0.01$, $dt = 0.0025$. The simulation domain is too small to show the 3D instability structure, so only the Weibel instability is visible and the result is two-dimensional. It allows to perform a 2D Fourier analysis along x and y axes, by averaging on the z axis. The result of such analysis for B_x field component is shown on fig. 3. The simulation process shows the filament formation on the linear stage of the instability growth, and their merging into a large-scale stable magnetic structure at later stages.

To observe the 3D nature of the instability a greater mesh has been chosen for further simulations, namely $512 \times 512 \times 240$ mesh size, mesh steps $dx = dy = 0.02$, $dz = 0.05$, time step $dt = 0.005$. The simulation has been performed on the Intel core i5-2400 @ 3.1GHz desktop computer, 1 time iteration step takes approximately 8 seconds.

The result for energy dependency on time are shown on fig. 4. To show the correspondence with the theoretical estimate of the increment, Fourier analysis had been carried out (fig. 5).

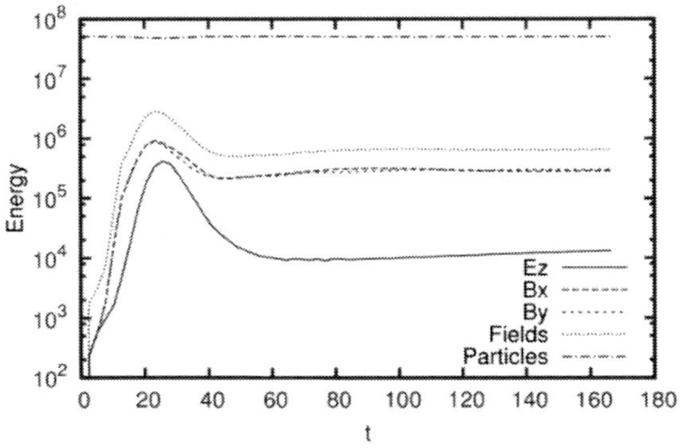

Figure 4. Energy dependency on time. Energy of the E_z field component which grows due to beam plasma instability, and weibel instabiity induced B_x and B_y component energy are shown. Full energy of fields and particles are also present on the figure. Their sum is conserved with 1.2% accuracy by the end of calculation.

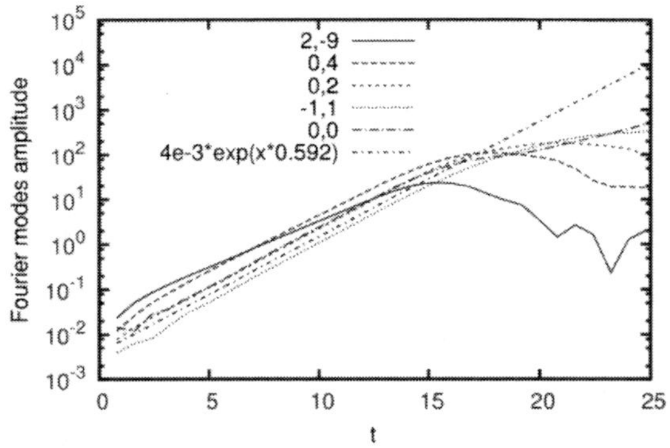

Figure 5. Fourier mode amplitude time dependency. Modes which were dominant on some time step were chosen for the plot. The straight curve shows theoretical estimate.

5. Results

We have developed a 3D3V kinetic code for plasma particle-in-cell simulation. The code is optimal for plasma instability simulations since it doesn't contain most common approximations: a full self-consistent Vlasov-Maxwell system is solved, the second-order numerical scheme is used both for field and PIC calculations, multiscale simulations are possible, parameter calibrations are not present. The focus of the development is on the most effective computational algorithms: the LRnLA algorithms and vectorization are implemented.

For the demonstration of the code possible applications the Weibel instability study has been performed. The instability growth on a linear stage shows good correspondence with analytical estimations, and the stable magnetic field structure at later stages agrees with theoretical expectations. High computational effectiveness allows the simulation to be performed in a reasonable time even on a desktop computer.

The combination of the high effectiveness and numerical scheme accuracy becomes a key benefit of the CFHall code. With it it becomes possible to perform numerical experiments in the problems which were not solved before.

Acknowledgments

The work is supported by RFBR grants 12-01-00708-a, 11-01-00282-a

References

[1] Justin M. Fox. *Advances in Fully-Kinetic PIC Simulations of a Near Vacuum Hall Thruster and Other Plasma Systems*, PhD thesis, Massachusetts Institute of Technology, 2007.

[2] Choueiri E.Y., *An overview of oscillations in Hall thrusters. In 31rd International Conference on Spacecraft Propulsion*, Cannes, France, ICSP-134. (2000)

[3] V. Levchenko, *Asynchronous parallel algorithms as a way to archive effectiveness of computations*, J. of Inf. Tech. and Comp. Systems (in russian) **1** (2005) p.68

[4] Y. Sigov, *Numerical experiment - the link between the past and the future of plasma physics* (in russian), (FIZMATLIT, Moscow, 2001)

[5] Morton, G. M. (1966), A computer Oriented Geodetic Data Base; and a New Technique in File Sequencing, Technical Report, Ottawa, Canada: IBM Ltd.

[6] Weibel E.S., *Spontaneously growing transverse waves in a plasma due to an anisotropic velocity distribution*, Phys. Rev. Lett., 2, 83-84, 1959.

[7] B. D. Fried. *Mechanism for Instability of Transverse Plasma Waves*. Physics of Fluids, 2:337, 1959.

[8] Ricardo A. Fonseca et al., *Three-dimensional Weibel instability in astrophysical scenarios*, Phys. Plasmas 10, 1979 (2003)

Journal of Physics: Conference Series **441** (2013) 012015 doi:10.1088/1742-6596/441/1/012015

Investigation of Electrode Erosion Mechanism of Multi-Phase AC Arc by High-Speed Video Camera

M Tanaka, T Ikeba, Y Liu, S Choi and T Watanabe
Dept. Environmental Chemistry and Engineering, Tokyo Institute of Technology Yokohama, Kanagawa 226-8502, Japan

E-mail: mtanaka@chemenv.titech.ac.jp

Abstract. A multi-phase AC arc has been applied to the glass melting technology. However, the electrode erosion is one of the most considerable issues to be solved. In order to investigate the erosion mechanism of the multi-phase AC arc, the combination of the high-speed video camera and the band-pass filters was introduced to measure the electrode temperature. Results indicated the tip temperature of the electrode surface in the 12-phase arc was lower than that in the 2-phase arc, while erosion rate in 12-phase arc was higher than that in the 2-phase arc. Furthermore, the dynamic behaviour of the vapours in the arc was investigated by using the same high-speed camera system. The tungsten electrode mainly evaporates at the anodic period during AC cycle. The oxygen concentration in the arc increases with larger number of the phases, resulting in the higher erosion rate in the 12-phase arc.

1. Introduction

The glass industry is a large global industry that annually produces more than 100 million tons of glass products such as sheet glass, container glass, fiber glass, optical glass, and so on. Most glass has been produced by typical Siemens-type melter fired in air with heavy oil or natural gas as the fuel. This type of melter has been used for more than 140 years because of its superior large-scale performance and continuous melting system. In the air-fuel fired furnace, the heat transfer from above burner flame to glass melt is so low that the conventional melting technology is energy intensive and time consuming, especially in the melting and the refining processes. With the rapid growth of glass usage and the increased energy and environment issues, it is crucial to develop a new glass melting technology.

An innovative in-flight glass melting technology with thermal plasmas was developed to solve the above problems [1-3]. The granulated raw material with small diameter is injected into thermal plasmas and the powders contact fully with the plasma and/or burner flame. The high heat-transfer and temperatures of the plasma will melt the raw materials quickly. In addition, the decomposed gases of carbonates are removed during the in-flight treatment to reduce the refining time considerably. Compared with the traditional glass production, the total vitrification time is evaluated as only 2-3 h at the same productivity as the fuel-fired melter.

Arc plasmas as an energy source with high energy efficiency have been applied to the in-flight glass production. In particular, the multi-phase AC arc is one of the most suitable heat sources for the in-flight glass melting because it possesses many advantages such as high energy efficiency, large plasma volume (about 100 mm in diameter), low velocity (5-20 m/s) [4, 5]. In spite of the recent experimental efforts [6, 7], the multi-phase AC arc remains to be explored. In particular, the electrode erosion is one of the most considerable issues for the practical use of the multi-phase arc in glass melting process, because it determines the electrode lifetime and the purity of the products.

The objective of the present study was to investigate the erosion mechanism of the electrode in the multi-phase AC arc. Combination of the two-colour pyrometry and the high-speed camera observation was applied to measure the temperature of the electrode on the millisecond time scale. Furthermore, the high-speed observations of the vapours in the arc were conducted to understand the dynamic behaviour of the metal vapour in the arc, which would be strongly related to the erosion phenomena. In the present work, the electrode temperature and the dynamic behaviour of the vapours in the arc were compared between the 2-phase and 12-phase arcs.

2. Experimental Procedure

2.1. Experimental setup

The schematic diagram of the experimental setup is shown in figure 1. It consisted of 12 electrodes, arc chamber, and AC power supply. The electrodes were symmetrically arranged by the angle of 30 degree and were divided into two layers, upper six and lower six electrodes. The electrodes were made of tungsten (98wt%) and thoria (2wt%) with diameter of 6 mm. City water was used to cool the electrodes at 3 slpm of water flow rate for each electrode. 99.99% argon was injected around the electrode at 5 slpm of gas flow rate to prevent them from the oxidation. Because the multi-phase arc discharge was generated under the atmospheric air except of above mentioned argon, main plasma forming gas was air. 24 Sets of arc welding transformers with single-phase AC (DAIHEN B-300, Japan) were used to generate the multi-phase AC arc discharge. The more detail about the power supply was explained in the previous report [6, 7]. In the present work, arc current was adjusted at AC 100 A for each electrode. The electrode erosion rates were evaluated from the weight difference of the electrode before and after experiments.

The applied voltage between each electrode and the neutral point of the coil of the transformer can be calculated by the following equation:

$$V_i^N = V_m^N \sin\left[\omega t - 2\pi(i-1)\big/12\right], (\mathrm{i} = 1, 2\ldots 12) \tag{1}$$

where V_i^N indicates the applied non-load voltage for each electrode number i and V_m indicates the amplitude of the non-load voltage (about 220 V). The voltage was applied to only electrode No. 1 and No. 7 for the 2-phase arc discharge. In the case of the 12-phase arc discharge, the voltage was applied to electrode No. 1, 2…12.

2.2. Temperature measurements by high-speed video camera

According to the Wien's approximation of Planck's law, the spectrum of thermal radiation from a blackbody is used to determine the electrode temperature at the applied wavelengths. A real body that emits less thermal radiation than the blackbody has surface emissivity ε less than 1. Since the emissivity is unknown for many applications, the grey body assumption $\varepsilon(\lambda_1) = \varepsilon(\lambda_2)$ is used where it is hypothesized that the surface emissivity is independent of wavelength.

$$TR_{\lambda 1} = K\varepsilon(\lambda_1)d^2\lambda_1^{-5}\left(e^{k2/\lambda_1 T} - 1\right)^{-1} \tag{2}$$

Based on these approximations, the electrode temperature T is obtained from the ratio of the radiation intensity,

$$T = \frac{K_2(\lambda_1 - \lambda_2)}{\lambda_1\lambda_2}\left[\ln\left(\frac{TR_{\lambda 1}}{TR_{\lambda 2}}\right) + 5\ln\left(\frac{\lambda_1}{\lambda_2}\right)\right]^{-1} \tag{3}$$

where K_2 is the second radiation constant in Plank's law which value is 1.439×10^{-2} m·K, $TR_{\lambda 1}$ and $TR_{\lambda 2}$ are the thermal radiations emitted by the electrode at λ_1 and λ_2, respectively.

Spectroscopic measurements (iHR550, Horiba Jobin Yvon) were conducted to determine appropriate wavelengths for two-colour radiations. High-speed video camera (FASTCAM SA-5, Photron) with the band-pass filter system was used to measure the radiation intensities as shown in figure 2. Electrode temperature was then evaluated from the radiation intensities at 785 nm and 880 nm. Typical frame rate and exposure time of the measurements were 10,000 fps and 20 μs, respectively. The voltage of each electrode was recorded at 1MHz by an oscilloscope (Scope Corder DL 850, Yokogawa) synchronized with the high-speed camera.

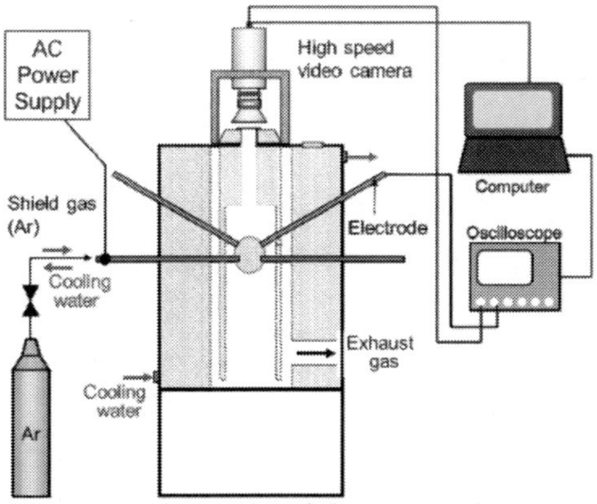

Figure 1. Schematic of the generating system of the multi-phase arc.

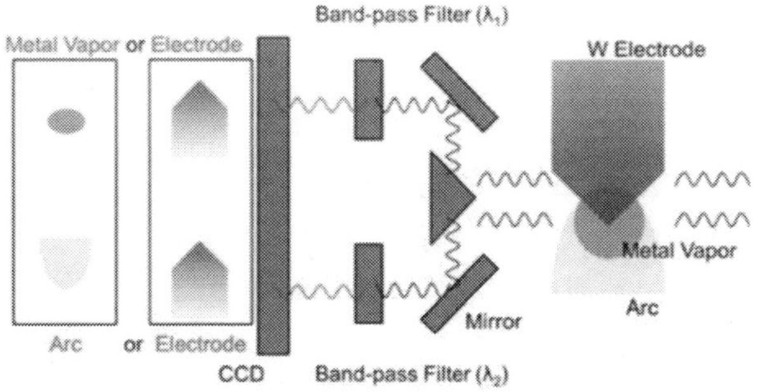

Figure 2. Schematic of the combination of the high-speed video camera and the band-pass filters.

2.3. High-speed observations of vapours in the arc

The dynamic behaviour of the vapours in the arc was investigated by using the same system with the temperature measurements. Emissions from atomic argon, oxygen, and tungsten vapours were observed at the wavelength of 738 nm, 777 nm, and 393 nm, respectively. Typical frame rate and the exposure time of the measurements were 10,000 fps and 100 μs. The voltage of each electrode was also measured at 1MHz synchronized with the high-speed camera.

3. Results and Discussion

3.1. Temperature estimation

Figure 3 shows the snapshot of the electrode obtained by the high-speed camera and the representative temperature distribution of the electrode for the 12-phase arc. The grey image indicates the arc. The temperature around the electrode tip was higher than the melting point of tungsten (3,695 K). The electrode tip temperature was estimated from the obtained temperature distributions.

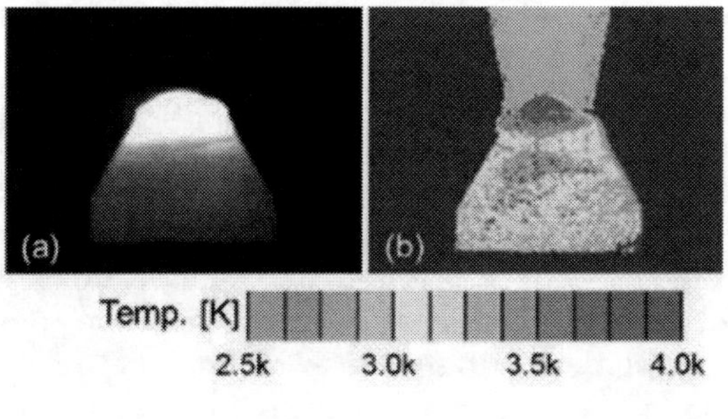

Figure 3. Snapshot of the electrode taken by the high-speed camera (a) and representative electrode temperature distribution in 12-phase AC arc (b).

Figure 4 shows the temperature variation of the electrode tip for the 2-phase and the 12-phase AC arcs in one AC cycle. Two peaks of the tip temperature were found in both cases for the 2-phase and the 12-phase arcs. These peaks originated in the peaks of the sinusoidal current waveform in AC cycle, resulting in the peak values of the heat transfer from the arc to the electrode. The temperature variation range of the 12-phase arc was smaller than that of the 2-phase arc. This could originate in the existence of the multiple arcs at the same time in the case of the multi-phase arc, while the only single arc exists in the case of the 2-phase arc.

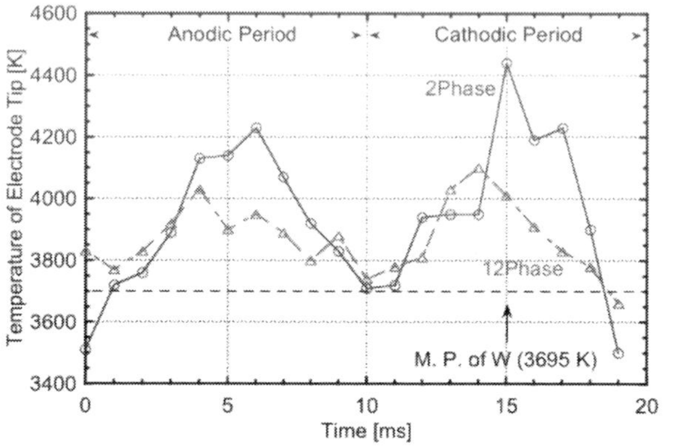

Figure 5. Temperature variation of the electrode tip for the 2-phase and the 12-phase arcs in one cycle.

Table 1 shows the comparison of the maximum temperature of the electrode tip and the erosion rate between the 2-phase and the 12-phase arcs. Maximum temperature in 12-phase arc was lower than that in the 2-phase arc, although the erosion rate for each electrode in the 12-phase arc was larger than that in the 2-phase arc. In general the temperature is one of the most crucial factors on the material evaporation, thus this relationship between the above maximum temperature and the erosion rate is unreasonable. This will be discussed in the next section.

Table 1. Maximum temperature of the electrode tip and the erosion rate per each electrode for the 2-phase and the 12-phase arcs.

	Maximum tip temperature in one cycle [10^3 K]	Erosion rate for each electrode [mg/min]
2-phase arc	4.4	0.31
12-phase arc	4.1	1.64

3.2. High-speed observations of the vapours in the arc

In order to understand the reason for the higher erosion rate in the 12-phase arc, high-speed observations of the vapours in the arc were conducted. Especially, the dynamic behaviour of the tungsten vapour was investigated to observe the evaporation of the electrode directly. Then, the oxygen entrainments into the arc were compared between the 2-phase and the 12-phase arcs. This is because the tungsten oxide formation on the electrode leads to sever erosion due to the lower melting point of tungsten oxide than metal tungsten [8, 9].

Figure 5 shows the arc current waveform synchronized with the high-speed observations. Figure 6 shows the snapshots of the high-speed video images for tungsten and argon during one cycle. The indicated time in figure 6, such as 1 ms, corresponds to the time in figure 5. The electrode was at the anodic period in the first half period, while that was at the cathodic period in the second half period. Importantly, the strong emissions from the tungsten vapours were observed in the 12-phase arc only at the anodic period. The tungsten electrode started to evaporate just after the peak top of the arc current at 5 ms in the anodic period. This is because the larger heat transfer to the electrode from the arc at the anodic period compared to the cathodic period, resulting from the electron condensation. These results indicate that the tungsten electrode was mainly evaporated at the anodic period.

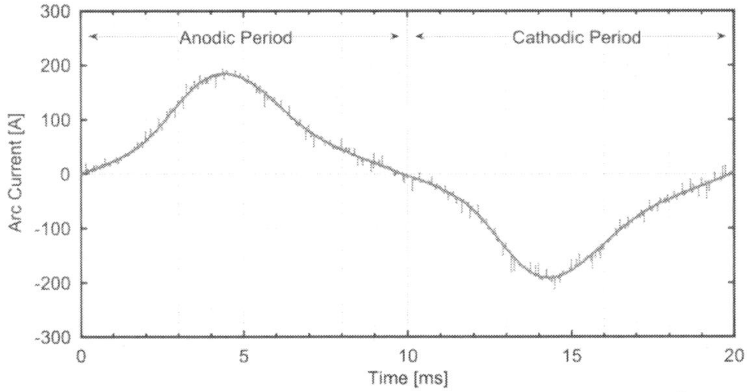

Figure 5. The waveforms of the arc current synchronized with the high-speed images shown in figure 5.

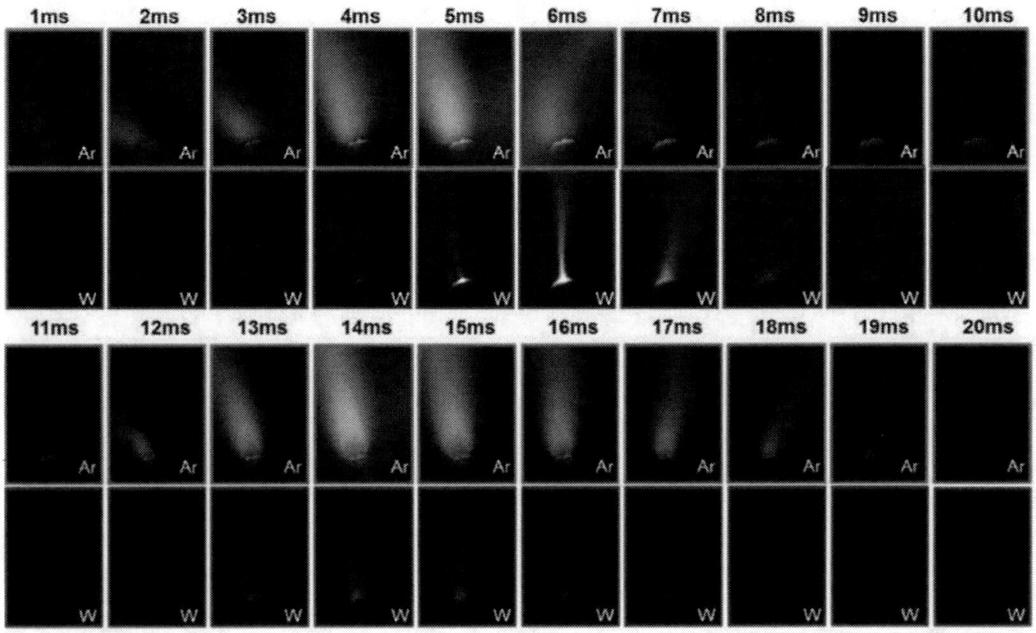

Figure 6. The snapshots of the high-speed video images for tungsten and argon during one cycle.

Figure 7 (a) and (b) shows the representative snapshots of the high-speed video images for argon and oxygen at anodic period. In order to evaluate the number density of oxygen qualitatively, the relative intensity of emissions from oxygen to that from argon was calculated. Figure 7 (c) shows the relative intensity distributions for the 2-phase and the 12-phase arcs. The relative intensity of oxygen to argon in the 12-phase is higher than that in the 2-phase arc. This result indicates that the oxygen concentration in the 12-phase arc is higher than that in the 2-phase arc. In fact, the arc-spot motion in the 12-phase is more widely distributed compared to the 2-phase arc [6]. Therefore, the ambient oxygen gas is more easily entrained with the increase of the number of the phases, resulting in the decrease of the melting point due to the oxidation of the tungsten. Consequently, the electrode erosion in the 12-phase arc was severe compared to the 2-phase arc. To reduce the oxygen entrainment must be important to apply the multi-phase AC arc to industrial field.

Further experimental studies enable us to understand the electrode erosion mechanism of the multi-phase AC arc and currently under investigation by using the high-speed camera system.

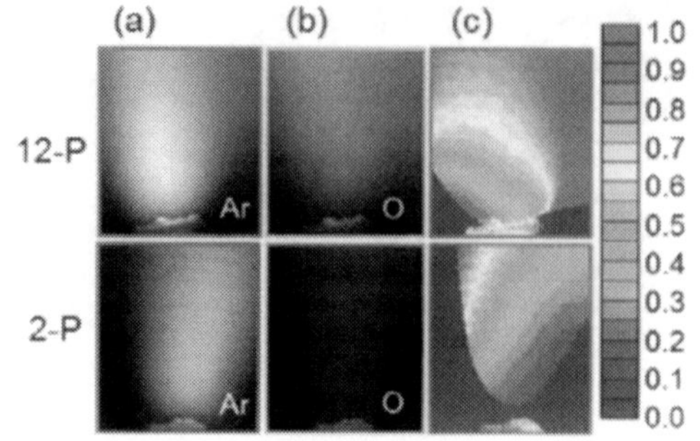

Figure 7. The representative snapshots of the high-speed video images for argon (a) and oxygen (b) and relative intensity distributions of the emissions from oxygen to that from argon (c).

4. Conclusions

The erosion mechanism of the tungsten electrode in the multi-phase AC arc was investigated by using the high-speed video camera combined with the band-pass filters. The obtained results are as follows.

(1) The electrode temperature was successfully measured on the millisecond time scale. Results indicated that the tip temperature in the 12-phase arc was lower than that in the 2-phase arc.

(2) The high-speed observations of the vapours in the arc were conducted to observe the metal vapour during the arc and the effect of the oxygen entrainments into the arc. The erosion of tungsten electrode was visually observed under anodic period with higher oxygen concentration.

(3) The ambient oxygen gas is more easily entrained into the arc with increasing the number of phases, resulting from the wider distributed arc-spot motion. Thus, the erosion rate increases with the number of the phases.

Acknowledgments

The financial support provided by the Strategic Development of Energy Conservation Technology Project of NEDO (New Energy and Industry Technology Development Organization, Japan) is gratefully acknowledged.

References

[1] Yao Y, Watanabe T, Yano T, Iseda T, Sakamoto O, Iwamoto M and Inoue S 2008 An innovative energy-saving in-flight melting technology and its application to glass production *Sci. Technol. Adv. Mater.* **9** 025013

[2] Yao Y, Yatsuda K, Watanabe T, Funabiki F and Yano T 2008 Investigation on in-flight melting behavior of granulated alkali-free glass raw material under different conditions with 12-phase AC arc *Chem. Eng. J.* **144** 317-23

[3] Watanabe T, Yatsuda K, Yao Y, Yano T and Matsuura T 2010 Innovative in-flight glass-melting technology using thermal plasmas *Pure Appl. Chem.* **82** 1337-51

[4] Liu Y, Tsuruoka Y, Tanaka M, Ichihashi T, Yano T and Watanabe T 2011 In-flight melting behavior of different glass raw materials by hybrid heating of twelve-phase AC arc with oxygen burner *Thin Solid Films* **519** 7005-8

[5] Tanaka M, Tsuruoka Y, Liu Y and Watanabe T 2011 Investigation on in-flight melting behaviour of granulated glass raw material by multi-phase AC arc plasma and hybrid plasma *IOP Conf. Ser.: Mater. Sci. Eng.* **18** 112010

[6] Tanaka M, Tsuruoka Y, Liu Y, Matsuura T and Watanabe T 2011 Investigation of multiphase AC arc behaviour by high-speed video observation *IEEE Trans. Plasma Sci.* **39** 2904-5

[7] Tanaka M, Tsuruoka Y, Liu Y, Matsuura T and Watanabe T 2011 Stability analysis of multi-phase AC arc discharge for in-flight glass melting *Current Appl. Phys.* **11** S35-9

[8] Ogawa Y, Matsuda J and Morita T 2004 Effect of ambient pressure on arc-electrode behaviour *Proc. ISOPE*, 171-8

[9] Ogawa Y 2011 High speed imaging technique Part 1 – high speed imaging of arc welding phenomena *Sci. Technol. Weld. Join.* **16**(1) 33-43

Journal of Physics: Conference Series **441** (2013) 012016 doi:10.1088/1742-6596/441/1/012016

Temperature Behavior in a Tandem Type of Modulated Induction Thermal Plasma for Materials Processings

Katsuya Kuraishi[1], **Mika Akao**[1], **Yasunori Tanaka**[1,2], **Yoshihiko Uesugi**[1,2], **Tatsuo Ishijima**[2]

[1]Faculty of Electrical and Computer Engineering, Kanazawa University, Kakuma, Kanazawa 920-1192, Japan

[2]Research Center for Sustainable Energy and Technology, Kanazawa University, Kakuma, Kanazawa 920-1192, Japan

E-mail: tanaka@ec.t.kanazawa-u.ac.jp

Abstract. A tandem type of modulated induction thermal plasma (Tandem-MITP) system has been developed using two rf power supplies and two coils for one plasma torch. This system was developed to control temperature and reaction fields temporally and spatially in a thermal plasma. The modulation of each coil current can change the thermal plasma temperature in the torch temporally and spatially. Time variation of Ar excitation temperature was estimated by the two-line method using Ar atomic lines. The Tandem-MITP could be established successfully, and the Ar excitation temperature in the thermal plasma could be controlled temporally and spatially by the developed system.

1. Introduction

We have so far developed a series of modulated induction thermal plasma system: a pulse modulated induction thermal plasma (PMITP) system [1], an arbitrary-waveform modulated induction thermal plasma (AMITP) system [2] and a feedback control type of modulated induction thermal plasma (FBC-MITP) system [3]. These systems can modulate the coil current amplitude to control the temperature of thermal plasmas in the time domain. Such a modulation of the coil current can produce periodical temperature field or non-equilibrium effects even in thermal plasmas. These unique features of the modulated induction thermal plasma lead to their various applications. One example of the PMITP application is TiO_2 nanoparticle synthesis using Ar-O_2 PMITP [4]. That result implied that the PMITP can control the size and phase of TiO_2 nanoparticles [4].

The contribution of this paper is to provide a new modulated induction thermal plasma system: a tandem type of modulated induction thermal plasma (Tandem-MITP) system. The tandem type of induction thermal plasma without modulation had been developed by Uesugi *et al* [5]. However, there is few report concerning a Tandem-MITP. The Tandem-MITP is expected to produce a modulated gas flow field and a modulated temperature field not only temporally but also spatially in one torch. Furthermore, injection of reactive gases to such a Tandem-MITP can provide more effective modulated reaction fields which will be useful for various materials processings. One expected example is in particle treatments or nanoparticle synthesis to heat up injected particles along particle fight at specified time and spatial position. For a fundamental study of a Tandem-MITP, we measured response of the effective power

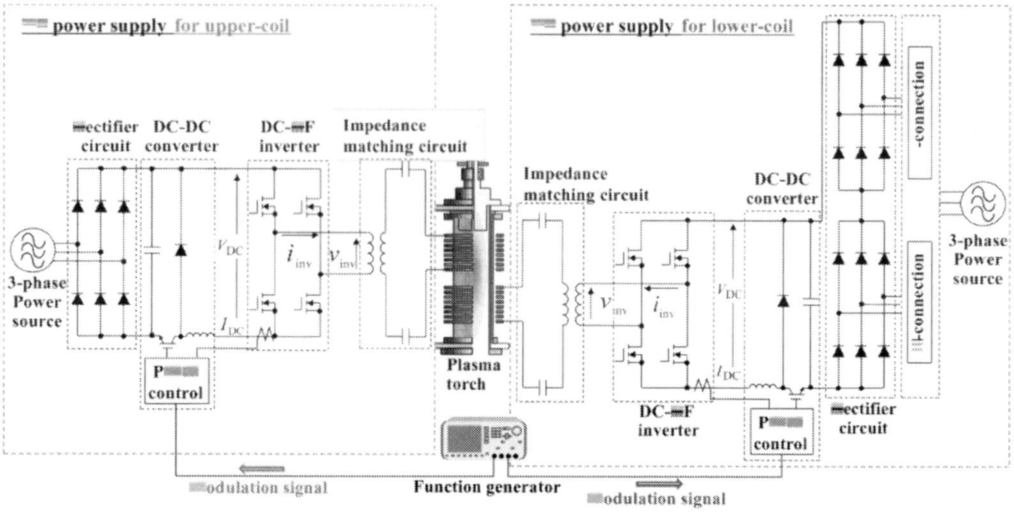

Figure 1. Schematic of the tandem type of modulated induction thermal plasma (Tandem-MITP) system.

from these two coils and Ar excitation temperature T_{ex}^{Ar} at different positions. Results indicated that a Tandem-MITP could be established successfully, and that T_{ex}^{Ar} in the thermal plasma could be controlled in more detail by the Tandem-MITP system spatially and temporally. This paper provides fundamental data for controlling temperature of a Tandem-MITP.

2. Experimental setup and conditions for Tandem-MITP

2.1. RF power supply for Tandem-MITP system

Figure 1 shows a schematic of the tandem type of modulated induction thermal plasma (Tandem-MITP) system. The tandem-MITP needs two rf power supplies: an rf power supply for the upper-coil and an rf power supply for the lower-coil. These two power supplies consist of four main parts: a rectifier circuit, an insulated gate bipolar transistor (IGBT) dc-dc converter (chopper) circuit, a metal-oxide semiconductor field-effect transistor(MOSFET) full-bridge inverter circuit, and an impedance-matching circuit with a matching transformer and an LC series circuit. The differences between them are the rectifier circuits with or without reduction of the current ripple, and the driving frequency for MOSFETs as described later. They can also modulate the coil current amplitude.

2.2. Plasma torch

Figure 2 illustrates a configuration of the plasma torch used in the experiment. The torch is composed of two coaxial quartz tubes with 430 mm in length, which is much longer than the conventional torches. The inner tube has an inside diameter of 70 mm. Between the inner and outer quartz tubes, cooling water flows from the bottom to the top side to keep the wall-temperature around 300 K. The sheath gas is supplied from the top of the plasma torch along the inner wall of the inner tube with swirl. This torch has two eight-turn induction coils: the upper-coil and the lower-coil. Distance between the upper-coil end and the lower-coil top is 50 mm. The two coils are connected with rf power supplies, respectively.

2.3. Experimental conditions

In order to measure a fundamental behavior of a Tandem-MITP, the following condition was set for experiment. The sheath gas was Ar, and its gas flow rate was set to 40 slpm for axial gas and 40 slpm for swirl gas. The pressure in the chamber was fixed at 70 torr. The driving frequency of the upper-coil current was fixed at 430 kHz, while and that of the lower-coil current was set to about 300 kHz. The

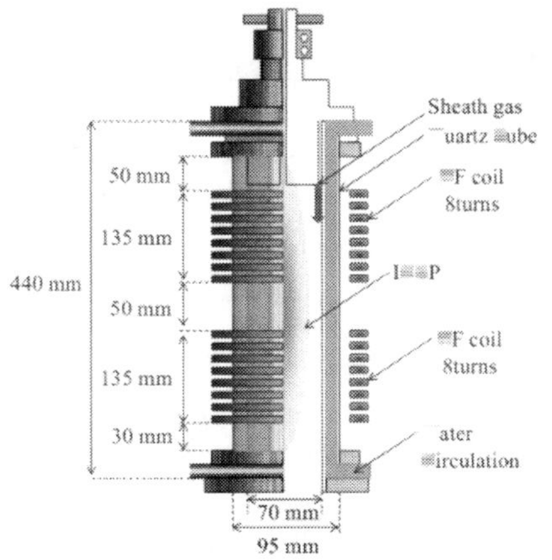

Figure 2. Configuration of plasma torch

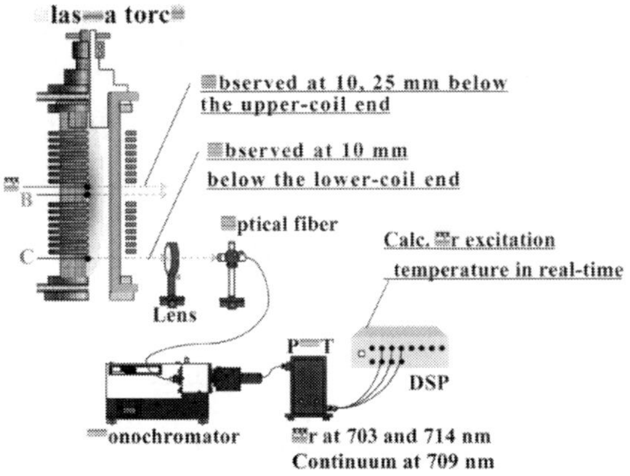

Figure 3. Spectroscopic observation system

difference between these driving frequencies was required to reduce the mutual electromagnetic coupling between two circuits. Meanwhile, generally the input power to the thermal plasma is proportional to the square value of the driving frequency. Thus, too lower frequency than 300 kHz was not adequate for the driving frequency of the lower-coil. As a modulation waveform for coil currents, a rectangular waveform was chosen for a fundamental study. The phase difference θ_d between two rectangular modulation signals for the upper-coil and the lower-coil was set to be $\pi/2$, π and $3\pi/2$. From this experiment, we also estimated the mutual influence between the two thermal plasmas produced by the upper and lower coils current.

2.4. Spectroscopic observation system

Figure 3 shows the spectroscopic observation system. Spectroscopic observation was carried out at three positions: at 10, 25 mm below the upper-coil end and at 10 mm below the lower-coil end at the center axis of the plasma torch, which are designated by A, B and C, respectively. The light radiated from each observation position is transmitted through an optical lens and an optical fiber bundle to the slit of the monochromator. Use of this system enables us to measure time evolution in the radiation intensities at different three wavelengths simultaneously. The wavelength resolution of the whole system including the monochromator and the optical fiber bundle is about 0.6 nm. In the present experiment, we measured temporal variations in the radiation intensities at 703.0 nm and 714.7 nm for Ar atomic lines and at 709 nm for continuum. These spectral lines are available because they satisfy the temperature estimation criteria [6]. Subtracting the measured radiation intensity at 709 nm from those at 703.0 nm and 714.7 nm yields the net radiation intensities of the Ar atomic lines at 703.0 nm and 714.7 nm. Using the net radiation intensities, Ar excitation temperature T_{ex}^{Ar} between the specified levels was estimated by the two-line method assuming Boltzmann distribution for the excited population. The T_{ex}^{Ar} was calculated in real-time in a Digital Signal Processor (DSP). Note that T_{ex}^{Ar} was estimated by the apparent intensities integrated along the line of sight without Abel inversion, and thus it may be underestimated. In spite of this, time evolution in the estimated T_{ex}^{Ar} was regarded as a parameter to understand thermal state of the Tandem-MITP formed. The above whole optical system was calibrated using a standard tungsten-halide lamp. This temperature determination includes relative errors $\Delta T/T$ ~0.06 around 10000 K, while $\Delta T/T$ ~0.5 around 4500 K due to intensity measurement errors.

3. Experimental results

3.1. Coil current and effective power

Various types of Tandem-MITPs were successfully established stably for the given conditions. To perceive responses of two rf power supplies for Tandem-MITPs, the instantaneous current $i_{inv}(t)$ and voltage $v_{inv}(t)$ for these two supplies were measured at the output terminal of each inverter circuit in Fig. 1. The effective power $P(t)$ was calculated by $i_{inv}(t)$ and $v_{inv}(t)$. Figures 4 (I) and (II) show time evolutions in (a) modulation signal for the upper-coil, (b) inverter output current for the upper-coil in root-mean-square (rms), (c) inverter output effective power for the upper-coil, (d) modulation signal for the lower-coil, (e) inverter output current for the lower-coil, and (f) inverter output effective power for the lower-coil. The root mean square values were estimated by integrating instantaneous value in each fundamental cycle. Panel (I) corresponds to the result for phase difference $\theta_d = \pi$, whereas panel (II) is the result for $\theta_d = 3\pi/2$.

As seen in Figs. 4(I)(a) and (I)(b), the rms value of the upper-coil current could be controlled to follow the modulation signal. Following to this current modulation, effective power $P(t)$ of the upper-coil is also modulated periodically from 2.5 kW to 12 kW. The effective power changes gradually following the modulation rectangular waveform with a time constant around 8 ms. This long time constant of 8 ms is attributed to thermal inertia ρC_p of the thermal plasma in the upper-coil, where ρ is the mass density and C_p is the specific heat at constant pressure. Meanwhile, the effective power of the lower-coil shows a rapid initial rise in corresponding rapid initial rise of the lower-coil current around t=25 ms. Here, t is the time referred to the horizontal time axis in this figure. After that, the effective power of the lower-coil increased until t=30 ms. This may arise from the following fact: The high-temperature plasma with high electrical conductivity and with low ρC_p still flows into the lower-coil region although the upper-coil current is rapidly decreased down around t=25 ms. Thus, the electric power from the lower-coil is rapidly elevated by the increased lower-coil current. On the other hand, from t= 30 ms to 50 ms, the effective power of the lower-coil decreases to 4.8 kW gradually although the lower-coil current remains 120 A_{rms}. This may be because the high-temperature plasma flowing from the upper-coil into the lower-coil decays. The above result implies that the input power to the plasma in the lower-coil is much affected not only by the lower-coil current but also by the upper- coil current. Whereas, the input power to the plasma in the upper-coil is influenced almost only by the upper-coil current.

Figure 4 (II) shows results for $\theta_d = 3\pi/2$. As illustrated in Figs. 4(II)(b)–(e), similar waveforms were obtained to Fig. 4(I). However, the waveform shape of the effective input power from the lower-coil in Fig. 4(II)(f) is quite different from that in Fig. 4(I)(f). In this case, the effective power of the lower-coil shows a rapid initial rise in corresponding rapid initial rise of the lower-coil current at t=37.5 ms. And then, the effective power of the lower-coil is kept to be about 5 kW from t=37.5 ms to 53 ms. After that, the effective power of the lower-coil rises from 5 kW to 10 kW from t=53 ms to t=62.5 ms. This rise up in the effective power of the lower-coil is clearly attributable to the flowing thermal plasma from the upper-coil.

The above-mentioned plasma behavior in the upper-coil and lower-coil region was also observed by a high-speed video camera. Figures 5 (I) and (II) depict the dynamic behavior of a Tandem-MITP for $\theta_d = \pi$ and $\theta_d = 3\pi/2$, respectively. The time t here is the same to that of Fig. 4. As seen in Figs. 4(I), a plasma with high intensity exists in the upper-coil region at t=4 ms. This plasma flows into the lower-coil region at t=12.5 ms. At t=30 ms, the plasma exists only in the lower-coil region on the contrary. In other words, the plasma seems to be transported from the upper to the lower coil regions. In case of $\theta_d = 3\pi/2$, from t=4 ms to t=12.5 ms, a plasma inside the lower-coil region has the increasing intensity, arising from the increasing input power to the lower-coil region. At t=30 ms, the plasma in the lower-coil region becomes to have the weak intensity. As seen, the developed system can controls the temporally and spatially modulated induction thermal plasmas.

3.2. Ar excitation temperature

As well known, the temperature is one of essential parameters to understand state of thermal plasmas because the thermal plasma is often under local thermodynamic equilibrium condition. In this work, the Ar excitation temperature T_{ex}^{Ar} was estimated by two-line method using the specified Ar atomic lines at 703.0 nm and 714.7 nm, assuming the excitation population follows the Boltzmann relation. Figure 6 depicts the time evolution in (a) modulation signal for upper-coil, (b) modulation signal for lower-coil,

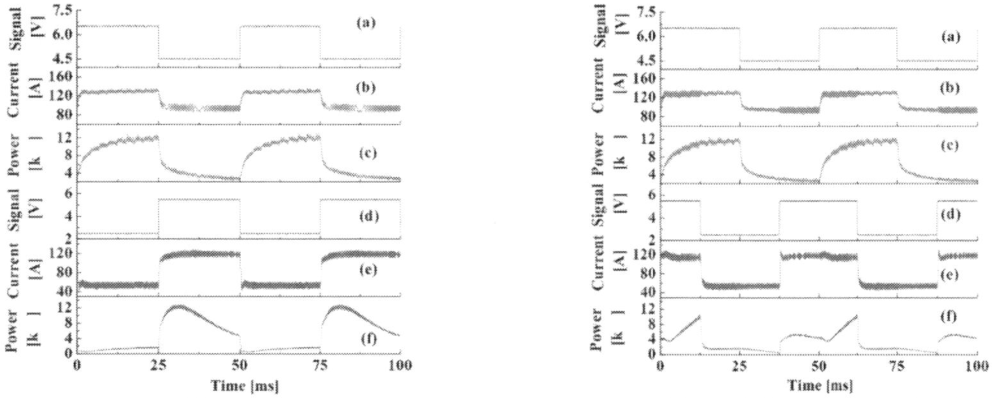

(I) Phase difference $\theta_d = \pi$ (II) Phase difference $\theta_d = 3\pi/2$

Figure 4. Time evolution in (a) modulation signal for upper-coil, (b) inverter output current for upper-coil in rms, (c) inverter output effective power for upper-coil, (d) modulation signal for lower-coil, (e) inverter output current for lower-coil (f) inverter output effective power for lower-coil.

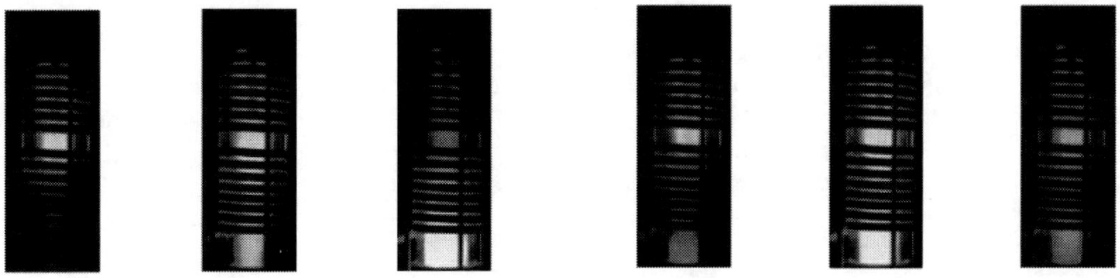

(a) t=4 ms (b) t=12.5 ms (c) t=30 ms (a) t=4 ms (b) t=12.5 ms (c) t=30 ms

(I) Phase difference $\theta_d = \pi$ (II) Phase difference $\theta_d = 3\pi/2$

Figure 5. High-speed video images of a Tandem-MITP.

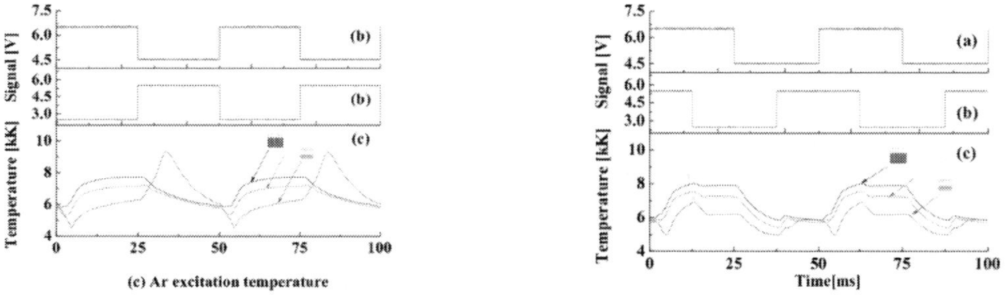

(I) Phase difference $\theta_d = \pi$ (II) Phase difference $\theta_d = 3\pi/2$

Figure 6. Time evolution in (a) modulation signal for upper-coil, (b) modulation signal for lower-coil, and (c) Ar excitation temperature T_{ex}^{Ar} in a Tandem-MITP.

and (c) Ar excitation temperature T_{ex}^{Ar}. As seen in Fig. 6(I)(c), the T_{ex}^{Ar} at positions A and B change periodically and similarly with a certain time constant according to the modulation signal for the upper-coil. The T_{ex}^{Ar} at A changes between 5800 K to 7700 K gradually, due to the thermal inertia ρC_p of the thermal plasma. On the other hand, the T_{ex}^{Ar} at position C shows quite different temporal variation against those at A or B. The T_{ex}^{Ar} at C rapidly decreases from 6000 K to 4500 K just after the lower-coil current down, for example around t=0 ms in Fig. 6(I)(c). After that, the T_{ex}^{Ar} gradually increases between

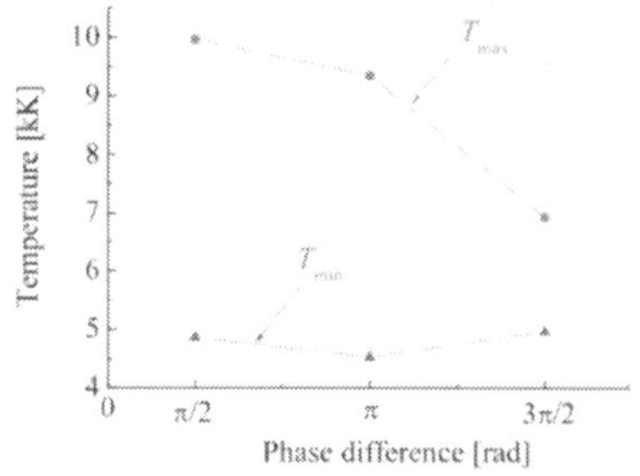

Figure 7. Maximum and minimum values of Ar excitation temperature measured at position C versus the phase difference between the modulation signals for the upper and lower coil currents.

t=4.7 ms and t=25 ms because of the flowing thermal plasma from the upper-coil. From t=25 ms to 37 ms, the T_{ex}^{Ar} sharply rises up from 6000 K to 9300 K. This rise up in the T_{ex}^{Ar} results from the rise up in the effective input power from the lower-coil to thermal plasma in the lower-coil region. From t=37 ms to 50 ms, the T_{ex}^{Ar} gradually decreases with time, since the thermal plasma flowing from the upper-coil into the lower-coil region decays as mentioned in the previous section. These results indicate that T_{ex}^{Ar} can be controlled by the lower-coil current as well as by the upper-coil current, which leads to the large modulation in T_{ex}^{Ar} from 4500 K to 9300 K.

In case of $\theta_d = 3\pi/2$ as indicated in Fig. 6(II), time evolutions in T_{ex}^{Ar} at A and B are similar to those in case of $\theta_d = \pi$. This means that T_{ex}^{Ar} at A and B has a weak influence from the lower-coil at this time. On the other hand, the T_{ex}^{Ar} at C changes with a lower amplitude compared to that for $\theta_d = \pi$.

Figure 7 plots the maximum and minimum values of T_{ex}^{Ar} measured at position C versus the phase difference. From Fig. 7, the maximum values of T_{ex}^{Ar} at C strongly depends on the phase difference. This result implies that phase difference between the upper-coil current and lower-coil current is also an important parameter for controlling the temperature in the Tandem-MITP.

4. Conclusions

In this paper, a tandem type of modulated induction thermal plasma (Tandem-MITP) system has been developed using two rf power supplies and two coils for one plasma torch. This system was developed to provide a method for controlling temperature and reaction fields both temporally and spatially in thermal plasmas. We measured time variations in the Ar excitation temperature estimated by the two-line method using Ar atomic lines. Results suggested that Ar excitation temperature in the Tandem-MITP could be controlled in more detail temporally and spatially with two coil currents. These features of the Tandem-MITP can be useful for various materials processings.

References

[1] Sakuta T, Tanaka Y, Hashimoto Y and Kazuki M 2000 *Trans. IEE of Japan* **120-A** 972.
[2] Tanaka Y, Morishita Y, Fushie S, Okunaga K and Uesugi Y 2007 *Appl. Phys. Lett.* **90** 071502.
[3] Tanaka Y, Uesaka Y, Tsubokawa Y and Uesugi Y 2010 *10-th Asia Pacific Cof. Plamsa Sci.& Technol.* OCC-08.
[4] Tanaka Y, Nagumo T, Sakai H, Uesugi Y, Sakai Y and Nakamura K 2010 *J. Phys. D. Appl. Phys.* **43** 265201.
[5] Uesugi T, Nakamura O, Yoshida T and Akashi K 1988 *J. Appl. Phys.* **64** 3874-3879.
[6] Tanaka Y, Yokomizu Y, Matsumura T, Kito Y 1995 *J. Phys. D: Appl. Phys.*, **28**, 2095-2103.

Journal of Physics: Conference Series **441** (2013) 012017 doi:10.1088/1742-6596/441/1/012017

Time Evolution in Radiation Intensities of C_2 and H Spectra in $Ar/CH_4/H_2$ Pulse Modulated Induction Thermal Plasmas for Diamond Film Deposition

Yosuke Haruta[a], Kenta Fujimoto[a], Sosuke Horita[a], Yasunori Tanaka[a,b], Yoshihiko Uesugi[a,b], Tatsuo Ishijima[b]

[a] Faculty of Electrical and Computer Engineering, Kanazawa University, Kakuma, Kanazawa 920-1192, Japan

[b] Research Center for Sustainable Energy and Technology, Kanazawa University, Kakuma, Kanazawa 920-1192, Japan

E-mail: `tanaka@ec.t.kanazawa-u.ac.jp`

Abstract. The present paper describes the first trial application of an $Ar/CH_4/H_2$ pulse modulated induction thermal plasma (PMITP) to polycrystalline diamond film deposition on a silicon substrate. The PMITP system has been originally developed by our group to control the heat flux and densities of chemical species in thermal plasma treatment. Polycrystalline diamond film was successfully deposited on a 25×25 mm^2 Si(100) substrate using the PMITP. Spectroscopic observation was carried out to measure the time evolution in the spectral intensity from the PMITP irradiated to the Si substrate. We found that the C_2 molecular spectra have strong intensities at 4–7 ms after rise-up of the coil current, while the H spectral line has much high intensity at 6–12 ms in a 15 ms modulation cycle. This implies that the PMITP produces repetitive irradiation from different chemical fields in one modulation cycle, which may effectively provide diamond film deposition.

1. Introduction

The inductively coupled thermal plasmas (ICTPs) have been widely adopted as effective heat and chemical species sources for various materials processings such as syntheses of diamond films [1], thermal barrier coatings [2], fullerene synthesis [3], and surface modifications, etc. The ICTP has some advantages like their high enthalpy and high radical density without any contamination.

To obtain a further effective radical source, we have previously developed a series of modulated induction thermal plasma system including a pulse-modulated induction thermal plasma (PMITP) system and an arbitrary-waveform-modulated induction thermal plasma (AMITP) system [4, 5]. The PMITP and AMITP are established by the coil current amplitude of the order of several hundreds amperes modulated almost into a rectangular or other waveforms. Such a millisecond modulation can perturb the thermal plasma markedly, and then it can change the temperature and radical densities as well as the gas flow field in the thermal plasma in time domain [4]. It also produces chemically non-equilibrium effects even in high-temperature and high-pressure plasmas [4, 6]. We found actually that the $Ar\text{-}N_2$ PMITP provides simultaneous control of the increased nitrogen radical flux and the decreased enthalpy flow by chemically non-equilibrium effects in application to surface modification [6]. Such a modulated induction thermal plasma has further possible other applications to be effectively adopted.

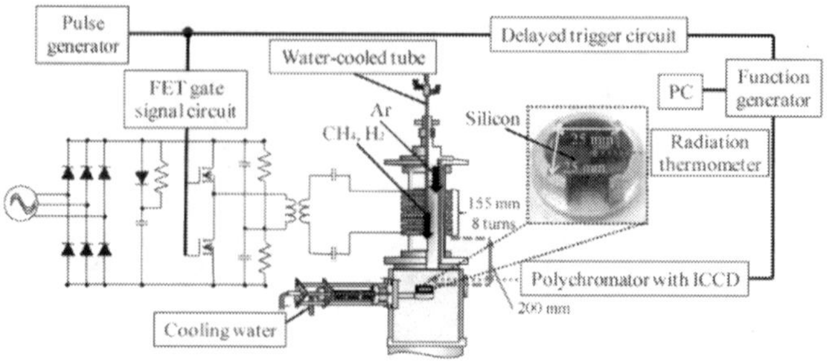

Figure 1. Schematic diagram of experimental system.

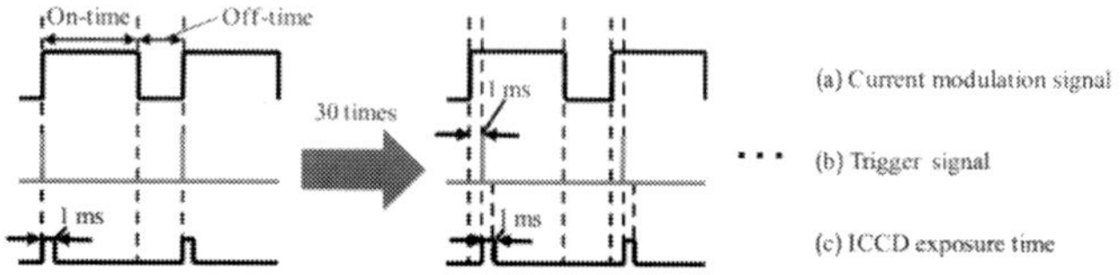

Figure 2. Timing chart in spectroscopic observation.

This paper describes the first results on application of the PMITP to polycrystalline diamond film deposition. A several kW-class Ar-CH_4-H_2 PMITP was adopted to fabricate polycrystalline diamond film on a Si (100) substrate. A polycrystalline diamond film was successfully deposited almost uniformly on a 25×25 mm^2 Si(100) substrate using the PMITP, whereas it could not be deposited using a non-modulated plasma with the same heat flux. Spectroscopic observation was carried out to measure the time evolution in the spectral intensity from the PMITP irradiated to the Si substrate. Strong C_2 molecular spectra were found during the first stage in the on-time and then strong H atomic line was observed during the successive stage in the on-time of the PMITP. This implies that a carbon film is deposited at the first stage, and then a deposited graphite film is chemically etched by hydrogen atom to leave a diamond film in one modulation cycle.

2. Experimental

2.1. Experimental arrangements

Figure 1 shows the PMITP system and also the spectroscopy observation system used in the present work. The rf power source for PMITP is operated with metal-oxide semiconductor field emission transistors (MOSFET). The rated power of the source and the fundamental rf frequency are 30 kW and 450 kHz, respectively. This power source can modulate the output electric current amplitude of several hundreds of amperes into a rectangular waveform. The rise time of the current amplitude modulation is 80 μs with a delay time less than 1μs. The modulated coil current has four modulation control parameters: the quantity 'On-time' which is the time duration with higher current level (HCL), and 'Off-time', the time duration with lower current level (LCL) as indicated in [7]. We also defined the shimmer current level 'SCL', as a ratio of LCL/HCL. The 100%SCL condition corresponds to non-modulation condition. In addition, the duty factor (DF) is set to a ratio of the on-time to one modulation cycle: DF= 'on-time' /('on-time' +'off-time').

The configuration of the plasma torch is the same to our previous work [7]. It has two coaxial quartz tubes. The inner diameter of the interior quartz tube is 70 mm; its length is 330 mm. Argon gas was supplied as a sheath gas along the inside quartz tube. Into the plasma torch, a water-cooled tube was inserted from the plasma torch head along the center axis. Through this tube, CH_4/H_2 gas mixture was injected to the plasma. The tip position of this water-cooled tube was set between 5-th and 6-th turns of the coil. This means that CH_4/H_2 gas mixture was directly injected into high-temperature region

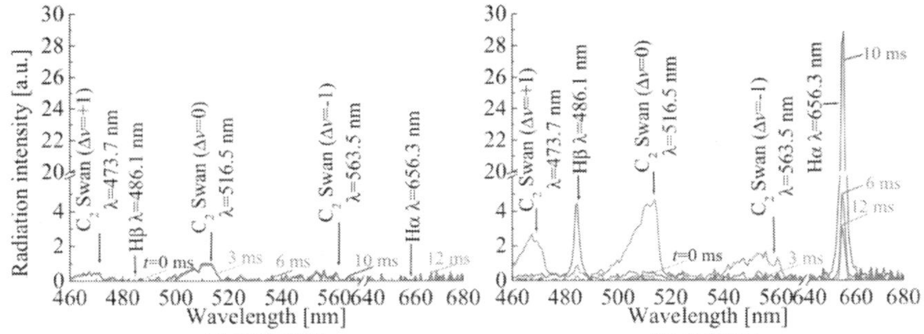

Figure 3. Radiation intensity observed at 5 mm above Si substrate.

in the thermal plasma. The direct injection of CH_4/H_2 gas mixture to the high temperature region can cause its complete decomposition. Downstream of the plasma torch, a Si (100) substrate with a size of 25 × 25 mm^2 was located. The distance between the substrate surface and the coil end was 200 mm. This substrate surface is directly irradiated by an Ar/CH_4/H_2 PMITP. The irradiation of the PMITP also increases the surface temperature. This surface temperature of the Si substrate was measured with a radiation thermometer. As pretreatment processes of the Si substrate surface, it was scratched with diamond powder of 5.6 μm in ethanol using an ultrasonic cleaning device.

Spectroscopic observation was carried out to measure the emission spectra from the PMITP. The observation position was 195 mm below the coil end, which corresponds to 5 mm above the substrate surface. The observation size was 2.5 mmϕ. To measure emission spectra at different timings against the coil current modulation, we used a delayed trigger circuit and a polychromator with an ICCD (HAMAMATSU: PMA-20). Figure 2 schematically illustrates the timing chart of (a) the pulse generator signal for the coil current modulation, (b) the delayed trigger signal for the ICCD, and (c) the ICCD exposure signal. First, the ICCD was operated at 0 ms just after transition from LCL to HCL. The exposure time of the ICCD was set 1.0 ms. At the same timing against the modulation signal, the same measurement was done 30 times to obtain the averaged value of the radiation intensity of spectra. After this measurement, the timing of the ICCD exposure was shifted by 1 ms using the delayed trigger circuit for the next timing measurement. Similarly, the timing of the ICCD exposure was shifted by every 1 ms in the modulation cycle to obtain time evolution in the averaged radiation intensities of emission spectra from the PMITP.

2.2. Experimental conditions

In the present work, the modulation cycle was fixed at 15 ms. The DF of the modulated coil current was set at 67%. Only the SCL was changed at six values from 45% to 100%. One purpose of the present work is to find the modulation effect on the diamond deposition. The surface temperature of Si substrate was thus controlled to be fixed to be 1,000(±20) °C for any modulation conditions. The fixed surface temperature was realized by carefully regulating a base input power, i.e. the time-averaged input power, for any modulation conditions. For example, the 100%SCL condition required 9.3 kW for input power to obtain the surface temperature of 1000 °C, while 14 kW in time-averaged power was needed for the 45%SCL condition. Irradiation time of Ar-CH_4-H_2 PMITP was set to six hours.

The flow rate of Ar sheath gas was fixed at 65 L min^{-1} (liters per minute). The center gas flow rate of CH_4 and H_2 gas mixture through the water-cooled tube was set at 0.03 L min^{-1}, and 3 L min^{-1}, respectively. The pressure inside the chamber was controlled at 180 Torr with an automatic feedback control.

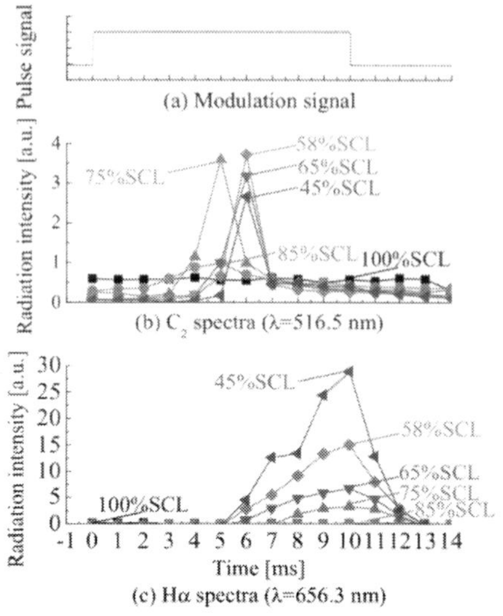

Figure 4. Time evolution in (a) modulation signal, radiation intensity of (b) $C_2(\Delta v = 0)$ (λ=516.5 nm), (c) Hα (λ=656.3 nm) in different coil current modulation.

3. Results and discussion

3.1. Ar-CH$_4$-H$_2$ PMITP spectroscopic observation at 5 mm above Si substrate

Figure 3 shows the emission spectra observed at 5 mm above the Si substrate at timings of t=0, 3, 6, 10 and 12 ms, for (a) 100%SCL (non-modulation condition), and (b) 45%SCL (modulation condition). The timing t=0 ms against the modulation signal corresponds to the timing just after transition from LCL to HCL. For 100%SCL, C_2 Swan system around wavelengths of 473.7 nm and 516.5 nm can be seen. However, those intensities are relatively low and they are almost unchanged with time because of non-modulation condition. On the other hand, for 45%SCL, i.e. a large modulation condition, C_2 Swan system around 473.7 nm and 516.6 nm has higher intensities especially at t=6 ms. In addition, H atomic lines at 486.1 nm and 656.3 nm have higher intensities at t=10 ms.

Figure 4 shows time evolution in (a) modulation signal, (b) the radiation intensity of $C_2(\Delta v = 0)$ (λ=516.5 nm), and (c) the radiation intensity of Hα (λ=656.3 nm) in different modulation conditions, i.e. in different SCLs. As seen, the non-modulated plasma with 100%SCL provides a certain constant intensity from C_2 molecular spectra, and a low intensity of H atomic line. On the other hand, for PMITP, reducing SCL produces changing intensities of the C_2 spectra and also the H atomic line. The radiation intensity of the C_2 molecular spectra increases with time from t=4 ms to t=7 ms, and it has a peak around t=5–6 ms almost irrespective of SCL condition. Concerning the H atomic line, it has a higher intensity from t=6 ms to t=12 ms for different SCLs. One interesting point is that it has a peak at t=10 ms almost independently of modulation conditions. The above measurement results implies that the PMITP irradiates a particle flux including C_2 around t=4–7 ms, and then H flux to the substrate around t=6–12 ms to the substrate. The irradiation of this alternative kinds of chemical species fluxes with C_2 molecules and H atom in one modulation cycle is one feature of the present method using the PMITP. Furthermore, this irradiation repeatedly occurs by modulation cycle.

According to the equilibrium composition calculation of C-H system, molecules C_2 is produced around 5000–6000 K, while H atom is done around 6000–8000 K by thermal dissociation. Thus, C_2 is produced at the boundary between the high-temperature plasma and the cold gas, whilst H is created at high-temperature region. On the other hand, the PMITP provides a modulated gas flow. Thus, this gas flow created by the PMITP transfers C_2 around the edge of the plasma to the substrate first. After that, H atoms in the high temperature region of the PMITP is transported to the substrate. The roles of C_2 flux and then H flux in one modulation cycle will be discussed later.

Next is the SCL dependence on these radiation intensities. Reducing SCL from 100% to 58% increases the radiation intensity of C_2 spectra. However, a further reduction in SCL to 45% decreases that intensity. This is because 45%SCL condition has a higher instantaneous input power to the PMITP during the on-time, which may lead to higher dissociation degree of C_2 molecules. This 45%SCL condition contrarily produces more H atoms by higher thermal dissociation of CH_4 and H_2.

3.2. SEM images and Raman spectroscopy

Figure 5 presents the SEM images of the Si substrate surface after Ar-CH_4-H_2 PMITP irradiation in different shimmer current levels SCLs: (a) 100%SCL (non-modulated plasma), and (b) 45%SCL (modulated plasma). These SEM images were taken at the center of each substrate. In the SEM image for 100%SCL, several particles of around 3 micron in size were sparsely deposited on the Si substrate. On the other hand, for 45%SCL condition, a film with polycrystalline diamond structure was deposited and grown on the Si substrate. From other experiments on 1 hour irradiation of Ar-CH_4-H_2 PMITP, we found that reducing SCL from 100% to 45% decreased the diamond particle size formed, and in addition that the particles were rather densely deposited on the substrate. The thickness of the deposited diamond film in Fig. 5(b) was estimated to be 5.6 μm for 45%SCL.

The deposited particles or film were analyzed using Raman spectroscopy. The upper panel in Fig. 6 shows the Raman shift spectra of the Si substrate surface after Ar-CH_4-H_2 PMITP irradiation for two conditions: (a) 100%SCL (non-modulation), and (b) 45%SCL. These were measured at the center of each substrate. It can be seen that 100%SCL condition obtains not only Raman peak of disordered or diamond (k=1334 cm^{-1}) but also strong peaks from the Si (k=520 cm^{-1}) and strong G band (k=1590 cm^{-1}, graphite). A slight shift of diamond Raman peak (k=1332 cm^{-1}) may be due to the internal stress of diamond [9]. On the other hand, an important point that we emphasize in this paper is that the deposited film under 45%SCL condition has only diamond Raman peak (k=1328 cm^{-1}), and there is no peaks from Si and graphite. This result indicates that use of PMITP can fabricate a pure polycrystalline diamond film on the Si substrate. Another interesting point is the uniformity of the deposited film. The lower panel in Fig. 6 depicts the Raman shift spectra measured at the edge of the Si substrate irradiated by PMITP (a) with 100%SCL (non-modulation), and (b) with 45%SCL. For 100%SCL, only Raman shift peak from Si were found. On the other hand, the 45%SCL condition offers only Raman shift peak from diamond, and it is almost the same to that measured at the center of the substrate. This means that almost uniform film was obtained by using the PMITP. From the spectral measurement and deposited film analysis, we infer the following phenomena: The PMITP supplies a higher density flux including C_2 etc to be irradiated to the substrate at timing t=4–7 ms in the modulation cycle. This flux fabricates a carbon film consisting of sp^2 hybridized carbons (e.g. graphite, graphene, fullerene etc.), and also sp^3 hybridized carbons (e.g. diamond) and more carbon-based particles. After that, at timing of t=6–12 ms, the PMITP transports a high density hydrogen radicals to the substrate. The high density hydrogen flux effectively chemically-etches carbons with weaker bonds, for example, through sp^2 hybrid orbitals from the substrate. Whereas, only sp^3 hybridized carbon film is left on the substrate to create a polycrystalline

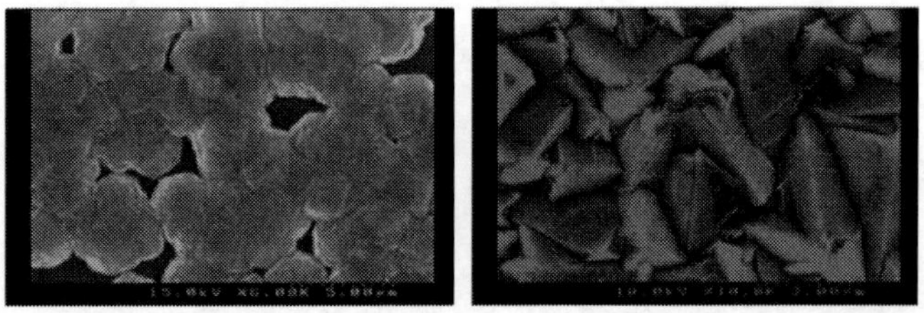

(a) 100%SCL (b) 45%SCL

Figure 5. SEM images of the Si substrate surface at the center after Ar-CH_4-H_2 PMITP irradiation (a) 100%SCL (non-modulated plasma), (b) 45%SCL.

Journal of Physics: Conference Series **441** (2013) 012017 doi:10.1088/1742-6596/441/1/012017

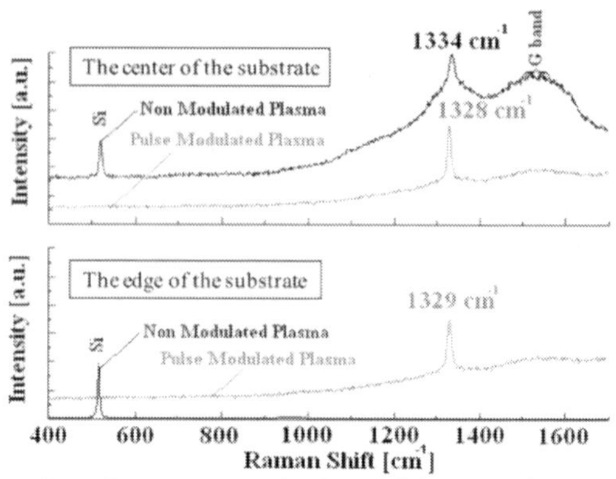

Figure 6. Raman shift measured at the center and edge of the Si substrate irradiated by non-modulated plasma or PMITP with Ar-CH_4-H_2 gas mixture.

diamond film.

4. Conclusions

This paper describes the application of the Ar-CH_4-H_2 pulse modulated induction thermal plasma (PMITP) to polycrystalline diamond film deposition on a Si substrate. The polycrystalline diamond film was successfully deposited on the Si substrate using the PMITP, while the non-modulated plasma could not deposit it under the same heat flux condition to the substrate. Spectroscopic observation was conducted at 5 mm above the Si substrate to measure time evolutions in the radiation intensities of both C_2 and H spectra. The results indicated that in one modulation cycle with the on-time/off-time of 10/5 ms, the intensity of C_2 has a peak at 4–7 ms after the rise-up of the coil current, and H atomic line has an intensity peak around 10 ms. This result implies that the PMITP provides a particle flux dominated by carbon-related chemical species and a flux dominated by H radicals, which are irradiated to the substrate, separately.

References

[1] Berghaus J O, Meunier J L and Gitzhofer F 2004 *Meas. Sci. Technol.* **15** 161
[2] Chen W R, Wu X, Marple B R and Patnaik P C 2005 *Surface & Coatings Technol.* **197** 109
[3] Wang C, *et al* . 2003 *Thin Solid Films* **425** 41
[4] Tanaka Y, Uesugi Y, Sakuta T 2007 *Plasma Sources Sci. & Technol.* **16** 281
[5] Tanaka Y, Morishita Y, Okunaga K, Fushie S and Uesugi Y 2007 *Appl. Phys. Lett.* **90** 071502
[6] Tanaka Y, Hayashi K, Nakamura T and Uesugi Y 2008 *J. Phys. D: Appl. Phys.* **41** 185203
[7] Ito T, Tanaka Y, Hayashi K, Uesugi Y 2009 *J. Plasma & Fusion Research Series* **8** 728
[8] Main R P 1969 *NASA* **CR-72570**
[9] Knight D S and White W B 1989 *J. Mater. Research* **4** 385

Journal of Physics: Conference Series **441** (2013) 012018 doi:10.1088/1742-6596/441/1/012018

Enhancement of Non-Equilibrium Atmospheric Pressure He Plasma Discharges by Using Silicon Diode for Alternating Current

Yujiro Sumiishi[1*], Yoshihiko Uesugi[1,2], Yasunori Tanaka[1,2], and Tatsuo Ishijima[2]

[1]Faculty of Electrical and Computer Engineering, Kanazawa University, Kakuma, Kanazawa, 920-1192, Japan
[2]Research Center for Sustainable Energy and Technology, Kanazawa University, Kakuma, Kanazawa, 920-1192, Japan

me111392@ec.t.kanazawa-u.ac.jp

Abstract. In this paper, we report that enhanced He dielectric barrier discharges (DBDs) using simple and inexpensive device, such as Silicon Diode for Alternating Current (SIDAC) and high voltage transformer at commercial frequency. The SIDAC is designed for direct interface with the ac power line as fast switching and pulse devices. Here, discharge characteristics of He DBDs using SIDACs connected in series are studied experimentally. It can be obtained by using 15 series SIDACs that a rapid voltage change at the SIDAC breakover is dv/dt ~20 kV/μsec and the pulsed DBD current with a duration of ~200 nsec reaches ~1 A, 100 times larger than that of the normal DBD without SIDACs. Emission intensity of He DBD is also increased strongly by series SIDACs to be ~100 times larger compared without the SIDACs.

1. Introduction

The non-equilibrium atmospheric pressure plasma discharges (APPDs) have various merits that easy handling of non-equilibrium plasmas and chemical reactivity. These favorable characteristics enable us to apply those to biological application, thin film fabrications with plasma CVD, surface modifications of materials, light source and display. Therefore, the non-equilibrium APPDs are one of the most useful discharges in the industry. Dielectric barrier discharges (DBDs) are one of the typical non-equilibrium APPDs. In order to obtain stable and efficient DBD plasmas it is usually required to use high voltage and high frequency power sources and/or high voltage pulse sources with high repetition rate. When large volume DBD plasmas are necessary for environmental and engineering applications, generation of such plasmas also requires expensive and complicated power sources. In this paper, we have been studied the enhancement effects of atmospheric He DBD plasmas by using a Silicon Diode for Alternating Current (SIDAC) as a simple high voltage power source with rapid voltage change.

Table 1. Basic properties of the SIDAC (Model No. K1V38(W))

Breakover voltage[V_{BO}]	360 ~ 400 V
Breakover current[I_{BO}]	0.5 mA
Hold current[I_H]	50 mA
OFF state voltage [V_{DRM}]	270 V
OFF state current [I_{DRM}]	10 μA
Switching Resistance[R_S]	0.1 kΩ
Thermal Resistance	15 °C/W

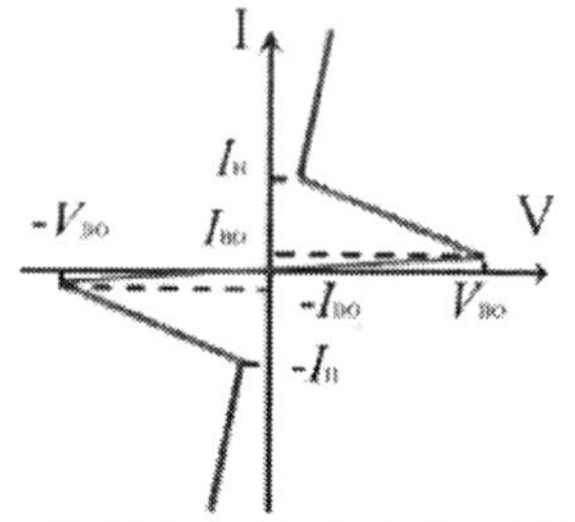

Figure 1. V-I characteristic of SIDAC

2. Experimental setup

2.1. What is the SIDAC?

The SIDAC is designed for direct interface with the ac power line. The operation of the SIDAC is functionally similar to that of a spark gap. Figure 1 shows V-I characteristic of SIDAC. The SIDAC remains non-conducting until the applied voltage meets or exceeds its rated breakover voltage (V_{BO}). Once entering this conductive state going through the negative dynamic resistance region, where conduction current is larger than the breakover current (I_{BO}), the SIDAC continues to conduct, regardless of voltage, until the conduction current falls below its rated holding current(I_H). At this point, the SIDAC returns to its initial nonconductive state to begin the cycle once again.

In general, the SIDAC is inexpensive and easy to use. In the present experiments, we used the SIDAC (Model No. K1V38(W)) made in Shindengen industry, which has the highest V_{BO}. Table 1 shows the basic properties of K1V38(W) used in the experiment. Although the V_{BO} is not enough high about 400 V for DBD plasmas, series connection of N SIDACs allows us to have much higher operation voltage given by $V=N\times V_{BO}$. For example, the switching voltage (effective breakover voltage) is increased to be 3,600 ~ 4,000 V with 10 series SIDACs.

2.2. DBD plasma reactor and power source

Schematic view of DBD plasma reactor used in the experiment is shown in Figure 2. An inside electrode is copper wire and outside one is aluminium sheet, and a quartz tube with 3 mm outer diameter and 0.8 mm thickness is used as dielectric barrier between two electrodes. The length between two electrodes is 10 mm.

The electric circuit using SIDAC and high voltage transformer is shown in Figure 3, and the experimental condition is shown in Table 2. The secondary voltage of the high voltage transformer is kept at 12 kV_{PP} at a frequency of 60 Hz. The SIDACs and DBD reactor are connected in series to apply SIDAC's high voltage switching to DBD. The number of the SIDACs in series is changed to be 4 patterns for 0, 5, 10 and 15. The SIDAC switching characteristic is checked using high voltage resistors and DBD plasmas as load. Rapid response of DBD plasmas at SIDAC switching phases are monitored electrically and optically. Time variation of emission intensity is observed by photo receiver (Model No. OE-200-UV, 190~1000 nm). The plasma emission is also measured by spectrometer and emission images are taken by digital camera.

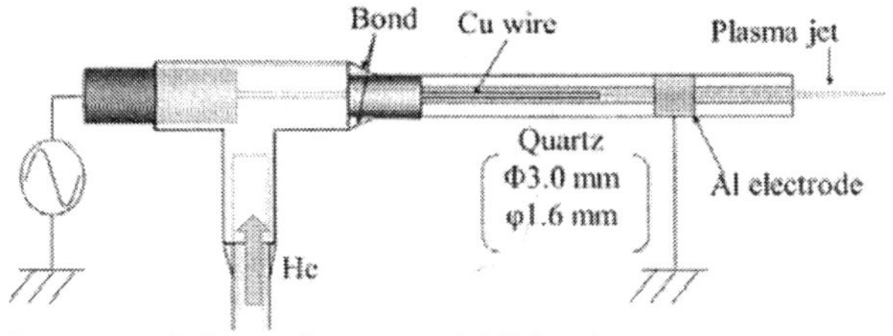

Figure 2. Schematic view of DBD plasma reactor

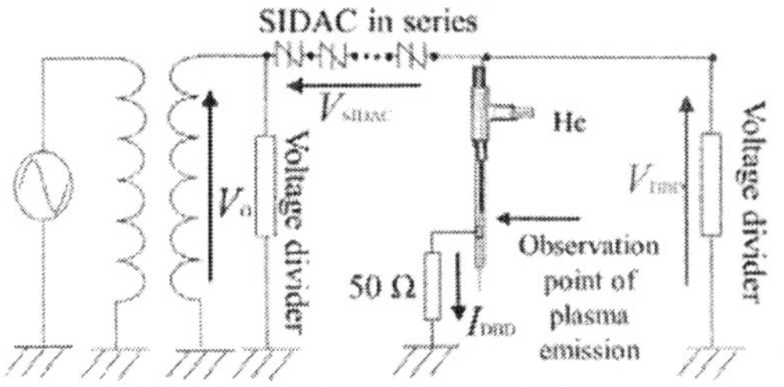

Figure 3. Experimental circuit

Table 2. Experimental condition

Output voltage [V_o]	12 kV_{pp}
Frequency	60 Hz
Gas	He
Gas flow	4 slpm
No. series SIDACs	0, 5, 10, 15

3. Experimental results

3.1. Characteristic features of the DBD plasma

Figures 4(a) and (b) show waveforms of transform secondary voltage(V_o), SIDAC applied voltage(V_{SIDAC}) and DBD voltage(V_{DBD}), DBD current including the charging of DBD(I_{DBD}) and emission intensity from DBD when 15 SIDACs were used. When V_o meets or exceeds V_{BO}, SIDAC turns into the conductive ON state from OFF state quickly and a high voltage pulse is applied to the DBD plasma reactor simultaneously. Typical switching time of series SIDACs observed in the experiment is about 200 ns. The voltage rise rate of dV_{DBD}/dt ~20 kV/μsec is obtained when 15 SIDACs are used. At the same time, small I_{DBD} of ~40 mA can be seen firstly. After this small DBD event, second large DBD pulse of 0.5~1 A height and ~200 ns width flows with a random delay time from the first DBD pulse. Corresponding to two DBD pulses visible light emission can be seen after ~20 μsec delay. DBD characteristics in positive and negative polarities are neary same in the present configuration.

When DBD is used as a load of the SIDAC circuit connected in series, the electric circuit should be governed by two switching characteristics, one is SIDAC switching and the other is DBD itself. The switching characteristics and generation of DBD at the positive switching phase are schematically shown in Figure 5. Firstly as shown in Figure 5(a), after previous DBD in the negative polarity V_o ~V_{SIDAC} goes up positively under the condition of $V_{SIDAC} < N \times V_{BO}$. In this phase, SIDAC is off state and no DBD current flows. The capacitor of dielectric barrier should be negatively charged and the circuit voltage from the transformer is dominantly applied to the SIDAC. When V_{SIDAC} exceeds $N \times V_{BO}$ SIDAC turns to be on state simultaneously and the voltage applied to DBD electrodes increases quickly as shown in Figure 5(b) and (c). At this phase, first small current pulse with a peak of several tens mA are generated. Since the SIDAC used in the present experiments requires $I_{DBD} > 50$ mA for the holding current to keep a continous ON-state, SIDAC forces to turn off the DBD current if $I_{DBD} <$ 50 mA transiently. The peak current of the first DBD pulse might be determined by the transient voltage applied to DBD at the SIDAC switching phase. So far the reason why the second DBD pulses observed in the experiments have such high current peak of ~ 1A is not clear, but it is speculated that the first DBD current is dominated by both DBD start-up and SIDAC transient switching characteristics. Probably, applied high electric field to DBD space by high speed SIDAC's switching is caused second large DBD pulse. It may come from the triggering effect of the remaining first DBD plasmas. Voltage is applied to the DBD space when SIDACs turn ON state, the discharge current of the first pulse discharge is eliminated because it cannot maintain a stable on-state. In the above state of the second pulse, SIDACs are on state from the voltage measurement data. It is considered that electron density associated with the discharge plasma by the first pulse attenuated as passage of time, but the second pulse discharge is formed stronger by small electronic discharges of first pulse. In the final stage shown in Figure 5(d), DBDs are terminated by charge-up effect of positively charged DBD barrier. Finally, the SIDACs return back to a non-conductive state and wait the following switching phase at the negative polarity.

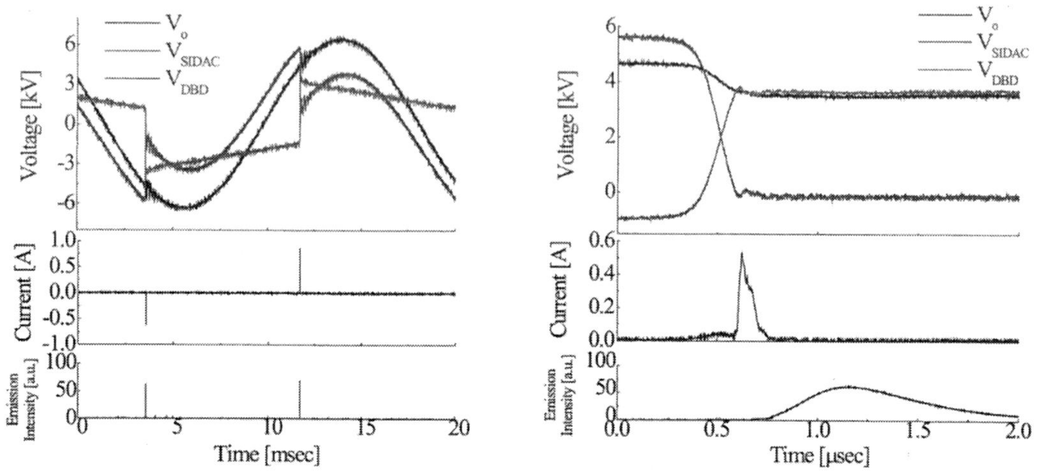

(a) One cycle (b) Enlarged view at the SIDAC OFF/ON transition

Figure 4. Waveforms of voltage, such as V_O, V_{SIDAC} and V_{DBD}, I_{DBD} and emission intensity when 15 SIDACs are used

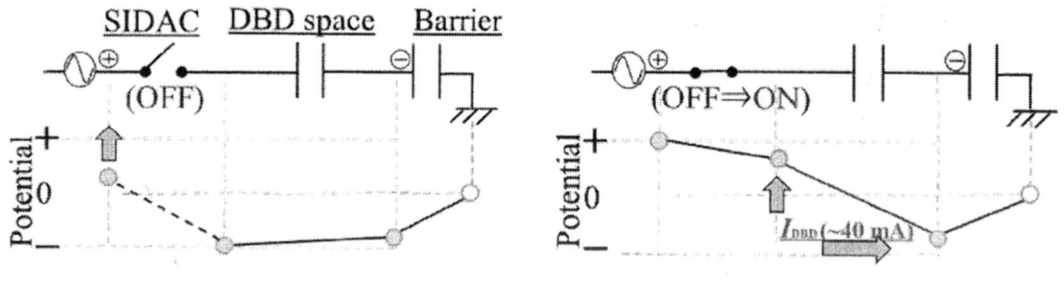

(a) Start from negative to positive polarity (b) SIDAC OFF to ON transiton

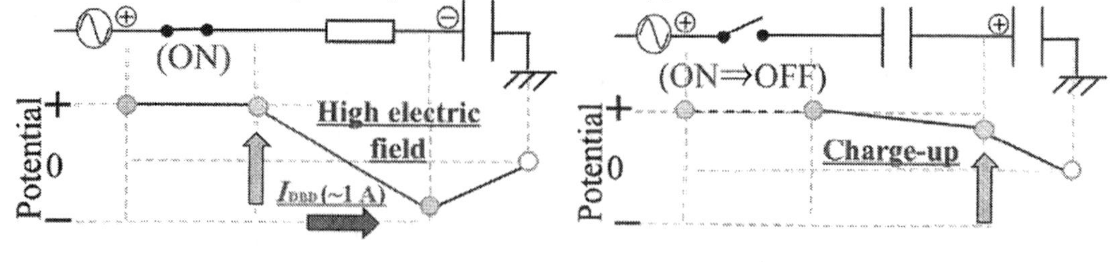

(c) High current pulse appear by high electric field (d) DBD stop with decreasing potential difference by charge up

Figure 5. Relationship between SIDAC switching at the positive phase and DBD

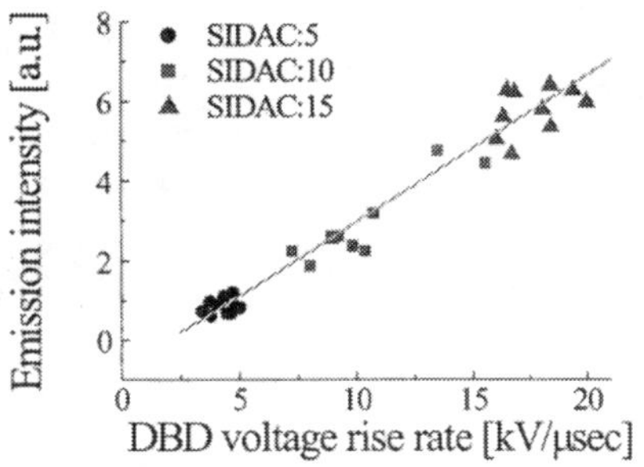

Figure 6. Relationship between emission intensity and dV_{DBD}/dt

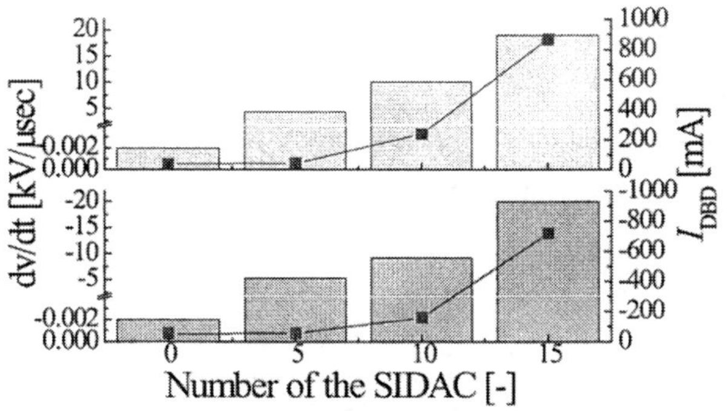

Figure 7. Relationship between dV_{DBD}/dt, I_{DBD} and number of the SIDAC

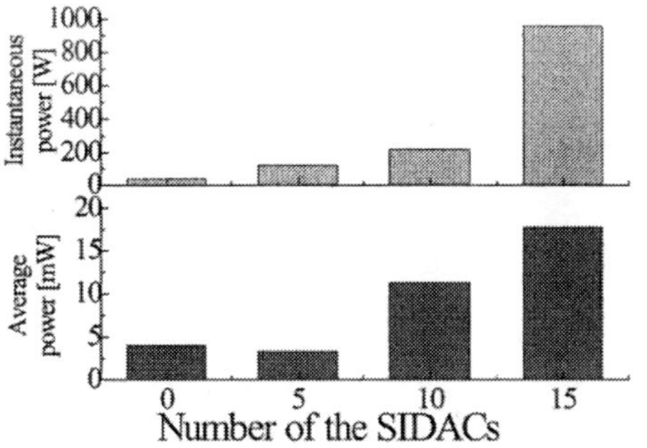

Figure 8. Instantaneous and average power against number of the SIDAC

Figure 6 shows relationship between the visible light emission intensity and dV_{DBD}/dt at the SIDAC OFF/ON switching in the positive polarity. The value of dV_{DBD}/dt increases with increasing the number of connected SIDACs in series, and the emission intensity also increases in proportion to dV_{DBD}/dt. Direct cause of this result is that discharge voltage which applied to DBD space is higher by large dV_{DBD}/dt. Figure 7 shows summary of the SIDACs characteristic by different polarity. It can be obtained by using 15 series SIDACs that a rapid voltage change at the SIDAC breakover is dV_{DBD}/dt ~20 kV/μsec and the pulsed DBD current reaches ~1 A, 50 times larger than that of the normal DBD without SIDACs. Figure 8 shows the instantaneous and average DBD power against number of SIDAC. Average power is also increased by series SIDACs to be ~5 times larger compared without the SIDACs. The reasons why average power decrease in series 5 SIDACs are the timing of the DBD is defined by breakover voltage of the connecting SIDAC, and greatly reduced the number of discharge than without SIDAC.

3.2. Spectroscopy of He plasmas

In this section, we show the result of He plasma emission spectroscopy. Pictures of He plasmas between two DBD electrodes are shown Figure 9. Shutter speed of CCD camera is set at 100 msec, so these pictures integrate the plasma emission for nearly 6 periodic cycles. The plasma emission doesn't show any drastic change at 5 SIDACs. However, when 10 or 15 SIDACs are connected in series, He plasma emission is increased strongly.

Spectroscopic results of He plasma emission are shown Figure 10. Shutter speed of CCD detector of a polychromator is set at 19 msec, and the number of averaging is 60 times. He(I) intensities of 558.5 nm and 667.8 nm show that increasing ~100 times compared that without SIDACs. Plasma emission does not show any spectra of electrode material impurities. Therefore the enhancement of visible emission shown in Figure 9 is caused by the increase of the densities of DBD plasma density.

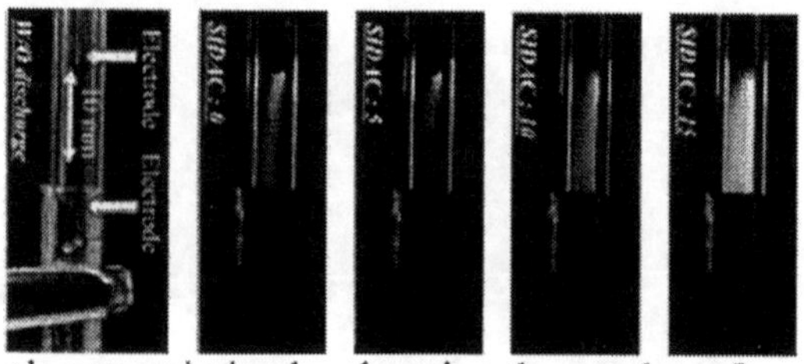

Figure 9. Pictures of He plasma emission by changing the number of connected SIDAC in series

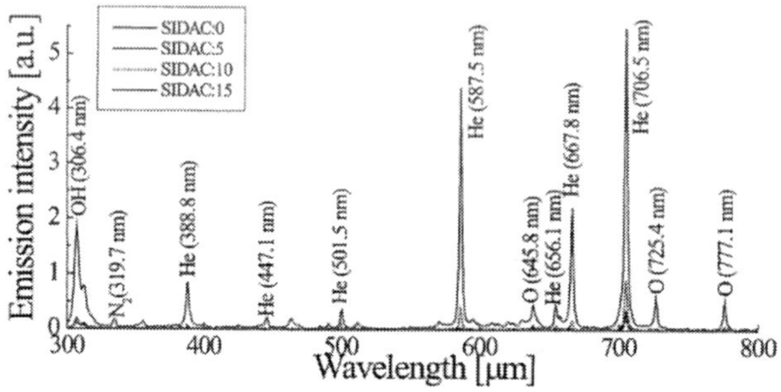

Figure 10. He plasma emission spectrum between two electrode

4. Summary

This paper has reported that enhancement effects of He APPDs by using Silicon Diode for Alternating Current (SIDAC) and high voltage transformer in series. Series connection of N SIDACs allows us to obtain high voltage pulse easily and simply, where the pulse operation voltage is given by $V=N\times V_{BO}$. For an example of the SIDAC application generation of DBD plasmas were tested and strong enhancement by SIDAC was observed. By using 15 series SIDACs a rapid voltage change of ~20 kV/μsec at SIDAC breakover was obtained in the several kV range and the large pulsed DBD current with a peak of ~ 1A and a duration of ~200 nsec was obtained. The emission intensity form He DBD plasmas has a linear relation with dV_{DBD}/dt. The increase of the emission intensity mainly comes from He(I) emission of 558.5 nm and 667.8 nm, which is ~100 times larger compared without using the SIDACs. As the results, it can be concluded that a simple circuit configuration of SIDACs connected in series and high voltage transformer can be applied to generate efficient atmospheric DBD plasmas with high radical density.

References

[1] S. Kanazawa, M. Kogoma, T. Moriwaki and S. Okazaki 1988 *J. Phys. D: Appl. Phys.* **21**, 838
[2] T. Nozaki and K. Okazaki 2006 *Pure Appl. Chem.*, **78**, No. 6, pp. 1157-1172
[3] D. B. Kim, H. Jung, B. Gweon, S. Y. Moon, J.K. Rhee and W. Choe 2011 *Phys. Plasmas* **18**, 043503
[4] T. Hoder, H. Hoft, M. Kettlitz, K.-D. Weltmann and R. Brandenburg 2012 *Phys. Plasmas* **19**, 070701
[5] K.-D. Weltmann, E. Kindel, R. Brandenburg, C. Meyer, R. Bussiahn, C. Wilke and T. von Woedtke 2009 *Phys. Plasmas* **49**, No.9, pp. 631-640
[6] Y. Sumiishi, Y. Uesugi, Y. Tanaka and T. Ishijima 2012 *IEEJ Annual Conf.* **1** p222 (in Japanese)

A Development of Atmospheric Pressure Plasma Equipment and Its Applications for Treatment of Ag Films Formed from Nano-Particle Ink

H Itoh[1], Y Kubota[1], Y Kashiwagi[1], K Takeda[2], K Ishikawa[2], H Kondo[2], M Sekine[2], H Toyoda[2] and M Hori[2]

[1]Technology Development Center, Tokyo Electron LTD., Japan

[2]Graduate School of Eng., Nagoya University, Japan

E-mail: hitoshi.itoh@tel.com

Abstract. We have developed an equipment of atmospheric pressure plasma with two micro-wave guide antennas, which have a discharge line with 41 slots. The antennas are set against a stage with a heater in a process chamber. A process gas, which is a 1 % H_2 gas diluted by Ar gas and its flow rate is 20 standard litter per minute (slm), flows into the micro-wave guide and goes to a process chamber through each slots. A micro-wave is introduced to the micro-wave guide and the atmospheric pressure plasma grows at each slots. We obtained the electron density of $1x10^{15}$ cm^{-3} and the H radical density of $1x10^{-16}$ cm^{-3} at the slot on the condition of a 10 GHz, 1.5 kWatt, pulsed micro-wave with 2.5 pulsed voltage, 4 kHz pulsed frequency, and a duty ratio of 0.16. We applied this system to improve the quality of the spin-coated Ag firm formed from Ag nano-particle ink. This Ag film showed a resistivity of 32 μΩ cm after annealing on the condition of 180 °C for 30 minutes recommended by the maker (The bulk resistivity of Ag is 1.6 μΩ cm). In order to make the annealing time shorter, we studied the effect of atmospheric plasma treatment of Ag film. We obtained the Ag film of the 5.7 μΩ cm resistivity after the atmospheric pressure plasma treatment under 180 °C for 5 min.

1. Introduction

Recently, the Atmospheric pressure plasma (AP-Plasma) has been investigated extensively. The AP-Plasma has a potential of large area, and high-speed surface treatment and it is useful for the many production lines. The AP-Plasma has high electron density and is able to produce high radical density of a process gas [1]. Then, we have developed a microwave AP-Plasma equipment with 1-meter-length plasma-head.

One of AP-Plasma applications is surface treatment of substrate for the printable electronics, because the AP-Plasma needs no expensive vacuum system and is suitable to apply an continues treatment of substrate surface, for example, hydrophilic treatment, hydrophobic treatment, and surface cleaning. We carried out to improve a quality of Ag film spin-coated using Ag nano-ink by our AP-Plasma equipment and the AP-Plasma produced high quality Ag film under simple treatment, lower process temperature, and shorter process time.

2. Experimental

Figure 1 shows our experimental system. The upper figure shows a horizontal view of our experimental system, and the middle figure shows a vertical view of our experimental system. This system has two microwave rectangle wave guides, and each wave guide has one slot line consisted of 41 slots. The length of the microwave antenna is 1-meter.

Journal of Physics: Conference Series **441** (2013) 012019 doi:10.1088/1742-6596/441/1/012019

The microwave antenna was combined to a treatment chamber to change and to maintain the uniform gas condition. This system has no dielectric plate between the antenna and treatment chamber. We can introduce a process gas into this rectangle wave guide antenna and introduce a process gas into the treatment chamber through the antenna. There is a movable stage to hold a substrate in the treatment chamber, and this stage has a hot plate to heat a substrate.

The power source of microwave produces pulsed microwaves and its frequency is 10 GHz. Maximum output power is 1.5 kW, and pulse width is the value between 5 and 50 μs, and its duty ratio is a value between 0 and 50 %.

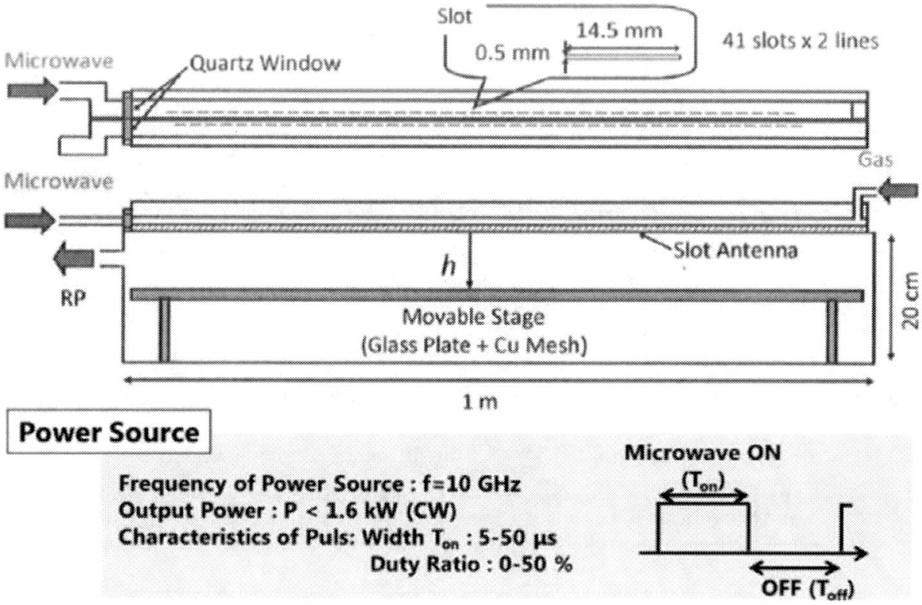

Figure 1. A schematic diagram of experimental apparatus.

Table 1 shows experimental conditions, measurement method of gas temperature (Tg), electron density (n_e), and hydrogen radical density ([H]) and detail of application of AP-Plasma for the improvement of the quality of Ag film deposited by spin coated method using Ag nano-ink. The process gas is the argon diluted 1 % hydrogen molecule. The flow rate of process gas is 20 slm or 50 slm, and its pressure is 1 atom in the treatment chamber. Substrate used for experiments are Si (100) chips of 15 mm×15 mm, and substrate temperature is a value between room temperature and 180 °C. Process time is a value between 5 min and 30 min.

In order to estimate the gas temperature and electron density, we measured the optical emission of N_2 and H_2 gases, respectively, introduced at a mixing ratio of less than 0.1 % in the main discharge gas. We measured optical emissions using an optical fiver and a spectrometer (SpectraPro 750, Acton Research Corporation, USA). The gas temperature was estimated by the curve fitting between experimental and simulated results of the emission of N_2 2nd Positive Band ($C^3\Pi_u - B^3\Pi_g$) [2]. The electron density was estimated by the curve fitting between experimental and simulation results of the optical emission spectra of H_β including the Doppler, pressure, instrumental, and Stark broadening effects [3]. H radical concentration was measured by VUVAS method [4].

Table1. List of experimental condition, measurement method of T_g, n_e and hydrogen radical concentration [H], and detail of application of the improvement of Ag film deposited by spin coated method using Ag nano-ink.

Experimental Condition	Process Gas	Pressure (aton)	Flow Rate (slm)	T_{sub} (°C)	Process Time (min)
	1 % H_2 in Ar	1	20, 50	15 – 180	5 – 30
Measurements of T_g, n_e and [H]	Gas Temperature (T_g)		Electron Density (n_e)		Hydrogen Radical Concentration
	Curve Fitting of N_2 2nd Positive Band		Curve Fitting of H_β Spectra Broadening		VUVAS
Application (Ag film)	Material	Material Maker	Deposition Method		Annealing
	Ag nano-ink	DIC Corporation	Spin Coated Method		Hot Plate in Air

We carried out to improve Ag film quality spin-coated using Ag nano-ink by our AP-Plasma equipment and the AP-Plasma produced high quality Ag film under simple treatment, lower process temperature, and shorter process time. We used Ag nano-ink (JAGT-05) produced by DIC Corporation. The Ag nano-ink consisted of Ag nano-particles with a mean diameter of 10 – 50 nm in the solvent of water or alcohol. Ag film was deposited by a spin coating method and was annealed using a hot plate in the air. It is believed that Ag nano-ink is a promising material to produce electrical lines and electro-rods of devices in the printable electronics which does not need expensive manufacturing tools.

3. Results

3.1. Dependence of Electron density and Gas Temperature on Process Parameters

Figure 2 shows the dependence of the electron density and the gas temperature on the process parameters. The left vertical axis is electron density, and the right vertical axis is gas temperature in each figure. The electron density is at around the value of 5×10^{15} cm^{-3}, and the gas temperature is at around the value of 1000 K.

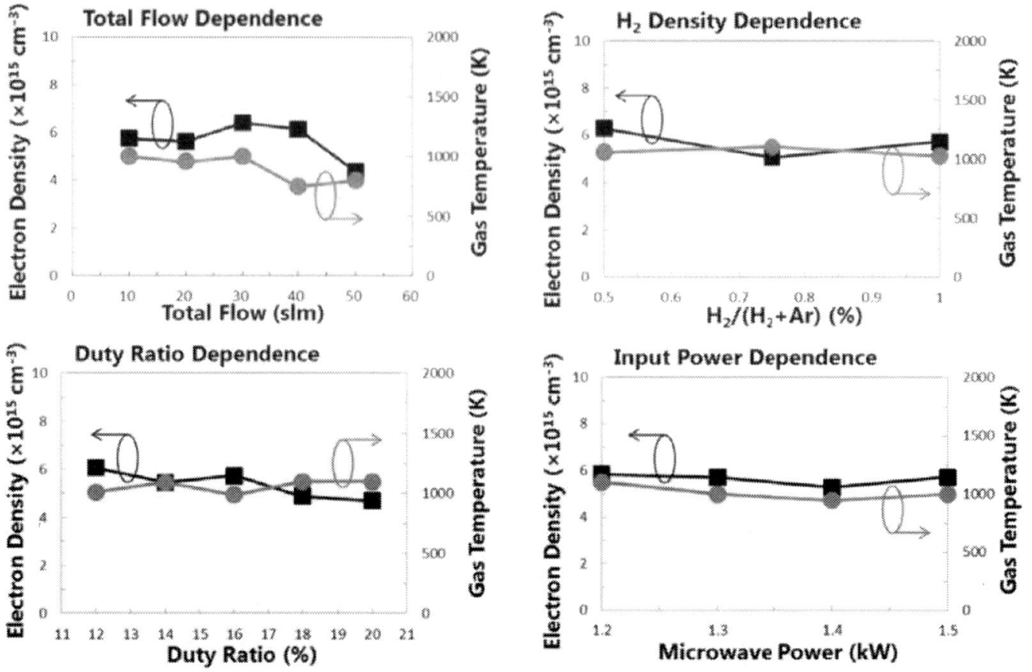

Figure 2. Dependence of electron density and gas temperature on process parameters.

3.2. Dependence of H density on process parameters

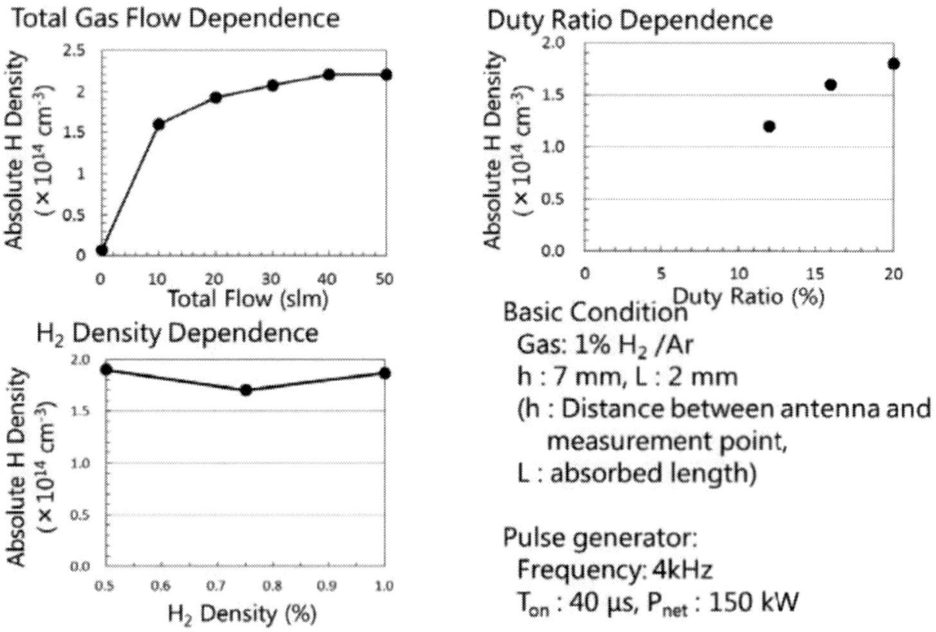

Figure 3. Dependence of H radical density on process parameters.

Figure 3 shows the dependence of the hydrogen (H) radical density on process parameters. The absolute density of H radicals increases rapidly with the increase of the total flow rate from 0 standard-litter-per-minute (slm) to 10 slm, and then the hydrogen radical absolute density increases slightly with the increase of total flow rate.

The dependence of the H radical absolute density shows no significant dependence on the hydrogen molecule (H_2) density and increases proportionally with the duty ratio.

3.3. Dependence of H radical density on the distance from slot antenna

H radicals are deactivated and recombined by many collisions in the atmospheric pressure condition. Figure 4 shows the dependence of H radical density on the distance between the slot antenna and the measuring point. The slope of the decay of the case of the 10 slm total gas flow is larger than that of the case of 50 slm. We can use, however, the high-density H radicals larger than the order of 10^{14} cm^{-3} in the region of the distance less than 5 mm from the slot antenna.

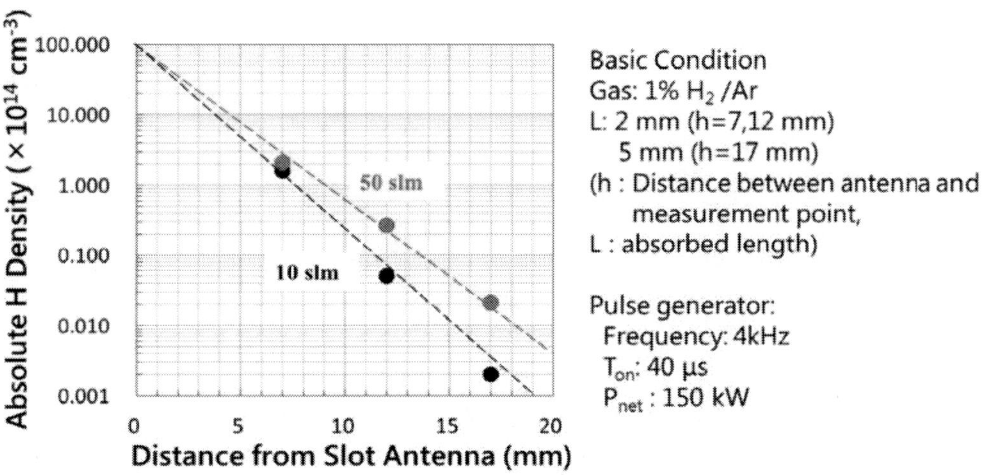

Figure 4. Dependence of H radical density on the distance from the slot antenna.

3.4. Ag Film Spin-Coated using Ag Nano-Ink: Treatment under the Condition recommended by the Ink-Maker

I show one of applications of our atmospheric pressure plasma to improve the film quality of Ag film spin-coated using Ag nano-ink. Figure 5 shows the characteristics of Ag film under the condition recommended by the ink-maker, DIC Corporation. Ag film is spin-coated using the amount of 1 ml Ag nano-ink under the coating condition of 2000 rpm and 10 seconds. And then, the film is under the thermal treatment of 180 °C and 30 minutes in the air.

Journal of Physics: Conference Series **441** (2013) 012019 doi:10.1088/1742-6596/441/1/012019

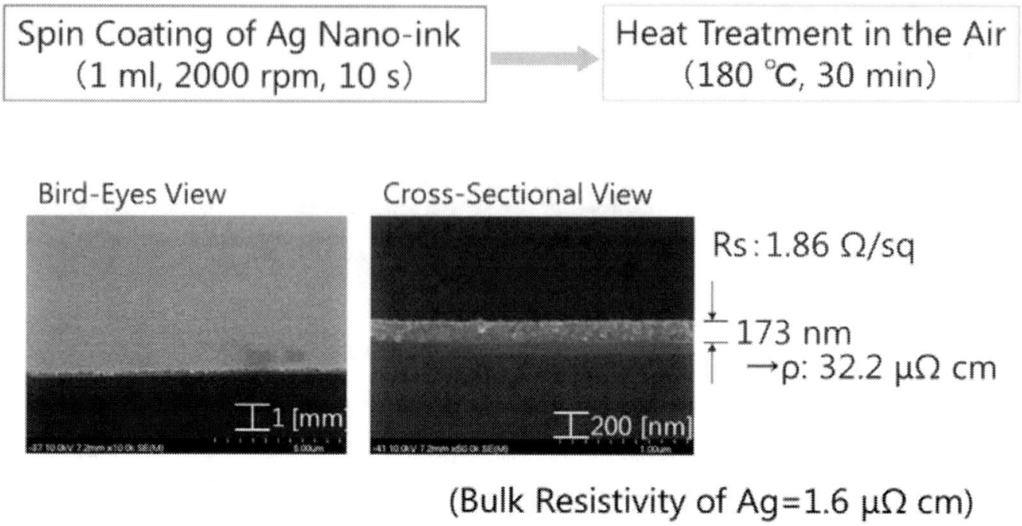

Figure 5. Ag film spin-coated using Ag nano-ink . The treatment is carried out under the condition recommended by the ink-maker.

The surface morphology of the Ag film is very smooth, as shown these two SEM micrographs in Figure 5. The resistivity of this Ag film is 32 μΩ cm. The bulk resistivity of Ag is 1.6 μΩ cm and then the resistivity of this film is about 20 times larger than that of the bulk value.

3.5. Heat and AP-Plasma Treatment in the H_2/Ar Atmospheric Pressure

Figure 6 shows the bird-eyes view and cross-sectional view of Ag film after treatment of only 180 °C heat treatment, only AP-Plasma treatment, and both heat and AP-Plasma treatment. The treatment time is 10 min.

All Ag films have a smooth surface and uniform thickness. However, the resistivity of either thermal only or AP-Plasma only treatment is very, very high. And the resistivity of Ag film after the both thermal and AP-Plasma treatment for 10 minutes is about 32 μΩ cm. This value is the same as the value after the ink-maker recommended treatment, that is, 30 minutes anneal time.

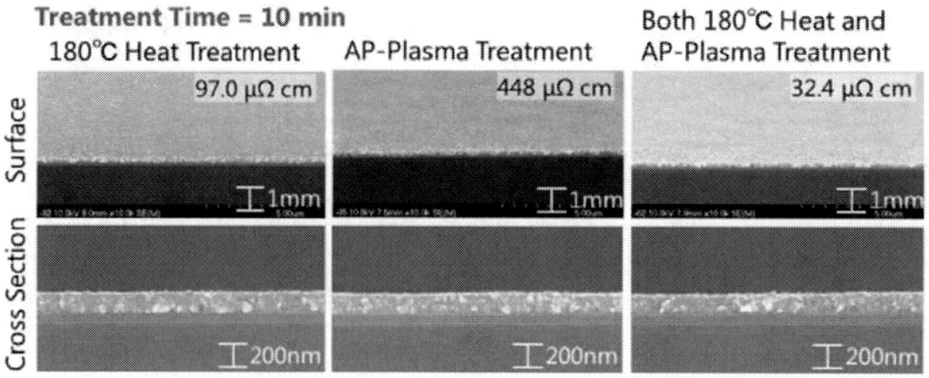

Figure 6. Heat and AP-Plasma treatment in the H_2/Ar atmospheric pressure.

3.6. Pre-Thermal Treatment to Reduce Ink Solvent

We found that the pre-thermal treatment after coating is effective to reduce the resistivity of the Ag film, as shown in Figure 7, which shows the effect of pre-thermal treatment. We carried out to do both AP-Plasma and 180 °C thermal treatment after 5 min pre-thermal treatment. In the case of 100 °C 5 min pre-thermal treatment, the resistivity of Ag film is 5.25 μΩ cm. Ag film has no grain boundary.

We believe that hydrogen radicals in the AP-Plasma of hydrogen and Ar mixture gas react on residuals of solvent in the Ag film and promote the grain growth of Ag film and then the Ag film produces a lower resistivity of 5.25 μΩ cm. Only the plasma treatment cannot reduce the residue of solvent effectively and both the plasma and heat treatments are useful to reduce the resistivity of Ag film.

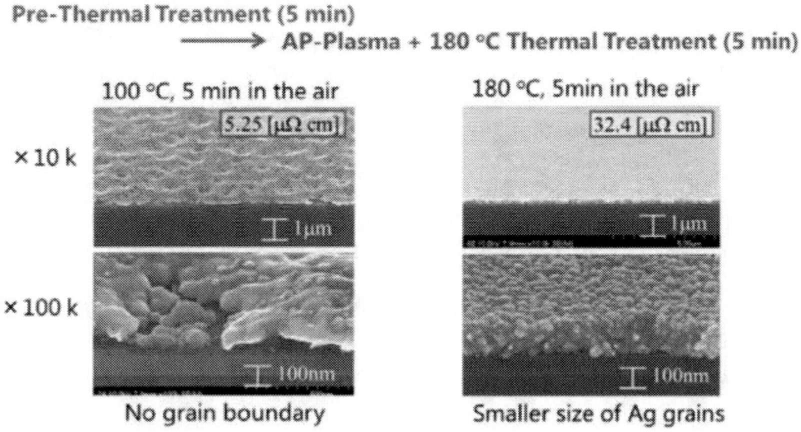

Figure 7. Pre-thermal treatment to reduce film resistivity.

4. Conclusions

We developed the microwave atmospheric pressure plasma system with 1-m-length antenna. This system yields the high electron density of 10^{15} cm^{-3} and produces high H radical densities of 10^{16} cm^{-3} at slots. We applied this system for the treatment of Ag film spin-coated using Ag nano-ink and obtained the high-quality Ag film with the resistivity of 5.25 μΩ cm for shorter process time.

References

[1] Nagatsu M, Xu Ge, Yamage M, Kanoh M and Sugai H 1996 *Jpn. J. Appl. Phys.* **35** L341
[2] Yuji T, Suzaki Y, Yamawaki T, Sakaue H and Akatsuka H 2007 *Jpn. J. Appl. Phys.* **46** 795
[3] Hiramatsu M, Furuhashi H and Goto T 1986 J. Appl. Phys. **60** 1946
[4] Iwasaki M, Matsudaira Y, Takeda K, Ito M, Miyamoto E, Yara T, Uehara T and Hori M 2008 *J. Appl. Phys.* **103** 023303

Atmospheric Pressure Plasma Generation using Dielectric Barrier Discharge Stacked by Insulator Coated Comb-Electrodes

Masaya Honda[1], Toru Sasaki[1], Tsukasa Aso[1], Takashi Kikuchi[1, 2] and Nob. Harada[1]

[1] Department of Electrical, Electronics and information Engineering, Nagaoka University of Technology, 1603-1 Kamitomioka, Nagaoka, Niigata 940-2188, Japan

[2] Department of Nuclear System Safety Engineering, Nagaoka University of Technology, 1603-1 Kamitomioka, Nagaoka, Niigata 940-2188, Japan

E-mail : honnda@stn.nagaokaut.ac.jp

Abstract. To generate volumetric atmospheric-pressure plasma, we proposed a dielectric barrier discharge system with stacked insulator-coated comb-electrodes. The large volumetric plasma was generated in each gap and surface of insulators by stacking electrodes. From the spectroscopic measurement of plasma, almost same shaped spectra were measured from 2 to 10 electrodes. In addition, the ozone concentration in plasma was estimated by the absorption spectroscopy. In this result, the constant ozone concentration was measured from 2 to 10 electrodes.

1. Introduction

Atmospheric-pressure plasma is used in various industrial applications such as material processing, environmental fields and medical fields [1-7]. The atmospheric-pressure plasma can be generated by capacitive, inductive and electromagnetic coupled methods [8-10]. One of the benefits of atmospheric-pressure plasma is that equipments for vacuum generation are not needed, exhibiting features such as the ease of creating non-equilibrium plasma, the generation of several active species, among others.

A dielectric barrier discharge (DBD) is one of the methods of an atmospheric-pressure plasma generation. The DBD consists of a pair of electrodes covered with insulator to prevent thermal damage, and a high-voltage power-supply. The DBD plasma can be easily obtained in atmospheric conditions using simple equipments and an arbitrary geometry [11-14]. Because of the concentration of different activated species between the electrodes, DBD is generally used in industrial applications such as ozonizer, surface modification, and decomposition of environmental pollutants, and so on [15-21].

Typical applications of atmospheric pressure plasma require the quantities of active species, that is, the density of active species and/or the volume of existing active species are needed to increase. Increasing the density of active species, we focus on the generation method for DBD plasma in atmosphere. To obtain large volumetric plasma, the large surface area as the electrodes and insulators is required. Thus, to obtain a relatively large volume of plasma by means of a compact device, we should produce the DBD plasma in the adjoining dielectric cavities [22]. However, the plasma parameters generated in the DBD plasma are difficult to control. To generate the large volumetric atmospheric-pressure plasma, we proposed a DBD system with stacked insulator-coated comb-electrodes and evaluating its plasma parameters.

2. Experimental Equipments

To generate the volumetric DBD plasma, we have developed a type of DBD device with stacked insulator-coated comb-electrodes. Figure 1 shows the insulator-coated comb-electrode. The electrode size is 25 mm × 45 mm. The comb-electrode is made of a tantalum of 0.8 mm of diameter, in alumina tube. Tantalum is used by reason of a high melting temperature and high corrosion resistance. The tantalum is covered by the alumina tube. The inner and outer diameters of the alumina tube are 1 mm and 2 mm, respectively. The relative dielectric constant of alumina is evaluated to be 6.9 at 45 kHz in frequency.

Figure 2 shows the schematic diagram of experimental setup. The electrodes were stacked, alternating the polarity as shown in Fig. 2. The gap distance between electrodes was fixed to 1 mm. A high-voltage power supply (IDX Co. ltd.) was used to generate the atmospheric-pressure plasma. The high-voltage power supply is an inverter system, it is contained by a boost transformer and parallel connection of two inverter circuits. The power supply can generate up to 20 kV_{pp} and 45 kHz in frequency. The voltage and current waveforms were measured using a high voltage probe (HV-P30 : IWATSU) and a current transformer (C.T. 2100 : PEARSON). To measure the active species, the spectrum emitted by the plasma was measured with a spectrometer (C12027-01 : HAMAMATSU).

The ozone, which is one of the active species, has been generated by this system. The ozone concentration in plasma was estimated by absorption spectroscopy of 254 nm wavelength generated by the mercury lamp (L937-01 : HAMAMATSU). The ozone concentration is estimated by the Beer-Lambert law, as follows [23-25];

$$C = -\frac{A}{aL}\log_{10}\left(\frac{I}{I_0}\right) \quad , \qquad (1)$$

where, C (g/m^3) is the ozone concentration, I / I_0 is the relative intensity, a (3000 ℓ/mol cm) is the

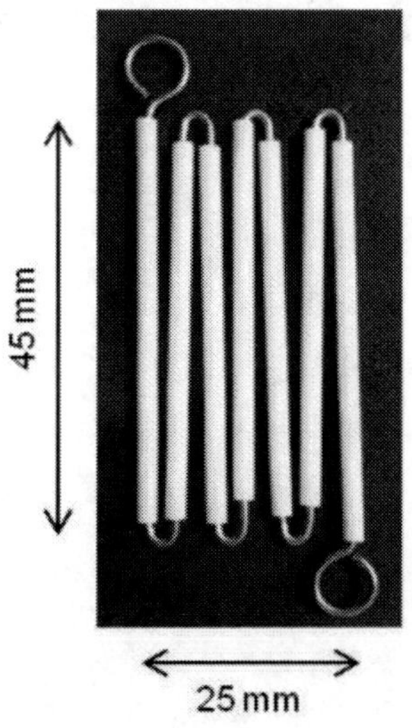

Figure 1. Insulator coated electrode

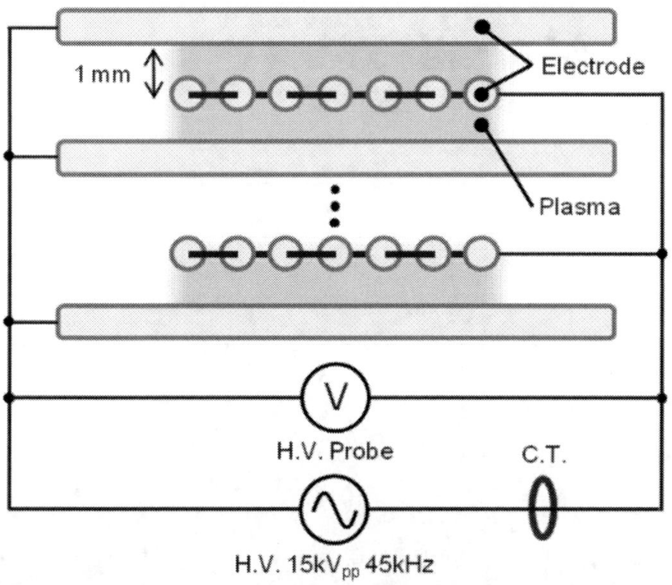

Figure 2. Experimental setup of atmospheric pressure plasma by using insulator-coated comb-electrodes

molar absorption coefficient of ozone, L (cm) is the light path length, and A (=48000) is the conversion coefficient of unit.

3. Experimental Results

To generate the DBD plasma in each gap of electrodes, the voltage of 15 kV_{pp} was applied. The DBD plasma was generated in changing from 2 to 10 electrodes. Figures 3 (a) and (b) show the photos of the generated DBD plasma in the case of stacking 10 electrodes. As shown in Fig. 3 (a), the DBD plasma was generated at the only overlapped area in alternately set electrodes, which was covered with the alumina. As shown in Fig. 3 (b), the DBD plasma was observed in each gap for stacking 10 electrodes, and it was also observed at the surface of the alumina tubes. The plasma volume was estimated by the visible emission region. In the case of stacking 2 electrodes, the plasma volume was estimated to be 6.25 x 10^2 mm^3. In the case of stacking 10 electrodes, the plasma volume was estimated to be 1.12 x 10^4 mm^3. Thus, the plasma volume was increased approximately 18 times from the comparison between the case of 2 electrodes and the case of 10 electrodes. From these results, the large volumetric plasma was obtained by using the stacking electrodes.

Figure 4 shows the typical voltage and current waveforms in the cases of stacking 2 and 10 electrodes, respectively. As shown in Figs. 4 (a) and (b), the applied peak voltage is fixed 15 kV_{pp} and the peak current is estimated to be 0.65 A_{pp} for stacking 2 electrodes, and 2.8 A_{pp} for the stacking 10 electrodes. In these results, the peak current for the stacking 10 electrodes is 4.3 times higher than that of the stacking 2 electrodes. This indicates that the peak current increases about 0.2 ~ 0.3 A_{pp} per electrode.

Figure 3 (a). Photo of plasma emission from top to bottom

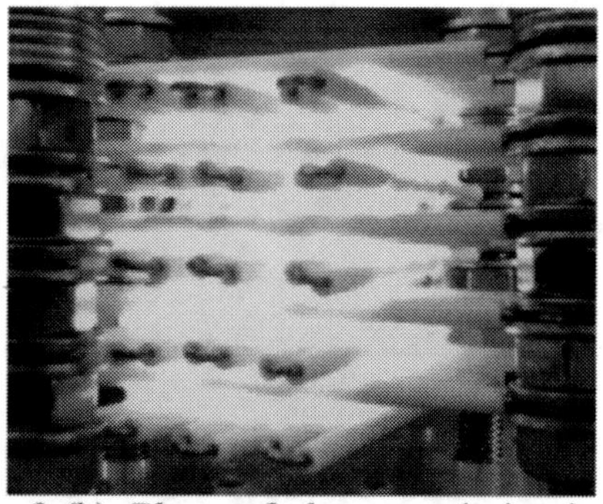

Figure 3 (b). Photo of plasma emission in each gap

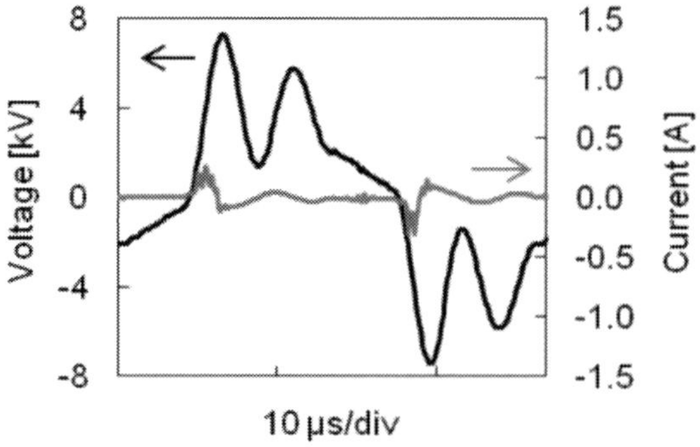

Figure 4 (a). Typical voltage and current waveforms in the case of stacking 2 electrodes

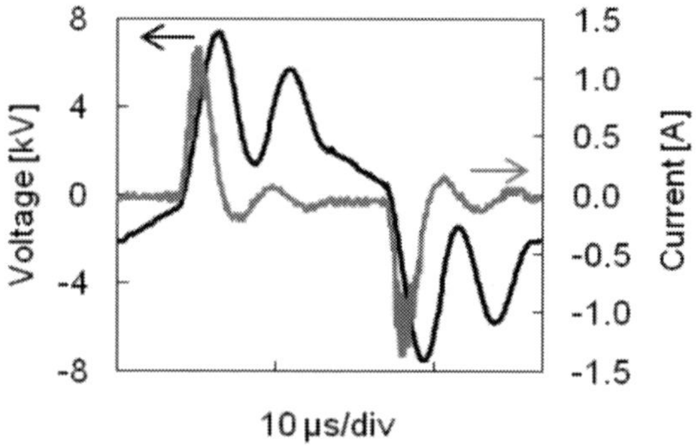

Figure 4 (b). Typical Voltage and current waveforms in the case of stacking 10 electrodes

Figure 5 shows the average input power used to generate the plasma as the function of electrode number. As shown in Fig. 5, the average input power is estimated to be 0.32 kW for the stacking 2 electrodes and 1.17 kW for the stacking 10 electrodes. The average input power to generate the DBD plasma requires 0.1 kW per electrode.

To evaluate the plasma spatial uniformity, the emission spectra from the plasma as a function of the number of electrodes were measured. Figure 6 shows the emission spectra from the generated plasma in the case of 2 and 10 electrodes. As shown in Fig. 6, nitrogen molecules, nitrogen ions, oxygen molecules, oxygen atoms and oxygen ions were observed from the emission spectra [26-27]. The spectra over 400 nm were not observed. It indicates that the UV emission was dominant. In addition, the emission spectrum was similar even in the different cases of number of stacked electrodes.

Figure 7 shows the generated ozone concentration at the central area of overlapped electrodes in the stacking direction. As shown in Fig. 7, the measured ozone concentration varied between 4.9 ~ 5.5 g/m^3 with increase of the number of electrodes. The amount of generated ozone was almost same for the different stacked cases, suggesting that the ozone generation could be increased by stacking more electrodes and by generating larger volumetric plasma.

Figure 8 shows the generated ozone concentration at each gap in the case of stacking 10 electrodes. Figure 8 shows the ozone concentration at each gap, where the upper gap is represented by gap 1, and the lower gap as gap 9, as represented in Fig. 3 (b). As shown in Fig. 8, the ozone concentration is estimated to be 3.1 ~ 3.5 g/m^3 in each gap. It means that the generated ozone concentration was almost same in each gap. In comparison with Figs. 7 and 8, the ozone concentration is different. This difference is attributed to the fact that the ozone concentration at central area of overlapped electrodes was higher than the concentration at edge area of overlapped electrodes. Therefore, the ozone concentration in each gap was lower than the concentration in electrodes stacked direction.

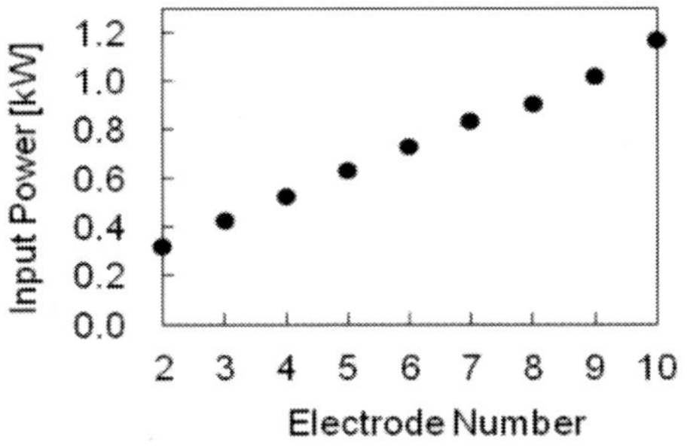

Figure 5. Input power as a function of electrode number

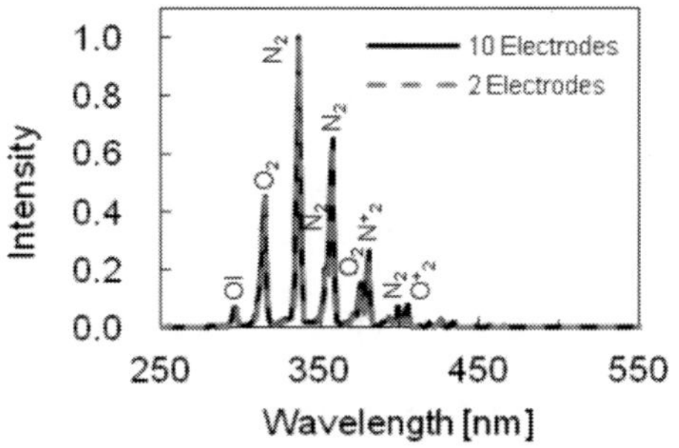

Figure 6. Plasma emission spectra in the case of stacked 2 and 10 electrodes

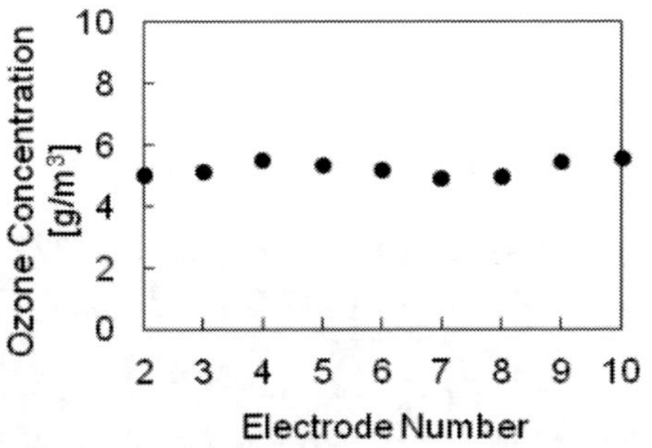

Figure 7. Ozone concentration in stacked direction of electrodes

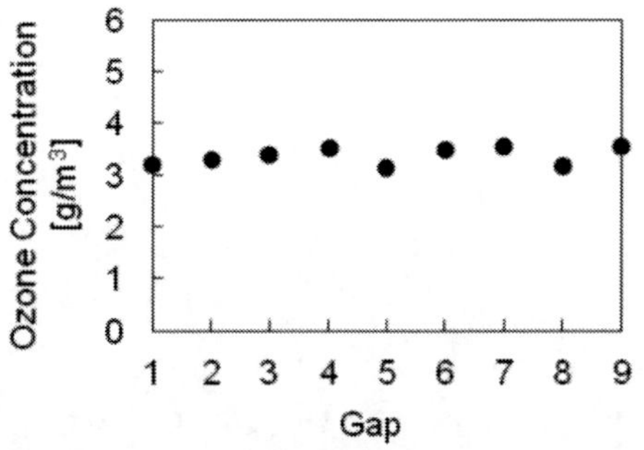

Figure 8. Ozone concentration at each gap

4. Conclusions

To generate large volumetric atmospheric-pressure plasmas, we have developed the DBD device consists of the insulator coated comb-electrode stack. The atmospheric-pressure plasma was generated by changing from 2 to 10 electrodes in each gap and the surface of the insulators. The large volumetric plasma was generated by stacking the electrodes. From the spectroscopic measurement of plasma, almost same spectra were observed in each gap from comparison of 2 to 10 electrodes. The results indicated that the generated plasma condition is independent of the number of electrodes. The absorption spectroscopy of ozone was measured at each gap direction and electrode stacked direction. The generated ozone concentration was almost same at each electrode gap position. The uniform ozone concentration in the direction of the stacked electrodes was measured by stacking 2 to 10 electrodes. In these results, large volumetric plasma and ozone generation could be increased by stacking more electrodes.

References

[1] Schutze A, Jeong J Y, Babayan S E, Park J, Selwyn G S and Hicks R F 1998 *IEEE Transactions on Plasma Science.* **26** 1685-94
[2] Jeong J Y, Park J, Henins I, Babayan S E, Tu V J, Selwyn G S, Ding G and Hicks R F 2000 *Journal of Physical Chemistry A.* **104** 8027-32
[3] Jeong J Y, Babayan S E, Tu V J, Park J, Henins I, Hicks R F and Selwyn G S 1998 *Plasma Sources Science and Technology.* **7** 282-285
[4] Takai K, Shimizu M, Mukaigawa S and Fujiwara T 2004 *IEEE Transactions on Plasma Science.* **32** 32-38
[5] Laroussi M 2002 *IEEE Transactions on Plasma Science.* **30** 1409-15
[6] Weltmann K D, Kindel E, Brandenburg R, Meyer C, Bussiahn R, Wilke C and Woedtke T V 2009 *Contributions to Plasma Physics.* **49** 631-640
[7] Weltmann K D, Kindel E, Woedtke T V, Hahnel M, Stieber M and Brandenburg R 2010 *Pure and Applied Chemistry.* **82** 1223-37
[8] Tendero C, Tixier C, Tristant P, Desmaison J and Leprince P 2006 *Specrtochimica Acta Part B.* **61** 2-30
[9] Park J, Henins I, Herrmann H W and Selwyn G S 2001 *Journal of Applied Physics.* **89** 20-28
[10] Razzak M A, Kondo K, Uesugi Y, Ohno N and Takemura S 2004 *Journal of Applied Physics.* **95** 427-432
[11] Sun L, Huang X, Zhang J, Zhang J and Shi J J 2010 *Physics of Plasmas.* **17** 1-4
[12] Belasri A and Harrache Z 2010 *Physics of Plasmas.* **17** 1-10
[13] Chu H Yand Huang B S 2011 *Physics of Plasmas.* **18** 1-5
[14] Dong L F, Xiao H, Fan W L, Yin Z Q and Zhao H T 2010 *Physics of Plasmas.* **17** 1-7
[15] Samaranayake W J M, Miyahara Y, Namihira T, Katsuki S, Hackam R and Akiyama H 2000 *IEEE Transactions on Dielectrics and Electrical Insulation.* **7** 849-854
[16] Kogelschatz U 2003 *Plasma Chemistry and Plasma Processing.* **23** 1-46
[17] Park S L, Moon J D, Lee S H, Shin S Y 2006 *Journal of Electrostatics.* **64** 275-282
[18] Liu C Z, Wu J Q, Ren L Q, Tong J, Li J Q, Cui N, Brown N M D and Meenan B J 2004 *Materials Chemistry and Physics.* **85** 340-346
[19] Liu C, Cui N, Brown N M D and Meenan B J 2004 *Surface and Coatings Technology.* **185** 311-320
[20] McLarnon C R and Mathur V K 2000 *Industrial Engineering Chemistry Research.* **39** 2779-87
[21] Ravi V, Mok Y S, Rajanikanth B S and Kang H C 2003 *Fuel Processing Technology.* **81** 187-199
[22] Honda M, Sasaki T, Kikuchi T and Harada N 2011 *Journal of JACT.* **16** 75-78 (in japanese)
[23] Hegeler F and Akiyama H 1997 *Japanese Journal of Applied of Physics.* **36** 5335-39

[24] Hakiai K, Takazaki D, Ihara S, Satoh S and Yamabe C 1999 *Japanese Journal of Applied of Physics.* **38** 221-224
[25] Wang D, Matsumoto T, Namihira T and Akiyama H 2010 *Journal of Advanced Oxidation Technologies.* **13** 71-78
[26] Paul H K 1972 *Journal of Physical and Chemical Reference Data.* **1** 423-520
[27] Alf L and Paul H K 1977 *Journal of Physical and Chemical Reference Data.* **6** 113-307

Journal of Physics: Conference Series **441** (2013) 012021 doi:10.1088/1742-6596/441/1/012021

Current-Voltage Characteristics of DC Discharge in Micro Gas Jet Injected into Vacuum Environment

K Matra, H Furuta and A Hatta
Department of Electronic and Photonic Systems Engineering,
Kochi University of Technology, Kochi, 782-8502, Japan

E-mail: 148001e@gs.kochi-tech.ac.jp, hatta.akimitsu@kochi-tech.ac.jp

Abstract. A current-voltage characteristic of direct current (DC) gas discharge operated in a micro gas jet injected into a secondary electron microscope (SEM) chamber is presented. Ar gas was injected through a 30 μm orifice gas nozzle (OGN) and was evacuated by an additional pump to keep the high vacuum environment. Gas discharges were ignited between the OGN as anode and a counter electrode of Si wafer. The discharge was self-pulsating in most of the cases while it was stable at lower pressure, larger gap length, and larger time averaged current. The self-pulsating discharge was oscillated by the RC circuit consisting of a stray capacitor and a large ballast resistor. The real time plots of voltage and current during the pulsating was investigated using a discharge model.

1. Introduction

Nowadays, micro plasma jet has attracted much attention because of its various merits, such as its ability to confine micro scale local plasma, to be generated in atmospheric and sub-atmospheric pressure [1-2]. Moreover, its low operating cost, non-thermal plasma, low power consumption, and easy handling make it attractive in many plasma applications. Some of its applications are surface modifications, analytical chemistry, nanostructure growth and biomedical treatment [3].

DC micro plasma in scanning electron microscope (SEM) using small orifice gas nozzle (OGN) anode has been developed by our group [4-6]. The usage of orifice type gas nozzle consisting of a small hole on a thin wall is a key issue for injection of localized high pressure micro gas jet into vacuum environment. Otherwise, by using a micro capillary type gas nozzle, the pressure of injected gas becomes too small to operate gas discharge due to the poor conductance of long capillary tube.

Micro plasma processing in SEM is expected to be used for local restoration of micro electronics devices such as large scaled integrated circuits (LSIs), fabrication of micro mechanical system devices (MEMS) and nanostructured material growth with SEM observation. In addition, there are advantages of plasma generation in SEM that break down voltage of plasma ignition can be reduced by electron beam and that the profile of injected gas can be observed directly [6].

Because the injected gas rapidly disperses in the vacuum environment, the discharge gas locally exists at the exit of the OGN. The discharge space is limited in the dense gas jet area which profile is determined by the local high pressure gas injection and the rapid dispersion without any wall or any strict boundary. The micro gas jet in vacuum will bring some interesting features of not only the rapid gas flow but also the self-limited discharge space with a large gradient of pressure as compared to conventional gas discharges.

In this paper, electrical characteristic of DC gas discharge operated in a micro gas jet injected in the SEM chamber is presented. The current and voltage characteristics were studied with variation of supplied pressure, electrode gap distance and source voltage. Because the discharge current sometimes oscillates the self-pulsating due to a large ballast resistance and a stray capacitance in the discharge circuit, the temporal wave forms of voltage and current were investigated with monitoring the time averaged values.

2. Experimental apparatus and procedure

The experiments in this research were conducted inside an SEM (Hitachi S-3000N) chamber as shown in figure 1. The anode of an OGN was supported by a 3-D micro-manipulator in the chamber. As the cathode, a cut of mirror polished Si wafer of low resistivity was placed on another 3-D micro-manipulator vertically and counter to the OGN. The gap distance between the electrodes was adjusted to 100, 200, or 300 μm by monitoring the scale in the SEM image, as shown in figure 2a.

The OGN was a standard 1/8" SS-316 tubing closed at one end with a laser drilled hole 30 μm diameter (Lenox Laser). Ar gas was supplied to the OGN through a mass flow controller (MFC). The Ar pressure applied to the orifice, monitored by a pressure gauge, was proportional to the gas flow rate of MFC. The conductance of the OGN as the ratio of gas flow rate to the applied pressure was about 0.075 sccm/kPa. A turbo molecular pump (PFEIFFER VACUUM, TMU071YP) was installed on a service port on the side wall of the chamber in addition to the original vacuum system. With injection of micro gas jet at 15 sccm flow rate which was the maximum in the present experiments, the chamber pressure was kept below 1 Pa and the SEM worked in the normal operating conditions.

Figure 2a shows a typical SEM image of the gas jet between the OGN and Si cathode at 15 sccm flow rate, 210 kPa orifice gas pressure (OGP) and with 300 μm gap. The gas molecules were ionized by the scanning electron beam at 5 keV in energy, and the produced electrons were detected in the same way as the detection of secondary emission from solid surfaces. The brightness of the image depends on total accumulated number density of gas molecules along with the beam line passing through the gas jet. Because the conductance of the straight part of OGN (3 mm inner diameter, D) was sufficiently larger than that of the small end orifice (30 μm, d), as depicted in figure 2b, the pressures inside of OGN and at the exit of orifice would be close to the applied pressure. When the discharge was ignited, the SEM image became fully white due to sufficient electrons from the discharge plasma.

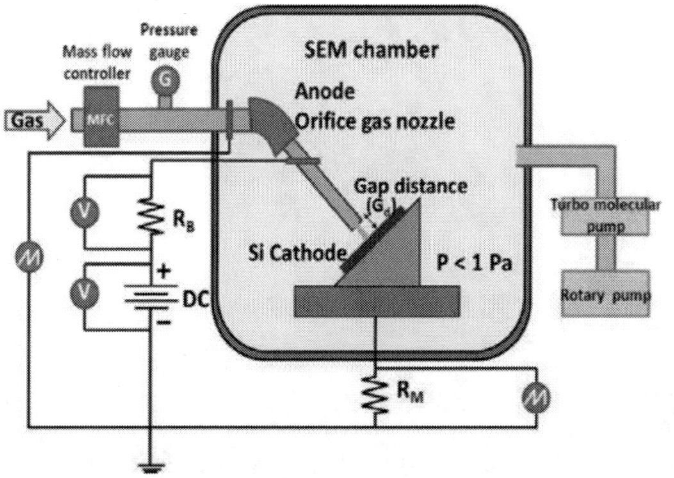

Figure 1. A schematic diagram of the micro gas jet set up in SEM observation chamber.

(a)

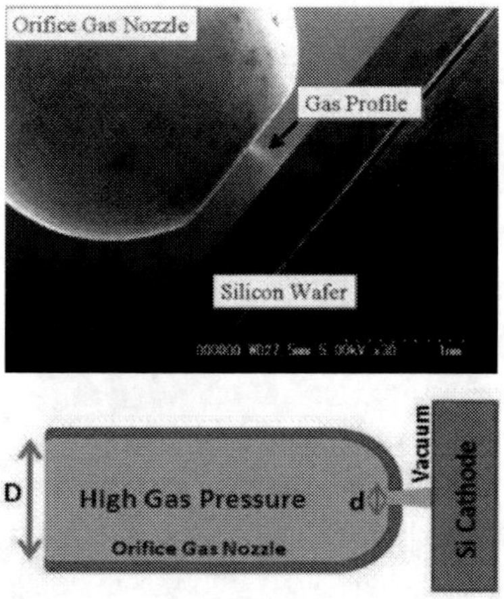

Figure 2. (a) A SEM image of the OGN anode and Si cathode with Ar micro gas jet and (b) a schematic drawing for OGN.

A positive voltage was applied to the OGN from a DC power source of 1 kV maximum through a 10 MΩ ballast resistor (R_B). The discharge voltage and current were monitored using a set of digital multi-meters (DMM). One was used for measuring the source voltage and the other was used for measuring the voltage drop across the ballast resister R_B which reflected the discharge current (I_d). The applied voltage to the OGN (discharge voltage) was obtained from the difference between the measured voltages. The discharge voltage and current sometimes pulsated due to oscillation in the circuit while the source voltage was kept constant. The monitored voltage and current by the multi-meters at a small sampling rate of 2 samples/s represented time averaged values when the discharge was automatically pulsating at frequencies around a several kHz. For measuring the real pulsating current, the counter electrode of Si substrate was grounded by a 50 Ω through terminator (R_M) inserted in BNC cable line, and for measuring the real discharge voltage of OGN, a high voltage probe was connected to the OGN. The current and voltage waveforms for the pulsating mode were stored in a digital oscilloscope (IWATSU-LeCroy, LT364).

3. Experimental results and discussion

3.1. Voltage-current characteristics of discharge in micro-gas jet in vacuum

Figure 3 shows the plots of voltage and current monitored by the multi-meters with variation of gap length (G_D) and the applied pressure (OGP). The obtained V-I curves look almost following the conventional characteristics of glow discharge, i.e., constant voltage with variation of current. By increasing the source voltage while the discharge voltage was almost constant, the discharge current limited by the ballast resistance gradually increased.

Figure 4 shows typical waveforms of voltage and current monitored by the oscilloscope at the experimental condition of 300 μm gap and 140 kPa pressure. There were two different modes of discharge observed, self-pulsating mode and continuous mode [7-10]. The discharge at 800 V (source voltage, V_s) was automatically pulsating even if the source voltage was kept constant. The discharge at 1,000 V was stable and continuous at the constant voltage and the constant current. In a previous work, it has been confirmed that the continuous discharge mode is suitable for local plasma processing like as micro sputter etching due to their stability [6].

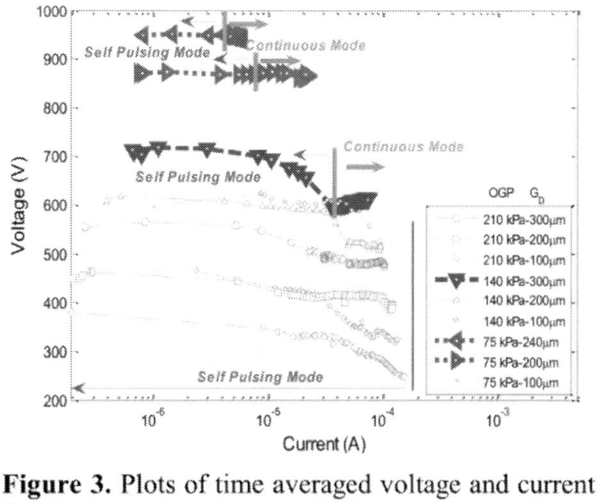

Figure 3. Plots of time averaged voltage and current for discharge in micro gas jet in vacuum monitored by the multi-meters.

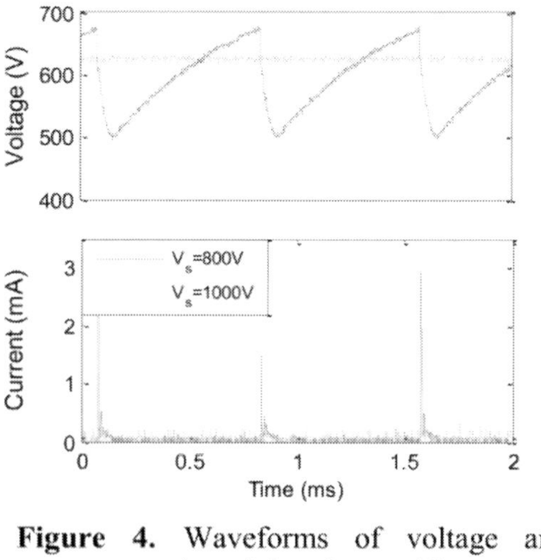

Figure 4. Waveforms of voltage and current monitored by the oscilloscope at 300 μm gap, 140 kPa pressure, and 800 V and 1,000V source voltages (V_s).

Figure 3 actually includes both the plots for the stable continuous discharge mode which represent real voltage and current, and also the hypothetical plots of time average for the pulsating discharge mode. The thick lines plotted in figure 3 show the results in which the discharge mode transited from the self-pulsating mode to the continuous mode when the monitored current exceeded critical current values as indicated. The minimum current to sustain continuous gas discharge (I_{min}) depended on the gas, pressure, electrode gap distance, electrode material and etc. All the plots on the thin lines show the apparent values on DMM for self-pulsating mode. The continuous mode was observed at the lower pressure (75 kPa) when the current exceeded 20 and 5.68 μA at the gap distance of 200 and 240 μm, respectively. At the medium pressure (140 kPa) the continuous mode appeared at the gap length 300 μm when the current exceeded 40 μA. At the higher pressures, 140 and 210 kPa, and at the shorter gap distance, the discharge was always pulsating even if the current was increased until the maximum source voltage.

As shown in figure 4, the current waveform at 800 V consists of sharp pulses with peak current ranging from 2-3 mA repeating with intervals about 0.75 ms. At 1,000 V, the discharge current was stable and continuous at about 70 μA as indicated in figure 3. The peak current in the pulsating mode was 30 to 40 times larger than that for the continuous mode. In the pulsating mode, the voltage dropped abruptly during the short pulses of discharge current and gradually increased during the intervals resulting in saw tooth wave. The pulsating frequency, mostly determined by the interval length, ranged in 0.5-5 kHz depending on the pressure, gap length, and also the source voltage.

3.2. Self-pulsating discharge oscillated in the circuit

Figure 5 shows typical waveforms of voltage and current during pulsating mode in the enlarged time scale for the experimental condition at 210 kPa pressure, 200 μm gap length and 800 V source voltage. Figure 6 shows the real plots of instantaneous current and voltage for the pulsating discharge at 210 kPa pressure, 200 μm gap length with variation of source voltage including the data shown in figure 5.

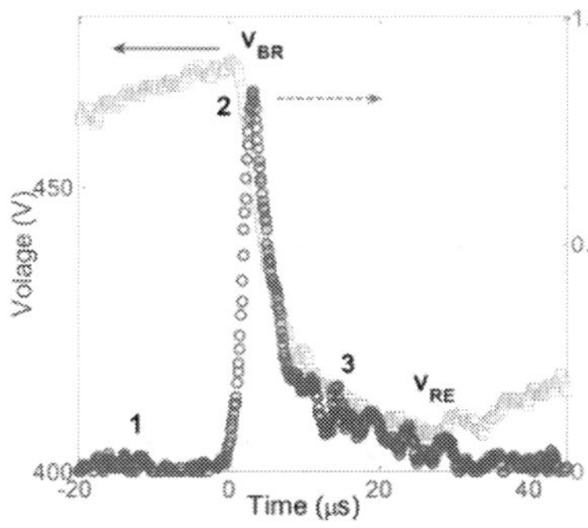

Figure 5. Typical waveforms of voltage and current during pulsating mode in the enlarged time scale at 210 kPa pressure, 200 μm gap, and 800V of source voltage.

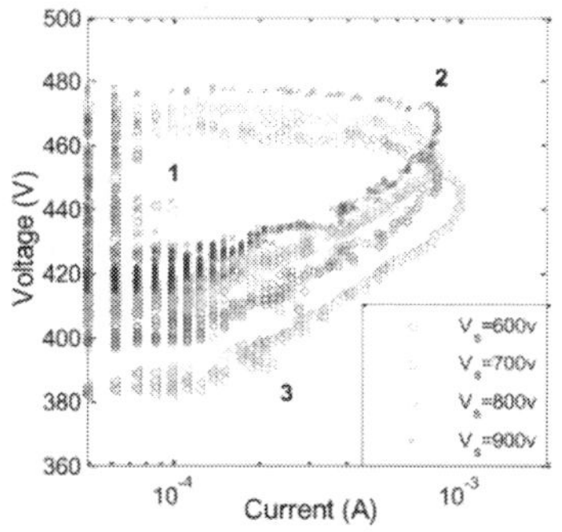

Figure 6. Real plots of instantaneous voltage and current for the pulsating mode at 210 kPa pressure, 200 μm gap, and variation of source voltage.

The self-pulsating mode is characterized by two threshold voltages. One is sustained voltage or recovery voltage as the minimum voltage to sustain plasma (V_{RE}) and the other is breakdown voltage as the maximum (V_{BR}) [12]. For the self-pulsating mode, the discharge voltage oscillated between these two voltages as indicated in figure 5 and 6. The discharge current appeared in the period of voltage drop from V_{BR} to V_{RE}. Here, the self-pulsating of the discharge can be divided into 3 phases as depicted in figure 5. The 1st phase is increasing voltage during the intervals of discharge, the 2nd phase is increasing of discharge current, and the 3rd phase is decreasing of discharge current with abrupt decrease of voltage to terminate until the voltage decreased to V_{RE} [13].

In the 2nd phase, in the present result, the discharge current abruptly increased to 0.83 mA within the rising time of 4.8 μs and the voltage dropped from 472 V to 453 V. In the 3rd phase, the discharge current decreased from the peak to around 50 μA within 16 μs together with decreasing of voltage from 453 V to V_{RE} (405V). After diminishing the discharge, the voltage increases to V_{BR} (472 V) again. This process was periodically repeated resulting in the self-pulsating discharge.

In figure 6, it should be noticed that, even at the same condition for gas discharge, the observed breakdown voltage V_{BR} and the minimum sustain voltage V_{RE} varied with variation of the source voltage. It is necessary to explain the reason why the voltage source affected the discharge characteristics.

For the minimum voltage, V_{RE}, the observed value was not the minimum sustain voltage for the discharge but just the turning voltage from decreasing phase to the recovery phase. By carefully looking at the transition from the 3rd phase to the next 1st phase, i.e., at the termination of the discharge, the voltage has started to recover before the current disappears completely. Because the current was continuously supplied through the ballast resistance also during the discharge, the minimum voltage V_{RE} appeared at the time when the discharge current decreased to the supplying current. When the discharge current was decreased to the supplying current, the actual voltage turned to increase before termination of discharge. When the source voltage was increased, due to increase of the supplying current through the ballast resistance, the decreasing of voltage was terminated faster and started to recover at the higher voltage.

For the breakdown voltage, it should be necessary to consider the pre-ionization before the distinctively observed current pulse. When the voltage approached the real breakdown voltage, ionization started to build up the plasma column and finally to ignite the main discharge. It will take a few μs for building up the plasma column. Due to the delay of building up the main discharge after the voltage became the breakdown voltage, the observed V_{BR} was exceeded the real breakdown voltage. When the source voltage was increased, the working voltage was increased more rapidly during the delay time resulting in observation of higher maximum voltage at the ignition of main discharge.

Figure 7 shows the voltage and current waveforms with variation of source voltage at the same discharge condition of self-pulsating mode, 210 kPa pressure and 100 μm gap. The pulsating frequency increased with increasing the source voltage. Figure 8 shows a model for the electrical circuit of discharge including the stray capacitor.

During the interval of discharge, the stray capacitor was charged through the large ballast resistor until the gas breakdown voltage. When the voltage exceeded the breakdown voltage, the gas discharge started to flow the electrical current with discharging the stray capacitor. When the current supplied from the ballast resistor was sufficient to sustain the continuous discharge ($I_d \geq I_{min}$), the continuous mode of discharge started. If the current supply from the power source through the ballast resistor was insufficient for sustaining the continuous discharge ($I_d < I_{min}$), the discharge was automatically terminated and the stray capacitor was charged again until the breakdown voltage. Therefore, I_{min}, the minimum current to sustain continuous discharge, was the threshold between the self-pulsing mode and continuous mode.

When the source voltage (V_s) was increased, the supplied current through the ballast resistor increased resulting in rapid recovery of voltage for the next discharge of ignition and higher frequency of pulsating. When the ballast resistor was decreased, the supplied current increased with resulting in higher frequency [14]. The trends of charging the stray capacitor simply follow the RC oscillation circuit model. In the present experiments, the stray capacitor was about 65 pF. It has been confirmed that the most of stray capacitor existed in the coaxial high voltage cable between the ballast resistor and the electrode.

The self-pulsating discharge also could be controlled by varying ballast resistor and stray capacitor [9, 14-15]. Due to the extreme peak power together with controllable pulse frequency, the self-pulsating mode is also applicable for micro plasma processing [14-16].

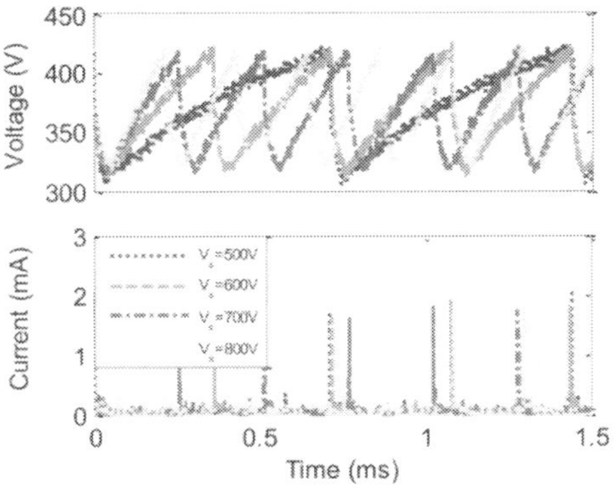

Figure 7. Voltage and current waveforms with variation of the applied voltage at 210 kPa pressure and 100 μm gap.

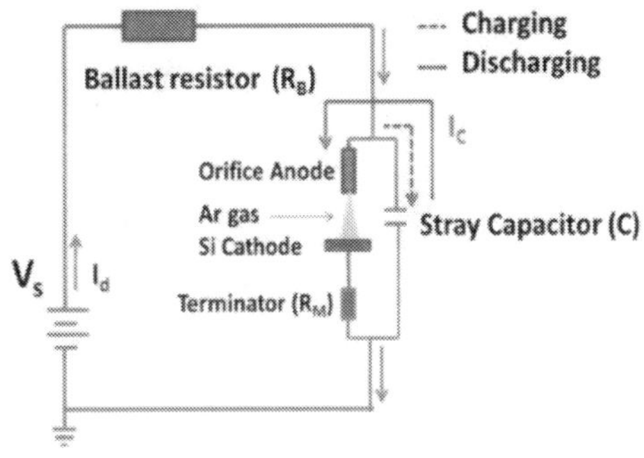

Figure 8. Model for the voltage oscillation due to charging and discharging of the stray capacitance through the ballast resistance.

4. Conclusion

Characteristic of current and voltage of discharge operated in micro gas jet injected into vacuum environment was investigated using the orifice of 30 μm diameter for anode and the Si substrate for cathode equipped in the SEM. The discharge showed self-pulsating mode and continuous mode. The continuous discharge occurred at lower pressure gas jet with longer gap distance at larger current. The self-pulsating of voltage and current was investigated using a model of discharge circuit including stray capacitor.

5. Acknowledgement

This work was supported by MEXT KAKENHI Grant Number 24110719 (a Grant-in-Aid for Scientific Research on Innovative Areas "Frontier Science of Interactions between Plasmas and Nano-interfaces").

6. References

[1] Yoshiki H 2006 *JJAP* **45** 5618
[2] Becker KH, Schoenbach KH, Eden JG 2006 *J Phys D* **39** 55
[3] Conrads H, Schmidt M. 2000 *Plasma Sources Sci Technol.* **9** 441
[4] Zou Q, Ishibashi T, Ohi H, Nakazawa K, Kanakusa H, Hatta A 2006 Microplasma discharge using carbon nanotubes for cathode *ISDEIV ' 06 International Symposium on Discharges and Electrical Insulation in Vacuum* **2** pp 841-844
[5] Matra K, Mitzobuchi Y, Furuta H, Hatta A 2011 Local sputtering by micro plasma jet in SEM *In: Proceedings of 11th international symposium on Sputtering & Plasma Process* pp 46-49
[6] Matra K, Mitzobuchi Y, Furuta H, Hatta 2012 *Vacuum* **87** 132
[7] Staack D, Farouk B, Gutsol A, Fridman A 2008 *Plasma Sources Sci Technol* **17** 025013
[8] Shirai N, Nakazawa M, Ibuka S, Ishii S 2009 *Jpn J Appl Phys* **48** 036002
[9] Yokoyama T, Hamada S, Ibuka S, Yasuoka K and Ishii S 2005 *J. Phys. D: Appl. Phys.* **38** 1684
[10] Arkhipenko V I, Kirillov A A, Safronau Y A, and Simonchik L V 2012 *The European Physical Journal* **60** 455
[11] Aubert X, Bauville G, Guillon J, Lacour B, Puech V and Rousseau A 2007 *Plasma Sources Sci. Technol.* **16** 23
[12] Ovidiu S and STOICAN J 2009 *Plasma Fusion Res. SERIES* **8** 804
[13] Lazzaroni C and Chabert P 2011 *Plasma Sources Sci. Technol.* **20** 055004
[14] Wu S, Lu X, Xiong Z, and Pan Y 2010 *Plasma Science, IEEE Transactions on* **38** 3404
[15] Lu X, Wu S, Chu Paul K, Liu D and Pan Y 2011 *Plasma Sources Sci. Technol.* **20** 065009
[16] Staack D, Farouk B, Gutsol A and Fridman A 2005 *Plasma Sources Sci. Technol.* **14** 700

Journal of Physics: Conference Series **441** (2013) 012022 doi:10.1088/1742-6596/441/1/012022

Characteristic of VHF plasma produced by balanced power feeding

K. Ogiwara[1], W. Chen[2], K. Uchino[2] and Y. Kawai[2]

[1] Graduate School of Information Science and Electrical Engineering, Kyushu University, 6-1 Kasugakoen, Kasuga, Fukuoka 816-8580, Japan

[2] Interdisciplinary Graduate School of Engineering Sciences, Kyushu University, 6-1 Kasugakoen, Kasuga, Fukuoka 816-8580, Japan

E-mail: ogiwara@asem.kyushu-u.ac.jp

Abstract. The characteristics of a VHF hydrogen plasma produced by a balanced power feeding method were examined using a two-dimensional hybrid model. The simulation results showed that the electron density peaks at a certain pressure inside the discharge electrodes and significantly decreases outside the electrodes for high gas pressure. In addition, the power absorption efficiency inside the electrodes was improved by increasing the gas pressure. On the other hand, the plasma was produced within the electrodes for low applied voltages.

1. Introduction

The production of a VHF discharge plasma with large area at high pressure is one of the key technologies in the fabrication of thin-film silicon solar cells because VHF discharge plasma provides high-rate film deposition with low electron temperature [1]. However, in this type of discharge, anomalous discharges occur outside the discharge electrodes. The anomalous discharges consume a large amount of the input power and cause unstable discharges inside the electrodes. Moreover, the anomalous discharges damage components around them, for example a power feeding cable, as the result, the film is contaminated with the sputtered materials. Thus, the occurrence of the anomalous discharges must be avoided for large area and qualified solar cells.

As one of the techniques to avoid the occurrence of the anomalous discharges, a balanced power feeding (BPF) method has been proposed [2]. Nishimiya et al. [2] demonstrated that this method successfully suppresses anomalous discharges and the electron density is higher than that without BPF.

We have studied the VHF plasma produced by the BPF method for effective application to the PECVD. In our previous work [3], we calculated the VHF hydrogen plasma produced by the BPF method, in which we focused on the effects of the VHF voltages applied to each electrode with same amplitude and opposite signs. We confirmed that the BPF method improved the VHF hydrogen plasma characteristics with short-gap parallel plate electrodes: the electron density outside the electrodes significantly decreased and the electron density and the absorbed power increased inside the electrodes compared to those in the conventional single-electrode power feeding (SPF) model. Thus, the advantage of the BPF method was shown in this preliminary work. In the present paper, we examined the characteristics of the VHF plasma produced by the BPF in detail, that is, the dependences on the gas pressure and on the amplitude of the VHF voltages which were applied to the both electrodes.

2. Description of the model

The simulation of a VHF plasma produced by the BPF method was performed by using the Plasma Hybrid Module (PHM) of PEGASUS software Inc.[4] The PHM has been developed on the basis of the Hybrid Plasma Equipment Model (HPEM) formed by Kushner and his colleagues [5]. The detailed description of the PHM is described in our previous paper [3]. Here, we briefly outline the computational procedure of the PHM. The density and the velocity of electrons, which are utilized for the calculation of the electron energy distribution functions (EEDFs) by Monte Carlo method, are calculated by the fluid model and the equation motion of electron. The density and the velocity of ions are determined by solving the equations of continuity in which the drift diffusion model is applied to the particle fluxes. For neutral particles, Navier-Stokes equation and the advective diffusion equation are derived. Some CCP models, which were demonstrated by Lymberopoulos and Economou [6] and by Rauf and Kushner [7], were calculated by using the PHM, and the simulation results of PHM showed good agreements in the previous researches.

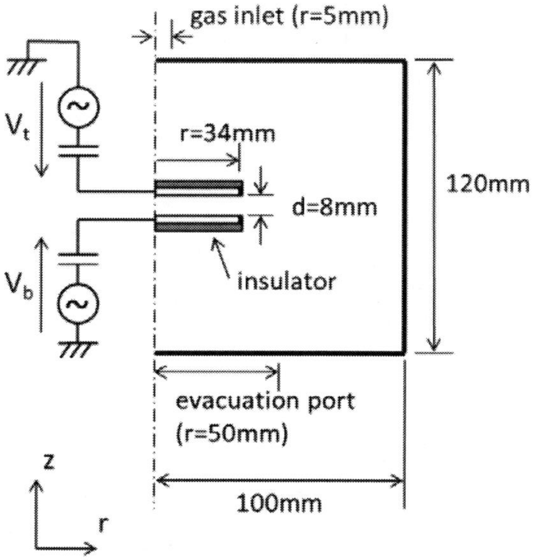

Figure 1. Schematic of the BPF model.

The BPF model we used here is a cylindrical coordinate system with axial symmetry, as shown in figure 1. A pair of parallel disk electrodes, whose radius is 34 mm, is situated in the center of a cylindrical chamber. The gap between the electrodes is fixed to 8 mm. The outside surfaces of the electrodes are covered with insulators, in order to suppress the emission of electric energy from them. In the BPF model, two VHF voltages V_t and V_b, whose amplitudes and phases are equal to and opposite from each other, are applied to the top and the bottom electrodes, respectively:

$$V_t = V_{rf}\sin(2\pi ft) + V_{dc,t},\ V_b = V_{rf}\sin(2\pi ft + \pi) + V_{dc,b}. \quad (1)$$

Here, V_{rf} and f are the amplitude and the frequency of the applied voltages, respectively. In this paper, f is fixed to 60 MHz. Self-bias voltages of each electrode $V_{dc,t}$ and $V_{dc,b}$ are included in the equations. The input power P_{elec} is defined as a time-averaged integral of the products of the applied voltages and the currents flow to the electrodes:

$$P_{elec} = P_t + P_b = \frac{1}{T_{rf}}\int_0^{T_{rf}} (V_t I_t(t) + V_b I_b(t))dt. \quad (2)$$

Here, T_{rf} is the time of a VHF cycle, and I_t and I_b are currents flow to the top and the bottom electrodes, respectively. The input power P_{elec} is equivalent to the total power absorbed in the calculating volume.

The hydrogen plasma in the model is consisted of electrons, H^+, H_2^+, H_3^+, H^-, H, H_2. The hydrogen gas is introduced with the flow rate of 300 sccm from the area around the center axis on the top face of the chamber. An evacuation port is located on the bottom face of the chamber. Here, the behavior of the leak currents is not considered in the model, so that, the effect of the applied VHF voltages in the BPF method on the VHF plasma with short-gap parallel electrodes was investigated.

3. Simulation results

1.1. Gas pressure dependence of BPF plasma

The gas pressure dependence ranging from 67 to 666 Pa was examined for V_{rf} = 100 V. In this operation, the input power was varied between 48 and 57 W. Figure 2 presents the pressure dependence of the electron density. As seen in figure 2 (a), the electron density outside the electrodes is significantly low when the gas pressure is higher than 200 Pa. This result indicates that the BPF method further suppresses the anomalous discharges for high pressures. On the other hand, the electron density inside the electrodes peaks at 400 Pa due to the electron trapping effect [8]. The same tendency was observed in the experiments of VHF hydrogen plasma produced by the BPF method [9]. When the gas pressure is higher than 400 Pa, the electron density in the center region between the electrodes becomes uniform with increasing the gas pressure, as seen in figure 2 (b).

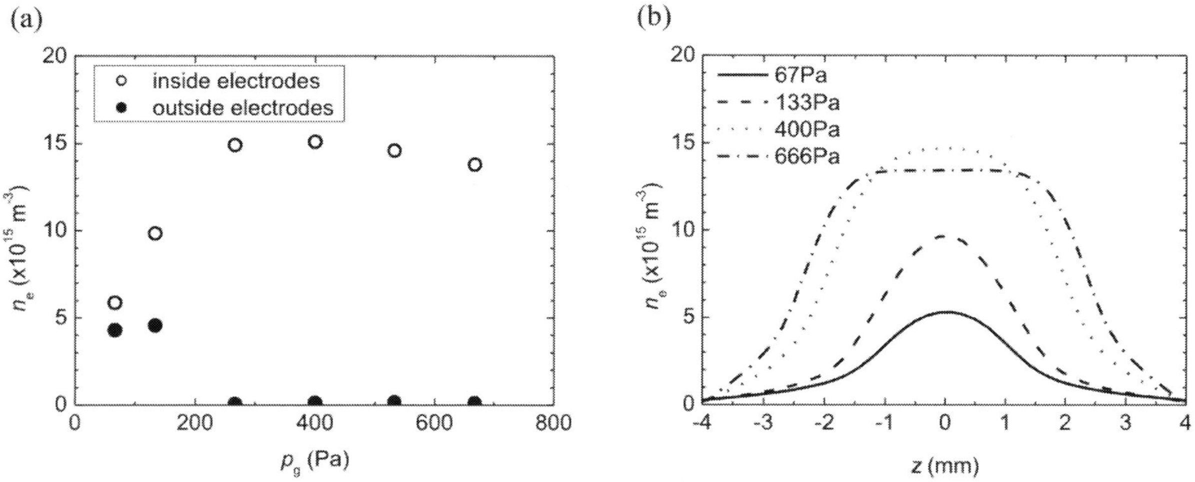

Figure 2. Characteristics of the electron density for different gas pressures at the applied voltage V_{rf} = 100 V: (a) pressure dependence of the peak electron density inside and outside the electrodes and (b) spatial distributions of the electron density between the electrodes in the *z*-direction at *r* = 20 mm.

Figure 3 shows the dependence of the absorbed power density P_{abs} inside the electrodes on the gas pressure. As shown in figure 3 (a), the power is more absorbed near the both electrodes due to the sheath heating [10]. As the gas pressure is increased, the absorbed power density increases in almost the whole region and takes a peak at $z = \pm 2.6$ mm. The enhancements of the power absorption in the sheath and in the middle region are due to the increases of stochastic heating proportionate to the increased electron density and of ohmic heating proportionate to the increased electron-neutral collision frequency, respectively. As shown in figure 3 (b), the total power absorbed inside the electrodes P_{in} also increases with increasing the gas pressure, and the absorption efficiency defined by P_{in} / P_{elec} reaches 87 % at p_g = 666 Pa. The reason for this tendency is because when the gas pressure is increased, the charged particles also increase inside the electrodes. These results indicate the BPF method is effective to supply the power into the process area for higher gas pressures.

The gas pressure dependence of the electron temperature is presented in figure 4. The electron temperature near the electrodes is about 2 times higher than that in the center region between the electrodes. From this result and figure 3 (a), the input power is used for heating of the electrons in the sheath region. The electron temperature decreases without changing its spatial profile when the gas pressure is increased. The low electron temperature is favorable for the production of the solar cell. The decrease of the electron temperature is due to the enhancement of the collisional quenching by neutral particles. On the other hand, the electron temperature outside the electrodes was almost constant around 0.8 eV.

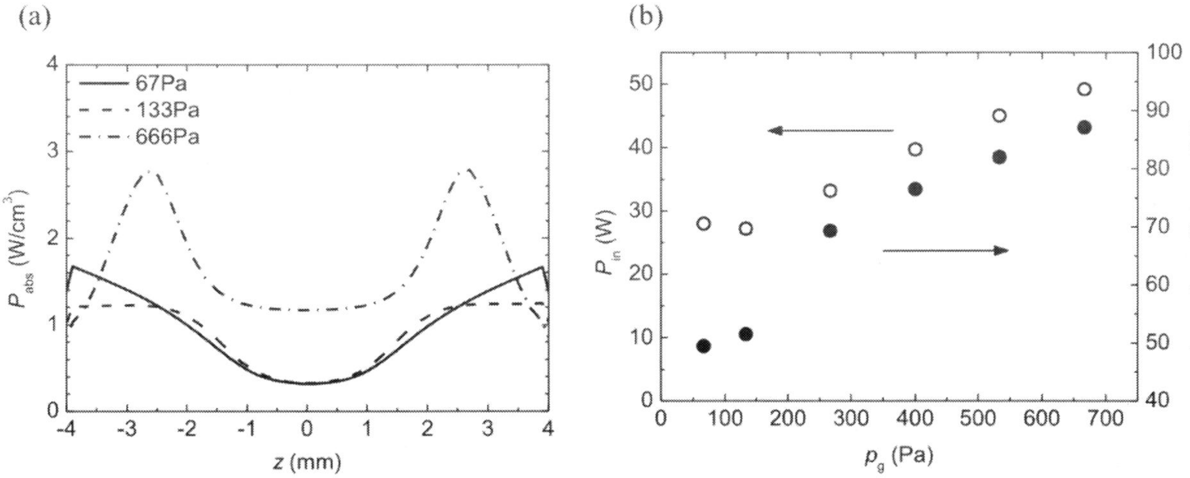

Figure 3. The characteristics of the power absorption for different gas pressures at the applied voltages of 100 V: (a) spatial distributions of the absorbed power density P_{abs} in the z direction at r = 20 mm and (b) the gas pressure dependence of the total power absorbed inside the electrodes P_{in} (open circles) and of the ratio of P_{in} to P_{elec} (closed circles).

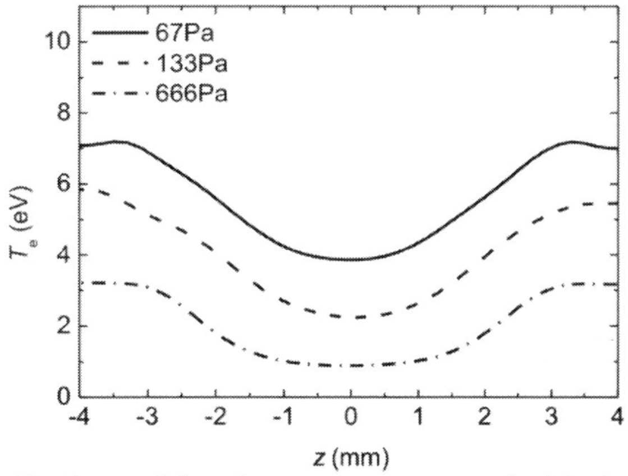

Figure 4. The spatial distributions of the electron temperature inside the electrode at r = 20 mm for different gas pressures. The applied VHF voltages are 100 V.

1.2. Applied voltage dependence of BPF plasma

The dependence of the BPF plasma on the applied voltages V_{rf} ranging from 80 to 150 V was obtained for the constant gas pressure of 133 Pa. In this operation, the input power was varied from 25 to 167 W with increasing the applied voltages. As shown in figure 5 (a), at the applied voltage of 80 V, the electron density inside the electrodes is one order of magnitude higher than that outside the electrodes. As the applied voltages are increased, the electron density increases both inside and outside the electrodes, and at V_{rf} = 150 V, the electron density outside the electrodes overcomes that inside the electrodes. However, it is noteworthy that the electrons outside the electrodes congregate adjacent to

Journal of Physics: Conference Series **441** (2013) 012022 doi:10.1088/1742-6596/441/1/012022

the reverse side of the electrodes. As seen in figure 5 (b), it is obvious that the electron density outside the electrodes in the BPF model rapidly decreases with the distance from the electrode compared to that in the SPF model, in which the same VHF voltage is applied to the top electrode and the bottom electrode is grounded. These results indicate that the supply of the VHF voltages with same amplitude and opposite sign to each electrode has an effect to limit discharge area in the vicinity of the electrodes.

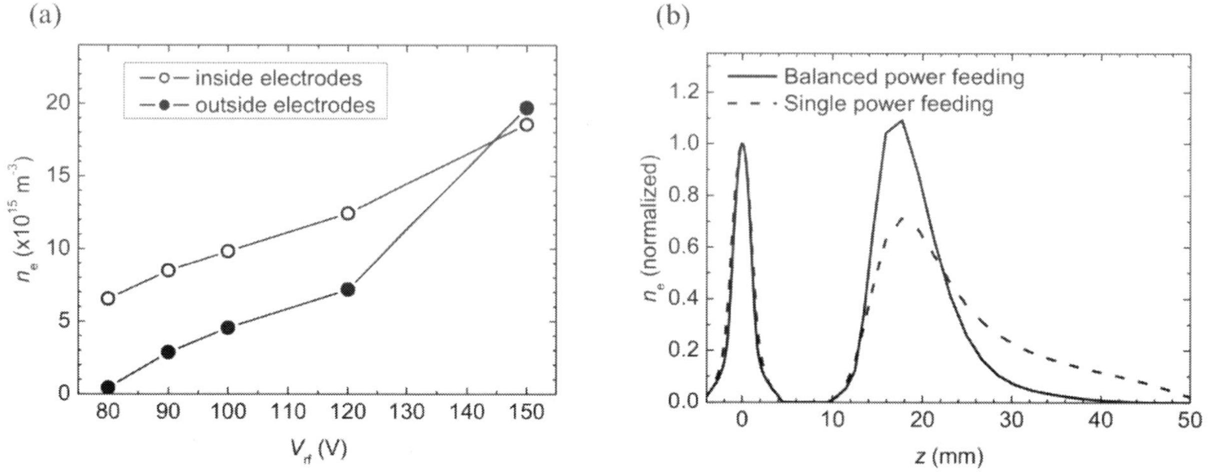

Figure 5. The characteristic of the electron density for the applied voltages for the gas pressure of 133 Pa: (a) the applied voltage dependence of the peak electron density inside the electrodes (open circles) and outside the electrodes (closed circles) and (b) spatial distributions of the normalized electron density in the BPF model (solid line) and the SPF model (dashed line) in z direction at r = 20 mm for the same applied VHF voltages of 150 V; $|z| < 4$ mm: inside the electrodes; $z > 10$ mm: outside the electrodes.

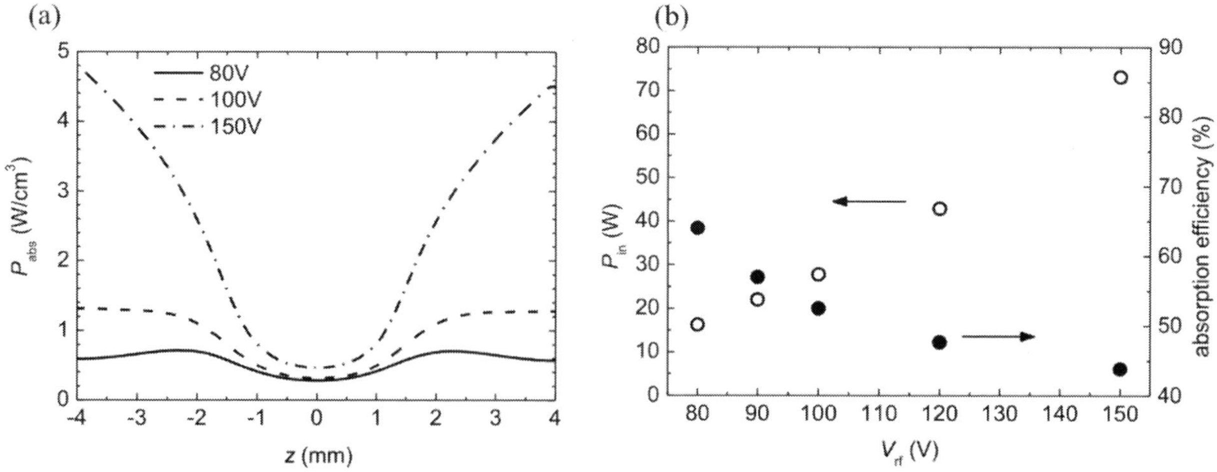

Figure 6. The characteristic of the power absorption for the applied voltages for the gas pressure of 133 Pa: (a) spatial distributions of the absorbed power density P_{abs} in z direction at r = 20 mm for the different applied voltages and (b) the applied voltage dependence of the total power absorbed inside the electrodes P_{in} (open circles) and the ratio of P_{in} to P_{elec} (closed circles).

As shown in figure 6 (a), the absorbed power density P_{abs} increases near the electrodes with increasing the applied voltages since the potential drop within the sheath becomes large, while the enhancement of P_{abs} in the center region between the electrodes is less than 0.2 W/cm^3. Figure 6 (b) indicates the total power absorbed inside the electrodes P_{in} increases as the applied voltage is increased.

However, the absorption efficiency decreases with increasing the applied voltages. This decrease relates to the increase of the electron density near the reverse faces of the both electrodes.

The electron temperature inside the electrodes increased from 5 to 7 eV near the electrodes when the applied VHF voltage was increased from 80 to 150 V for the constant gas pressure of 133 Pa. The decrease of the electron temperature in the center between the electrodes with increasing the applied voltages was less than 0.4 eV. The increase of the electron temperature near the reverse side of the electrodes was small compared to that inside the electrodes.

We found that the overloaded VHF voltages influence the expansion of the discharge area outside the electrodes through the escape of the power which is not absorbed inside the electrodes. In addition, the electron temperature near the electrodes increased with increasing the applied voltages. From these results, the overload of the VHF voltages, that is, the oversupply of the power causes the degradation of the film quality.

4. Conclusion

In this paper, we examined the dependences of a hydrogen plasma produced by the balanced power feeding method on the gas pressure and the applied VHF voltage by using the Plasma Hybrid Module of the software PEGASUS. The electron density outside the electrodes decreased drastically when the gas pressure is higher than 200 Pa. On the other hand, inside the electrodes, the electron density peaked at 400 Pa, and the power absorption increased and the electron temperature decreased with increasing the gas pressure. For the low applied VHF voltages, the electron density inside the electrodes was a decade higher than that outside the electrodes. However, the electron density outside the electrodes rose with increasing the applied voltages and overcame that inside the electrodes at 150 V. The power was more absorbed inside the electrodes for larger applied voltages, however, the absorption efficiency was reduced by increasing the applied voltages. It is concluded that the balanced power feeding method is suitable for producing VHF plasma at high pressures and at low applied voltages, leading to the fabrication of the microcrystalline silicon solar cell with qualified film deposition.

References

[1] Kondo M, Fukawa M, Guo L and Matsuda A 2000 *J. Non-Cryst. Solids* **266-269** 84
[2] Nishimiya T, Takeuchi Y, Yamauchi Y, Takatsuka H, Shioya T, Muta H and Kawai Y 2008 *Thin Solid Films* **516** 4430
[3] Ogiwara K, Chen W, Uchino K and Kawai Y Spatial characteristics of VHF hydrogen plasma produced by a balanced power feeding in Torr regime (submitted to *Thin Solid Films*)
[4] PEGASUS Software Inc., http://www.psinc.co.jp/english/index.html
[5] Kushner M J 2009 *J. Phys. D: Appl. Phys.* **42** 194013
[6] Lymberopoulos D P and Economou D J 1995 *J. Res. Natl. Inst. Stand. Technol.* **100** 473
[7] Rauf S and Kushner M J 1997 *J. Appl. Phys.* **82** 2805
[8] Brown S C 1966 *Introduction to Electrical Discharges in Gases* (New York: John Wiley & Sons) Chap. 10
[9] Yamauchi Y, Takeuchi Y, Takatsuka H, Yamashita H, Muta H and Kawai Y 2008 *Contrib. Plasma Phys.* **48** 326
[10] Lieberman M A and Lichtenberg A J 2005 *Principles of Plasma Discharges and Materials Processing* (Hoboken: John Wiley & Sons, 2nd ed.) Chap.11

Journal of Physics: Conference Series **441** (2013) 012023 doi:10.1088/1742-6596/441/1/012023

The generation characteristics of dielectric barrier glow discharge plasma in air

Guangliang Sun, Wenzheng Liu, Chuanhui Li and Rongrong Zhang
School of Electrical Engineering, Beijing Jiaotong University, China

E-mail: 11121659@bjtu.edu.cn

Abstract. This paper focused on how to suppress filamentous discharge and achieve glow discharge in air. In this study, glow discharge plasma was generated under low pressure by the plate electrode using dielectric barrier. Pulse supply and sinusoidal supply were adopted in this study and their influence on glow discharge were analyzed. Discharge characteristics under different air gap between the electrodes were studied. In addition, we studied the influence of dielectric barrier conditions on the glow discharge. The results show, pulse discharge, shortening the discharge gap and suitable dielectric barrier conditions, can effectively prevent the occurrence of filamentous discharge which contributes to achieve the glow discharge in air.

1. Introduction

Low temperature plasma which is chemically active contains many active particles and has light-emitting property and low temperature [1, 2]. It has a very broad prospect for industrial applications: ozone synthesis, high-power UV and VUV source, industrial waste gas treatment, wastewater treatment [3].

In general, the glow plasma is thought to be stable only in a low pressure, but it is possible to stabilize such a plasma at atmospheric pressure if three simple requirements are fulfilled: (i) use of a source frequency of over 1kHz, (ii) Dielectric barrier discharge (DBD), (iii) use of inert gas which increases costs[4]. Dielectric barrier discharge can weaken γ role and suppress the secondary electron's emission so that stabilize glow discharge. However, DBD at atmospheric pressure in air appears as filamentous discharge whose inhomogeneity and concentration of energy density limit the prospects of its application in many industrial areas [5, 6]. The difficulty to achieve atmospheric pressure glow discharge (APGD) in air is how to suppress the occurrence of filamentous. This paper discusses how to achieve stable glow discharge in air from three aspects: power waveform, air gap, conditions of dielectric barrier in order to realizing APGD finally.

2. Experimental devices

The structure of experimental system is shown in Figure 1. We have two power supplies: pulse supply creating pulse wave of 30 kHz and 0 ~ ± 5kV whose pulse width is 1 μs, sinusoidal supply creating sine wave of 20 kHz and 0 ~ ± 10kV. The electrode is in a reaction chamber which creates a low pressure environment. The discharge voltage is measured by a high voltage probe applied across the electrode and the discharge current is measured by a measurement resistor R_0 connected to the electrode. Z_{eq} represents the equivalent resistance of the supply.

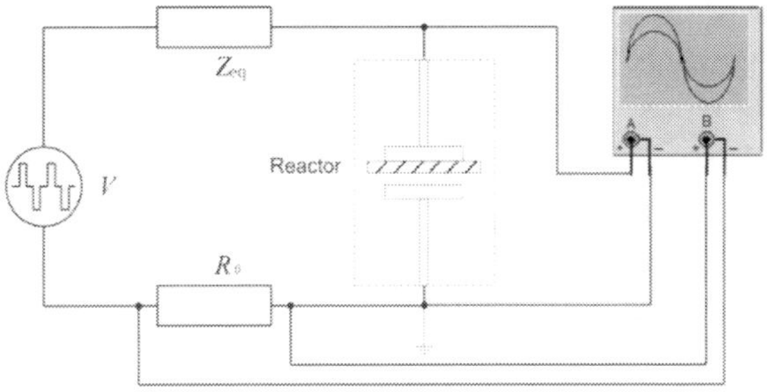

Figure 1. The structure of experimental system

The electrode is a pair of plate electrodes made of stainless steel with thickness of 5mm and diameter of 60mm. Figure 2 is a schematic view of the electrode in different blocking mode.

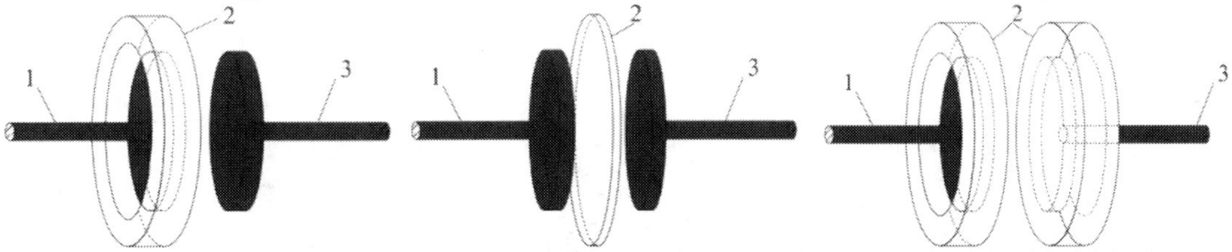

a. single-sided block b. middle block c. double-sided block

Figure 2. The electrode structure in different blocking mode
1-high voltage electrode 2-blocking dielectric 3-grounding electrode

3. The impact of discharge conditions on glow discharge

3.1 Supplies

The experiment was performed at 0.2×10^5Pa and the discharge gap was 2mm. We adopted polytetrafluoroethylene (PTFE) material with thickness of 1mm in single-sided block mode. We respectively conducted pulse power discharge and sinusoidal power discharge. The experiment result was shown in Figure 3 and Figure 4.

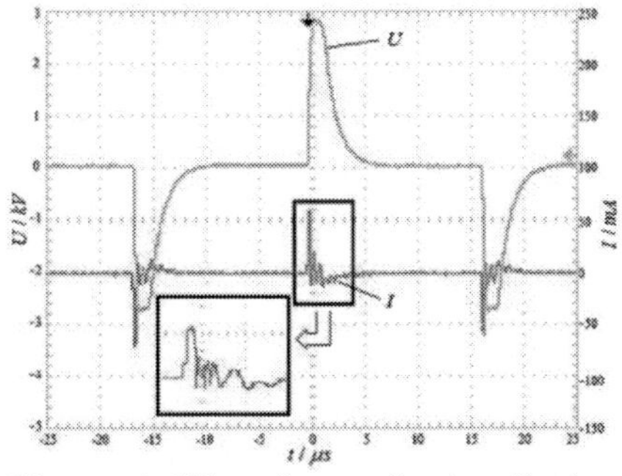

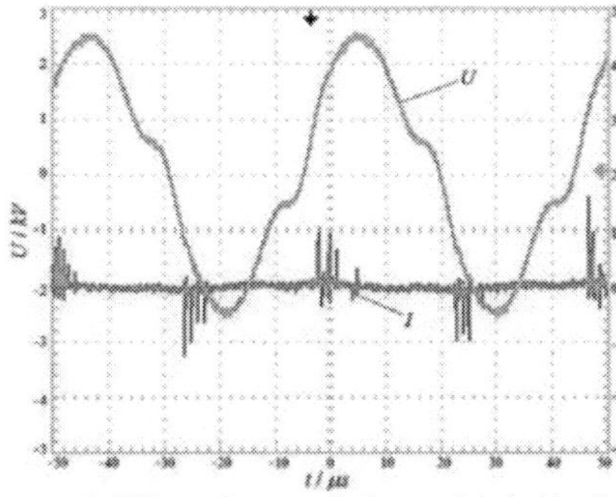

Figure 3. Waveform of pulse discharge **Figure 4.** Waveform of sinusoidal discharge

As we can see from the discharge waveform: pulse discharge generated at 2.8kV and only one maximum crest appeared in each half cycle of current's waveform, so it belonged to typical glow discharge; sinusoidal discharge generated at 2.4kV and multiple short pulses appeared in each half cycle, so glow discharge had been transformed to filamentous discharge[9].

Compared to sinusoidal discharge, pulse discharge can inhibit the development of the electron avalanche because it reduces the duration of the action of electric field which accelerates charged particles. Electron with smaller mass is accelerated, but ion with larger mass which mainly determine the temperature of plasma is not accelerated. Therefore the γ role (secondary electrons emit from the cathode when struck by positive ions) is weakened, the emission of secondary electron is reduced, and then the filamentous discharge is inhibited.

3.2 Air gap
Using pulse supply, the experiment adopted PTFE with thickness of 1mm in single-sided block mode. The distance (d) of air gap ranged from 1mm to 3mm. The breakdown voltage under different air gaps and pressures is shown in Figure 5 and the discharge current when applying 3kV is shown in Figure 6. Figure 7 is the discharge pictures at 0.2×10^5Pa under three different air gaps.

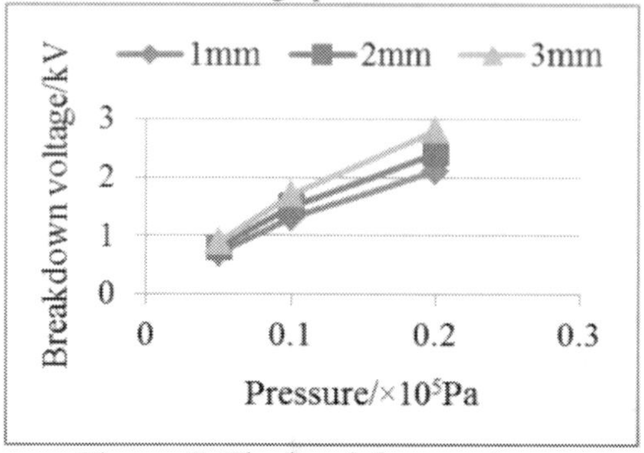

Figure 5. The breakdown voltage

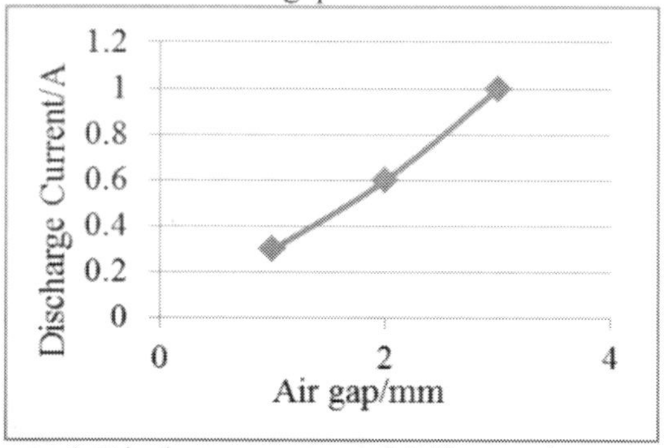

Figure 6. Discharge current of different air gaps

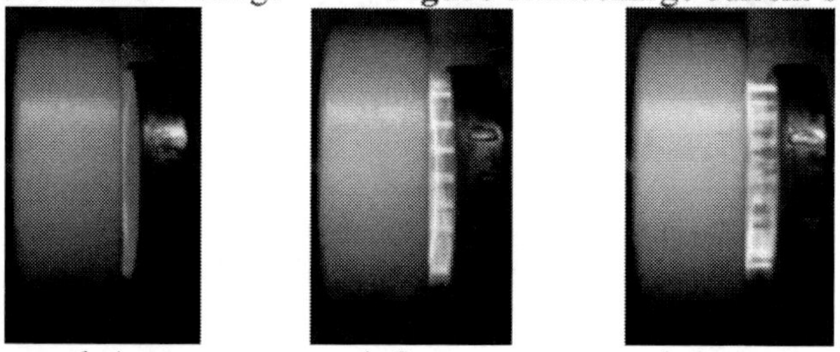

Figure 7. Discharge pictures at 0.2×10^5Pa under three different air gaps

As we can see from the results, the discharge voltage gradually increased with the rise of air pressure. With the air gap's decrease, the breakdown voltage and discharge current were both gradually reduced. At 0.2×10^5Pa, filamentous discharge appeared evidently between the air gaps of 3mm and was transformed into glow discharge while the air gap decreased to 1mm.

The electric field strength of air gap can be expressed by equation (1):

$$E_g = \frac{V\varepsilon_d}{l_d\varepsilon_g + l_g\varepsilon_d} \qquad (1)$$

Where, V represents the voltage across the electrodes, ε_d and ε_g respectively represents the dielectric constant of the material and the air gap, l_d and l_g respectively represents the thickness of the air gap and the barrier dielectric. The discharge voltage rises with the pressure's growth which significantly increases the difficulty of discharge under atmospheric pressure [7, 8, and 11]. Against this matter, we can decrease the difficulty of discharge through narrowing the air gap which decreases the discharge voltage when the electric field strength of discharging in air is constant.

Additionally, the discharge channel is shortened when narrowing the air gap and therefore the electron avalanche is restricted. The density of charged particles generated by the electron avalanche is reduced and the discharge current decreases, so that the filamentous discharge is suppressed.

4. Discharge characteristics under different conditions of dielectric barrier

4.1 Material
This experiment used three kinds of dielectric material shown in Figure 8. Keeping other conditions constant, we only changed the dielectric material. The experimental result is shown in Figure 9.

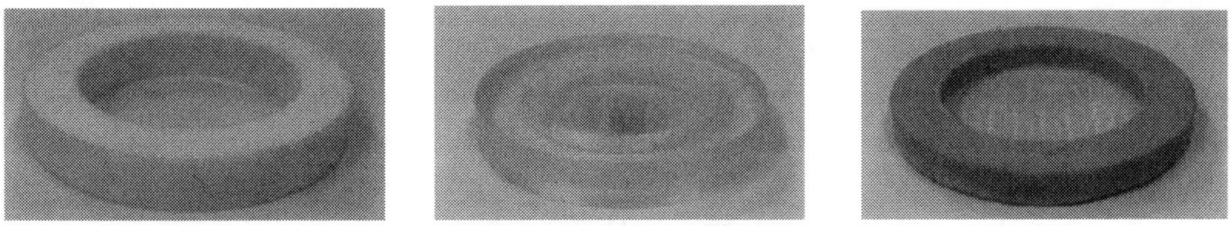

(a) PTFE (b) Poly methyl methacrylate(PMMA) (c) Epoxy

Figure 8. Three kinds of dielectric material

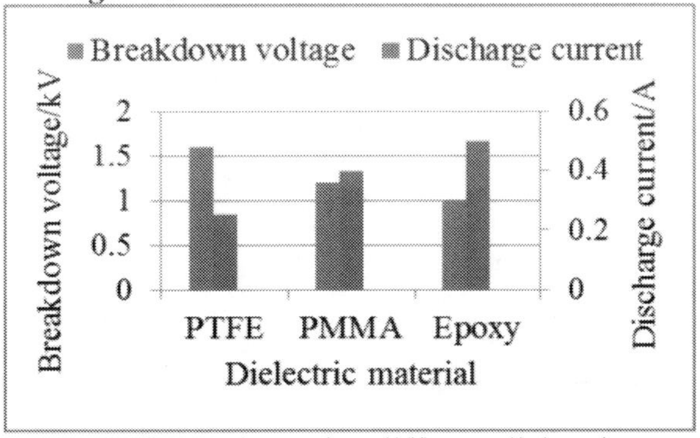

Figure 9. Discharge data using different dielectric materials

As the experimental results show that, the breakdown voltage in descending order: PTFE> PMMA>Epoxy, the discharge current in descending order: Epoxy>PMMA>PTFE.

The discharge is more difficult when using PTFE as the dielectric. But the filamentous discharge is avoided. PTFE is an electret whose surface is easier to accumulate electric charge and it can significantly reduce the electric field strength of the air gap [5, 10]. So the current is the smallest and the glow discharge can occur through least electrons at the initial stage and will be stable.

4.2 Thickness

In this experiment, we adopted PTFE with different thicknesses: 0.5 mm, 1 mm, 1.5 mm, 2 mm, as the barrier dielectric, keeping other conditions unchanged. Figure 10 is the experimental result.

The experiment results indicated that, the barrier dielectric with the thickness of 2mm possessed a greater breakdown voltage and exhibited a lower discharge current than that of 0.5mm and 1mm.

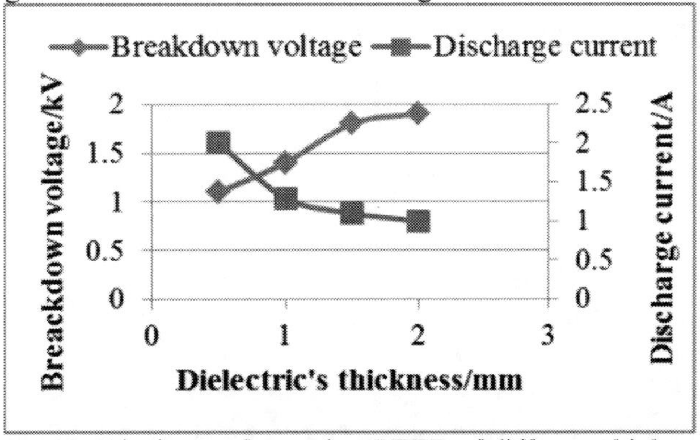

Figure 10. Discharge data using PTFE of different thicknesses

As the thickness increases, the equivalent capacitive reactance of the dielectric will be greater. Therefore the discharge current is reduced, which is conductive to suppress the filamentous discharge. However, as we can see from the equation (1), the value of voltage required to reach the discharge field strength become higher, so it's more difficult to achieve glow discharge in air. As the thickness decreases, it's easier to achieve glow discharge conversely. But the dielectric's equivalent capacitive reactance decreases and the discharge current increases which is not conducive to suppress the filamentous discharge. Therefore, a suitable thickness of the dielectric materials is of great significance to generate stable glow discharge plasma.

4.3 Blocking modes

This experiment kept the distance of air gap 1.5mm and PTFE was adopted as the barrier dielectric with thickness of 2 mm. Discharge data is shown in Figure 11.

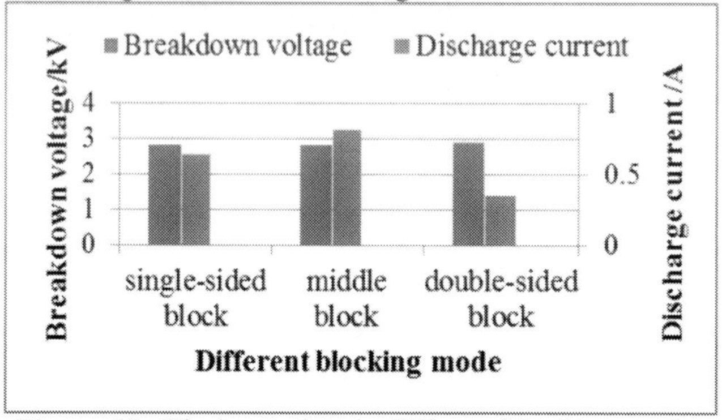

Figure 11. Discharge data in different blocking modes

The experiment results indicated that, the breakdown voltage in three modes was almost the same. The discharge current was the largest in the middle block mode and the smallest in the double-sided block.

Equation 1 shows that, the total thickness of air gap and barrier is the same in the three blocking modes, so the electric field strength in the air gaps is the same and then the breakdown voltage is almost the same.

As for the middle block mode, γ role exists in the whole cycle, a large number of secondary electrons appear in the gap and so the discharge current is the largest. Compared with single-sided block mode, double-sided block can completely eliminate γ role and secondary electron doesn't appear which contributing to suppressing filamentous discharge and achieving stable glow discharge.

5. Conclusions

This paper mainly studies the influence of power waveform, air gap and conditions of dielectric barrier on glow discharge. The results show:

1. Pulse discharge shorten the time of electric field and only accelerate electron, thereby suppressing the development of the electron avalanche, reducing the emission of secondary electrons, and then prevent the occurrence of filamentous discharge.
2. Narrowing the air gap can shorten the discharge channel, decrease the discharge voltage and current which contributing to achieving the glow discharge.
3. The electric charges are more easily accumulated on the surface of the dielectric barrier material when electret is selected. So the electric field strength of the air gap is reduced and the filamentous discharge is avoided.
4. If thicker dielectric is used the discharge voltage will increase, but the increase of current also can be avoided. So the suitable thickness contributes to achieving stable glow discharge.
5. Dielectric barrier in double-sided mode can completely avoid the emission of secondary electron which helps to achieve the glow discharge.

References

[1] Yi Luo, Zhi Fang, Yuchang Qiu. Material properties' influence to the characteristics of dielectric barrier discharge [J]. Insulating materials. NO.4, 45.

[2] Jie Xiangqian, Fang Zhi, Hao Yang. Research progress of evenly dielectric barrier discharge in air [J]. Journal of Vacuum Science and Technology, Vol 29 (6), 2009,649.

[3] Lijiang Pan. Research of air purification using pulsed discharge plasma [D]. Beijing: Beijing Jiaotong University. 2012.
[4] Satiko Okazaki, Masuhiro Kogoma, Makoto Uehara, Yoshihisa Kimura. Appearance of stable glow discharge in air, argon, oxygen and nitrogen at atmospheric pressure using a 50 Hz source [J]. Appl. Phys. 1993.
[5] Michael A. Force Berman, Allen • J. Lidengboge plasma discharge principle and material processing [M] The first edition of Beijing Science Press. 2007.
[6] Xuzhou Deng, Yuan Zhang, Lin Nan. Research of characteristics about dielectric barrier glow discharge at sub-atmospheric pressure [J]. China Illuminating Engineering Journal. 19 (2), 2008.
[7] Yun Yang, Guanjun Zhang. Research of influencing factors of dielectric barrier discharge in air [J]. High Voltage Engineering, 2007, 33.
[8] Kogelschatz U. Dielectric-barrier discharges: their history, discharge physics, and industrial applications [J]. Plasma Chemistry and Plasma Processing, 2003, 23 (1):1-46.
[9] Cheng Liu Li, Wenzheng Liu. One way to achieve atmospheric pressure glow discharge [J] Technology Innovation Herald, 2009, NO.02.
[10] F Massines, A Rabehi, P Decomps, et al. Experimental and theoretical study of glow discharge at atmospheric pressure controlled by dielectric barrier [J]. Appl. Phys. 1998, 83 (6): 2950-2957.
[11] Xinxin Wang, Mingzhe Lu, Yikang Pu. The possibility of uniform glow discharge in the air condition [J]. Acta Physica Sinica, 2002, 51 (12): 2778-2785.

Journal of Physics: Conference Series **441** (2013) 012024 doi:10.1088/1742-6596/441/1/012024

Deposition of Functional Membranes by Through-Substrate Plane-Plate Dielectric-Barrier Discharge

H. Kobayashi[1], H. Kakiuchi[2], K. Yasutake[2], T. Suzuki[3], M. Noborisaka[3], and N. Negishi[1]

[1]Central Research Laboratory, Hitachi, Ltd., 1-280 Higashi-koigakubo Kokubunji-shi, Tokyo, 185-8601 Japan

[2]Department of Precision Science and Technology, Graduate School of Engineering, Osaka University, 2-1 Yamadaoka Suita, Osaka, 565-0871, Japan

[3]Center for Science of Environment, Resource and Energy, Keio University, 3-14-1 Hiyoshi Kohoku-ku, Yokohama, 223-8522, Japan

E-mail: hiroyuki.kobayashi.sy@hitachi.com

Abstract. A novel roll-to-roll deposition method, namely, "through-substrate plane-plate dielectric-barrier discharge," was developed. This method uses a dielectric-barrier-discharge plasma plate, which has antenna and ground lines in a dielectric material, as an electrode. A film substrate was placed on the electrode so that an electric field, which passes through the substrate, generates a plasma above surface of the substrate. This method can avoid unexpected deposition on the electrode, which causes dust-particle contamination. In a preliminary experiment, deposition rates of Si, SiN, and Diamond-Like-Carbon were about 1, 1.5 and 5 μm/min, respectively. When there was a space between the plasma plate and the substrate film, plasma was generated on the back side of the substrate film, and then the deposition on the surface of the substrate film was prevented. This back-side discharge was suppressed by supplying nitrogen gas to the back-side of the substrate film.

1. Introduction

Low-cost mass-production of gas barrier films is required for fabrication of flexible devises such as electronic papers, photovoltaic (PV) and organic light-emitting diode (OLED) devises. Atmospheric plasma has been widely studied for its use in fabrication of gas barrier films due to its high deposition rate of a few micrometers per minute [1-5]. As an atmospheric-plasma source, dielectric-barrier discharge (DBD) [6-13], inductively coupled plasma (ICP) discharge [14], DC arc discharge [15], and microwave-excited plasma discharge [16] have been developed. The DBD plasma has been widely studied for a large-scale atmospheric plasma by parallel-plate configuration. However, when it is used for the deposition processes, several issues should be solved. For example, formation of unexpected deposition on the electrode should be suppressed because the deposition layer, which peels off the electrode, causes dust-particle contamination. In addition, a process gas should be uniformly fed in a large-size plasma to increase uniformity of deposition layers on the films. Moreover, low-temperature processes are required from the viewpoint of less heat resistance of substrate films such as polyethylene-terephthalate (PET) film. Thus, many types of new DBD plasma technique have been proposed. A plane-plate DBD plasma source that antenna and ground lines are formed in a dielectric material has been used in a plasma display panel and generates large-scale plasma [10-12].

In the present study, a novel roll-to-roll deposition method, namely, "through-substrate plane-plate dielectric-barrier discharge" was developed by using the DBD plasma plate. This method eludes unexpected deposition on the electrode.

2. Experimental

A schematic view of the through-substrate plane-plate dielectric-barrier discharge method is shown in Fig. 1, and the plane-plate DBD plasma source (DBD plasma plate) is shown in Fig. 2. As for the plasma plate, a plurality of antenna lines and ground lines were placed alternatively on a dielectric substrate plate and covered with a dielectric layer. Plasma is generated at the surface of the dielectric layer above the gap between the antenna and ground lines. A substrate film is placed on the surface of the plasma plate so that the substrate film is considered as a part of the dielectric layer of the plasma plate. Electric field, which passes through the substrate film, generates plasma above the surface of the substrate film. The plasma does not touch the plasma plate, so an unexpected deposition layer does not form on the plasma plate. The antenna and ground lines are placed on the back side of the substrate film, thus an optimal gas supply system can be freely designed in the region above the substrate film to improve the uniformity of the deposition layer.

The width of a gap between the antenna and ground lines (D in Fig. 2) was 1 mm. The thickness of the dielectric layer (T in Fig. 1) was 0.05 mm. The widths of the antenna and ground lines (W in Fig. 2) were 0.2 or 1 mm. The lengths of the antenna and ground lines (L in Fig. 2) were 100 mm. Alumina was used for the substrate plate and the dielectric layer of the plasma plate. A 0.05-mm-thick PET film was used as the substrate film.

The experimental setup is shown schematically in Fig. 3. The plasma plate was set in a chamber. A DC power-supply was connected to an inverter circuit, which generated a high-frequency discharge

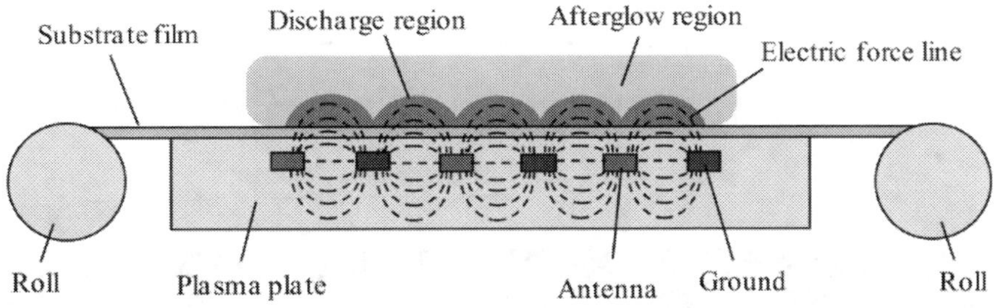

Fig. 1 Schematic view of a roll-to-roll deposition method by through-substrate plane-plate dielectric-barrier discharge.

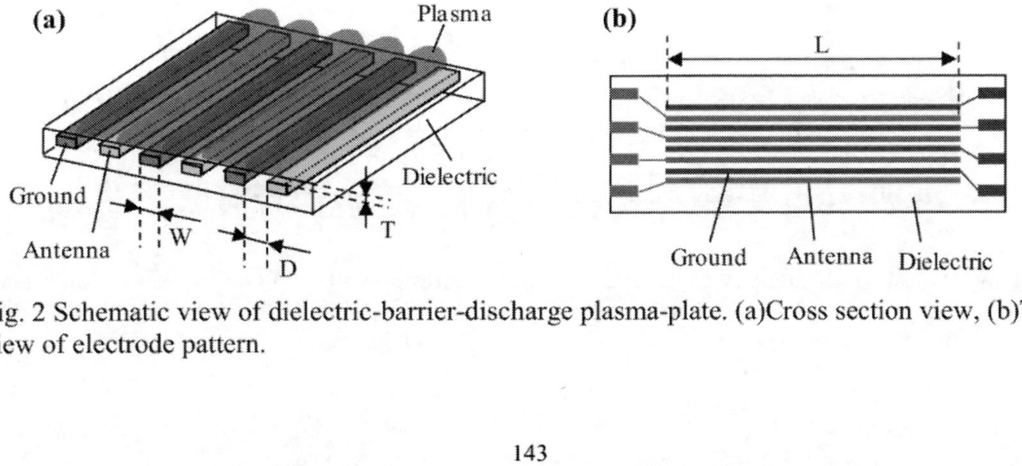

Fig. 2 Schematic view of dielectric-barrier-discharge plasma-plate. (a)Cross section view, (b)Top view of electrode pattern.

Journal of Physics: Conference Series **441** (2013) 012024 doi:10.1088/1742-6596/441/1/012024

power of about 20 kHz and 0-2 kV (Hereafter, discharge voltage denotes half value of the peak-to-peak voltage.) The electric power consumed by plasma discharge was about 75% of that supplied to the inverter circuit. A PET film was fixed on the plasma plate or wound by a roll set in the chamber. He/SiH_4, $He/SiH_4/NH_3/H_2$, and Ar/C_2H_2 were used as the process gases for deposition of Si, SiN, and Diamond-Like-Carbon (DLC) layers, respectively. The chamber was connected to a vacuum pump via a valve. The process pressure was maintained at 1 atm by changing the opening degree of the valve.

3. Results and discussion

Photographs of during and after deposition are shown in Figs. 4(a) and (b), respectively. A PET film was fixed on the plasma plate by a polyimide tape, as shown in Fig. 4(c). Width of both the antenna and ground lines was 0.2 mm. He/SiH_4 was supplied as the process gas with SiH_4 mixture ratio of 0.1%. The flow rate of the process gas was 10 L/min. The discharge voltage was 1.3 kV. Discharge power supplied to the plasma was only about 0.1 W/cm^2, so the plasma light was weak. A little bright plasma-light was locally observed, for example, at position Z in Fig. 4(a). At that position, plasma was generated on the back side of the PET film because there was a discharge space between the film and the plasma plate. At position Z, the deposition-layer was thin as shown as Z' in Fig. 4(b). It is clear that suppressing the back-side discharge is important for improving uniformity of the deposition-layer thickness.

The measured deposition rates of Si, SiN, and DLC layers (averaged across the plasma plate) are plotted in Figs. 5(a) and (b). The width of the antenna and ground lines (W) was 0.2 mm for a Si layer deposition. The discharge voltages were 1.3 and 1.5 kV, respectively, for SiH_4 mixture ratios of 0.1 and 0.3%. Discharge powers consumed by plasma were 0.1 and 0.13 W/cm^2 for SiH_4 mixture ratios of

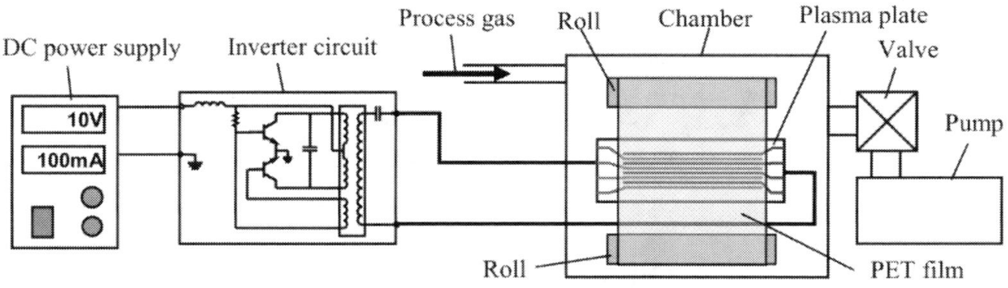

Fig. 3 Schematic view of experimental setup.

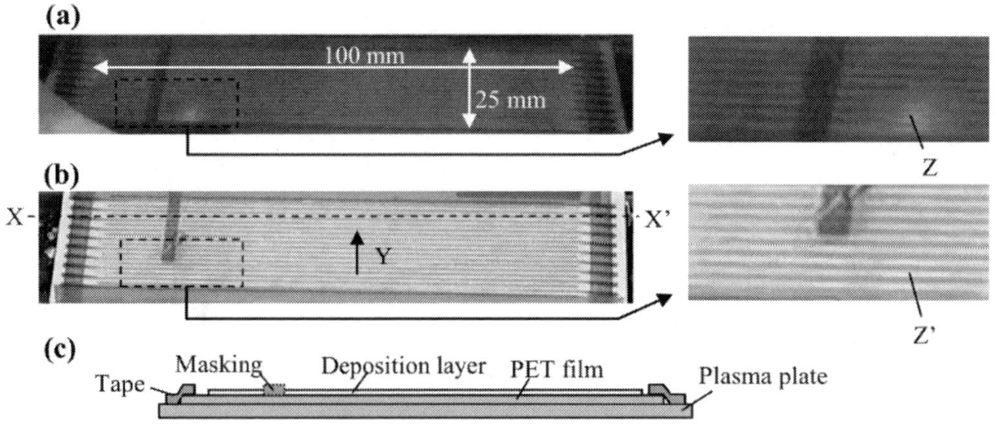

Fig. 4 Example of during plasma discharge and deposition results. (a) During deposition process, (b) After deposition process, (c) Schematic view of the sample for cross section of X-X' in (b).

Journal of Physics: Conference Series **441** (2013) 012024 doi:10.1088/1742-6596/441/1/012024

0.1 and 0.3%, respectively. For a SiN layer deposition, W of the plasma plate was 1 mm. Mixture ratios of SiH_4, NH_3, and H_2 were 0.1, 1, and 1%, respectively. The discharge voltage was 1.4 kV and a discharge power consumed by plasma was about 0.3 W/cm^2. Flow rate of the processes gas was 10 L/min and process pressures were 1 atm for both Si and SiN deposition. For DLC layer deposition, W of the plasma plate was 1 mm. The discharge voltage was 1.5 kV and discharge power was about 0.7 W/cm^2. Flow rate of the process gas was 3 L/min, and process pressure was 1 atm. Deposition rates were 1 and 1.5 μm/min for Si and SiN at SiH_4 mixture ratio of 0.3% and 0.1%, respectively. Deposition rate of DLC increased with increasing C_2H_2 mixture ratio, reaching about 5 μm/min at C_2H_2 mixture ratio of 3%. The order of the magnitude of these deposition rates is the same as that attained with other atmospheric-plasma methods [1-5]. Thus, the present deposition method is useful for high deposition-rate processes.

Distribution of the deposition-layer thickness across the film measured by laser microscope is shown in Fig. 6. The direction of the scan is marked as Y in Fig. 4(b). SiH_4 mixture ratio was 0.3%. The process time was 5 min. Deposition rate was high above the antenna and ground lines (A in Fig. 5) and between the antenna and ground lines (C in Fig. 6). The thickness of the deposition layer at position B (between A and C), which was measured with a scanning electron microscope (SEM) was at least about 0.5 μm.

SEM images of the surface of the Si deposition layers are shown in Figs. 7(a), (b), and (c), which correspond to positions A, B, and C in Fig. 6, respectively. Many particles were buried in the deposition layer. The surface of the deposition layer shown in Fig. 7(c) was coarser than that in Figs. 7(a) and (b). Moreover, many particles that were not buried in the deposition layer were visible at position C.

The distribution of electric-force strength above the film is shown in Fig. 8. It was calculated by roughly assuming that the relative permittivities of the dielectric layer, the PET film, and the gas phase were 6, 3, and 1, respectively. The voltages of the antenna and ground lines were 1.3 and 0 kV, respectively. Positions A, B, and C in Fig. 8 correspond to positions A, B, and C in Fig 6. The electric

Fig. 5 Deposition rate of (a) Si and SiN, (b) DLC layer.

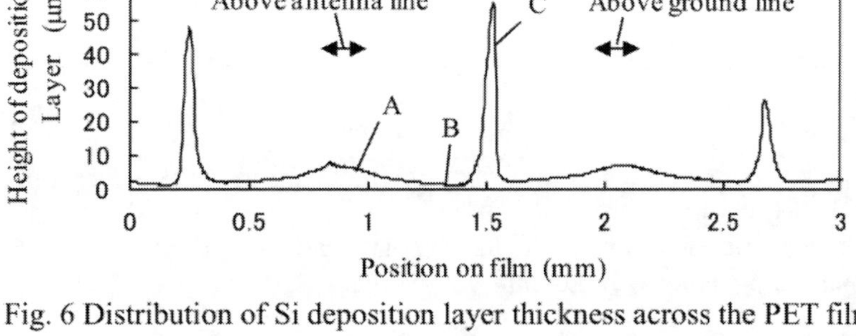

Fig. 6 Distribution of Si deposition layer thickness across the PET film.

Journal of Physics: Conference Series **441** (2013) 012024 doi:10.1088/1742-6596/441/1/012024

force at position A is two-times stronger than that at position C. It is clear that the surface of the deposition layer became rough at the position, where electric field strength was low. This result indicates that the agglomeration reaction was faster than the decomposition reaction of precursors, where electric field was weak. In this experiment, the process gas was not intentionary supplied to the plasma to increase the gas flow velosity for the purpose of supressing particle formation in the plasma. A study on reducing particle formation by increasing gas flow in the plasma is left for future work.

Measures to avoid the back-side discharge are described as follows. Minimum voltage required for the discharge was measured under three conditions. The discharge gas was He at 1 atm. Figure 9(a) denotes that the dielectric layer thickness (T) was 0.05 mm and PET film was not placed on the plasma plate. The discharge voltage was 0.6 kV. It can be assumed that the condition shown in Fig. 9(a) is the same as the condition shown in Fig. 9(a'), which indicates that there was sufficient space at the back side of the PET film. Figure 9(b) shows that the discharge voltage was 0.7 kV when T of the

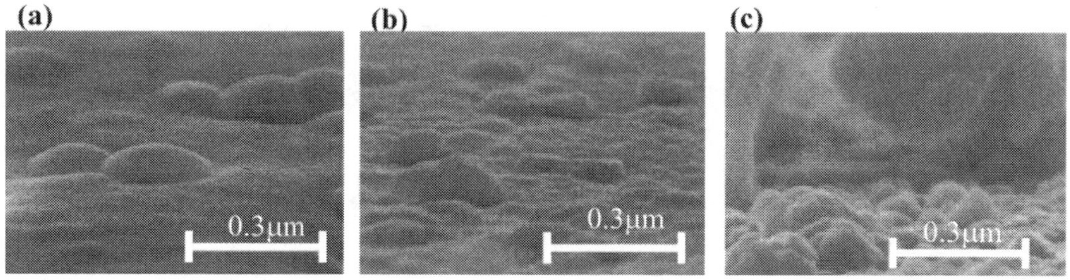

Fig. 7 SEM images of surface of Si deposition layer. Figs. (a), (b), and (c) correspond to A, B, and C in Fig. 6.

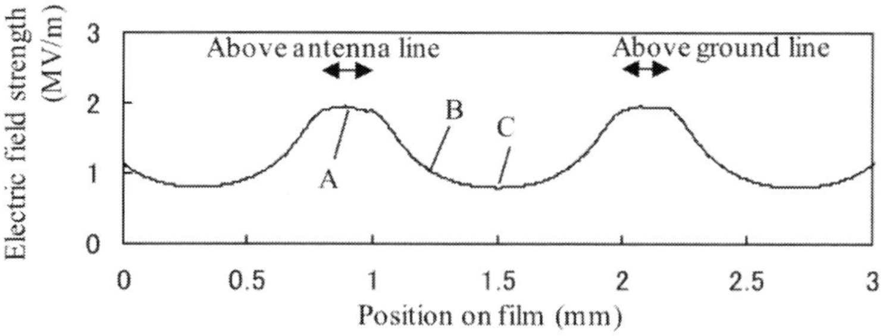

Fig. 8 Electric field strength above the film.

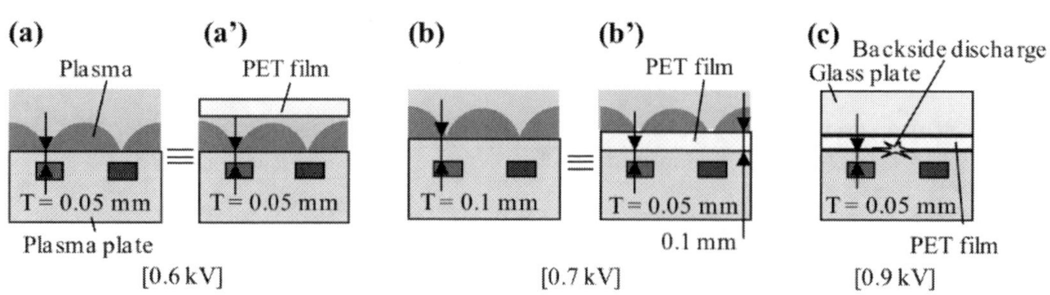

Fig. 9 Minimum discharge voltage of (a) 0.6 kV for T of 0.05 mm, (b) 0.7 kV for T of 0.1 mm, and (c) 0.9 kV for T of 0.05 mm with a PET film and a glass plate. (a) and (b) can be assumed to be equal to (a'); the PET film float above the plasma plate and (b'); about 0.1-mm-thick PET film was placed on the plasma plate for T of 0.05mm, respectively.

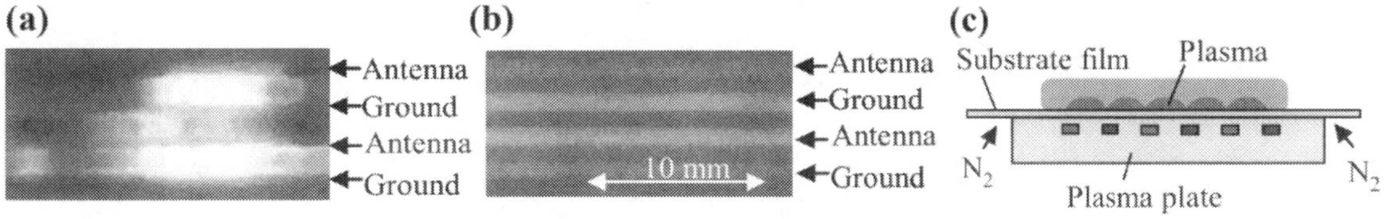

Fig. 10 (a) Photograph of back-side discharge, (b) Plasma on the surface of the PET film, (c) Schematic view of N_2 gas supply to the back side of the film.

plasma plate was 0.1 mm and no PET film was placed on the plasma plate. It can be roughly assumed that the condition shown in Fig. 9(b) is the same as the condition shown in Fig. 9(b'), which indicates that the dielectric layer thickness T was 0.05 mm for the relative permittivity of 6-10 (alumina) and about 0.1 mm-thick PET film for the relative permittivity of 3 was placed on the plasma plate. According to the comparison of Figs. 9(a') and (b'), plasma is generated at the back side of the substrate film rather than surface of the substrate film when the substrate film floats above the plasma plate. When T of the plasma plate was 0.05 mm and a glass plate was placed on the PET film to reduce the distance between the PET film and the plasma plate, as shown in Fig. 9(c), the voltage required for the back-side discharge was 0.9 kV. The comparison of Figs. 9(b') and (c) reveals that the back-side discharge can be prevented by reducing the space between the plasma plate and the substrate film.

There is another measure to suppress the back-side discharge. The discharge voltage of the process gas, which is sufficiently diluted by a rare gas, is lower than that of N_2 gas. About 0.5 and 2 kV were required for plasma discharge for He and N_2 at 1 atm, respectively, when T and D of the plasma plate were respectively 0.05 and 0.2 mm. This result indicates that discharge voltage required for N_2 is higher than that for He by a few times. A photograph of back-side discharge is shown in Fig. 10(a). The process gas was Ar/C_2H_2 (1%) at 1 atm. A bright light from the plasma was observed between the antenna and ground lines. On the contrary, N_2 gas was supplied near the region at the back side of the film by using another gas-supply line as shown in Fig 10(c), so that the N_2 gas was filled the gap between the PET film and the plasma plate in case that the PET film floated above the plasma plate. In this case, as shown in Fig. 10(b), the back-side discharge disappeared, and a weak plasma light was emitted from the region above the antenna and ground lines. This plasma-light emission indicates that the plasma was generated above the PET film. It is clear that the back-side discharge can be prevented by supplying N_2 gas to the back of the substrate film.

4. Summary

Deposition of Si, SiN and DLC layers by through-substrate plane-plate dielectric-barrier discharge was investigated. Deposition rates of the Si, SiN and DLC layers were about 1, 1.5 and 5 μm/min, respectively, which are the same order of the magnitude as those attained by other atmospheric-plasma methods. This method can avoid formation of an unexpected deposition layer on the plasma source, which causes dust-particle contamination. Deposition of particles formed in the plasma was observed because no intentional gas flow was generated in the plasma. In case that there is a space between the substrate film and the plasma plate, the plasma is generated on the back side of the substrate film, and the deposition on the surface of the substrate film is disturbed. The back-side discharge can be suppressed by reducing the space between the substrate film and the plasma plate. And also supplying N_2 gas to the back of the substrate film prevents the back-side discharge, when the process gas is sufficiently diluted by rare gas.

References

[1] K. Yasutake, H. Ohmi, H. Kakiuchi, T. Wakamiya, and H. Watanabe, Jpn. J. Appl. Phys., 45 (2006) 3592.
[2] T. Suzuki and H. Kodama, Diamond Relat. Mater., 18 (2009) 990.

[3] H. Kodama, A. Shirakura, A. Hotta T. Suzuki, Surf. Coat. Technol., 201 (2006) 913.
[4] H. Kakiuchi, H. Ohmi, K. Nakamura, Y. Yamaguchi, and K. Yasutake, Plasma Chem. Plasma Process., 30 (2010) 579.
[5] H. Kakiuchi, H. Ohmi, T. Yamada, K. Yokoyama, K. Okamura, and K. Yasutake, Plasma Chem. Plasma Process., 32 (2012) 533.
[6] B. Eliasson, M. Hirth, and U. Kogelschatz, J. Phys. D: Appl. Phys., 20 (1987) 1421.
[7] F. Massines, A. Rabehi, P. Decomps, R. B. Gadri, P. Segur, and C. Mayoux, J.Appl. Phys. 83 (1998) 2950.
[8] O. Sakai and K. Tachibana, J. Phys.: Conference Series 86 (2007) 012015.
[9] H. Kobayashi, T. Tandou, H. Nagaishi, K. Suzuki, and N. Negishi, Jpn. J. Appl. Phys., 51 (2012) 08HC04.
[10] S. Sato, H. Yamamoto, Y. Shirouchi, T. Iemori, N. Nakayama, and I. Morita, IEEE Trans. Electron Devices, 23 (1976) 328.
[11] T. Shinoda, M. Wakitani, T. Nanto, N. Awaji, and S. Kanagu, IEEE Trans. Electron Devices, 47 (2000) 77.
[12] K. Suzuki, N. Uemura, S. Ho, and M. Shiiki, AIP Conf. Proc., 636 (2002) 75.
[13] T. Oda, T. Takahashi, H. Nakano, and S. Masuda, IEEE. Trans. Industry Applications, 29 (1993) 787.
[14] H. Yabuta, H. Miyahara, M. Watanabe, E. Hotta, and A. Okino, J. Anal. At. Spectrom., 17 (2002) 1090.
[15] K. Yamakawa, M. Hori, T. Goto, S. Den, T. Katagiri, and H. Kano, J. Appl. Phys., 98 (2005) 043311.
[16] O. V. Penkov, H. J. Lee, V. Y. Plaksin, R. Mansur, and J. H. Kim, Thin Solid Films, 518 (2010) 6160.

GaN MOSFET with Boron Trichloride-Based Dry Recess Process

[1,2]Y Jiang, [1,2]Q P Wang, [1]K Tamai, [3]T Miyashita, [3]S Motoyama, [2]D J Wang, [1]J P Ao, and [1]Y Ohno

[1]Institute of Technology and Science, University of Tokushima, Japan
[2]School of Electronic Science and Technology, Faculty of Electronic Information and Electrical Engineering, Dalian University of Technology, China
[3]Research and Development Department, SAMCO Inc., Japan

E-mail: jpao@ee.tokushima-u.ac.jp

Abstract The dry recessed-gate GaN metal-oxide-semiconductor field-effect transistors (MOSFETs) on AlGaN/GaN heterostructure using boron trichloride (BCl_3) as etching gas were fabricated and characterized. Etching with different etching power was conducted. Devices with silicon tetrachloride ($SiCl_4$) etching gas were also prepared for comparison. Field-effect mobility and interface state density were extracted from current-voltage (*I-V*) characteristics. GaN MOSFETs on AlGaN/GaN heterostructure with BCl_3 based dry recess achieved a high maximum electron mobility of 141.5 $cm^2V^{-1}s^{-1}$ and a low interface state density.

1. Introduction

Gallium nitride (GaN) has attracted much attention in the field of high-power and high-efficiency applications due to its remarkable material properties such as wide bandgap, high breakdown field, and high electron saturation velocity. For power electronics application, the enhancement-mode (E-mode) plays a vital role of realizing safe operation and reducing power consumption owing to the normally-off operation [1-3]. In order to realize E-mode GaN MOSFET, source and drain ohmic contact must be primarily produced. To avoid activation at high temperature required by the ion-implanted source and drain regions, AlGaN/GaN heterostructure with high electron density of polarization-induced two-dimensional electron gas (2DEG) could be utilized as source and drain contacts. To accomplish this structure, dry recess process is necessary to remove the AlGaN layer for forming the channel region in the GaN layer. The avoidance of etching damage is required to obtain a high-quality channel with high carrier mobility and low interface states. Inductively coupled plasma (ICP) system is used to etch the gate. Low ICP and bias power could decrease the etching rate which is benefit to reduce etching damage [4-6]. Low power might also get an irregular recess profile and even failure in the MOSFETs operation. In our previous work [4, 7], we achieved a precise control of ICP etching rate of 1.2 nm/min and determined the optimum ICP condition of GaN MOSFETs on AlGaN/GaN heterostructure with the etching gas of $SiCl_4$. However, the devices etched with $SiCl_4$ gas had a negative threshold voltage due to the possible silicon (Si) contamination.

In this paper, we tried to use BCl_3 instead of $SiCl_4$ as the gate recess etching gas in order to reduce the influence of Si contamination on threshold voltage. Then we discussed the influence of etching gas and conditions on the performance of dry recessed-gate GaN MOSFETs on AlGaN/GaN heterostructure, which were characterized by extracting the field-effect mobility and interface state density.

2. Experiments

The device structure is a ring-type MOSFET on AlGaN/GaN heterostructure grown on a sapphire substrate, as shown in Fig. 1. The epitaxial structure includes a buffer layer, a 2 μm i-GaN layer, a 20 nm n-AlGaN layer, and a 10 nm n-GaN layer. The composition ratio of the aluminum in AlGaN layer is 25%. The n-AlGaN layer and n-GaN layer are doped with Si of a concentration of 1.0×10^{19} cm^{-3}. The sheet resistance of the wafer is around 500 Ω/□. To form the MOSFET structure, a 2DEG layer is used as the ohmic contact layer, and a surface etched i-GaN layer is used as the channel layer.

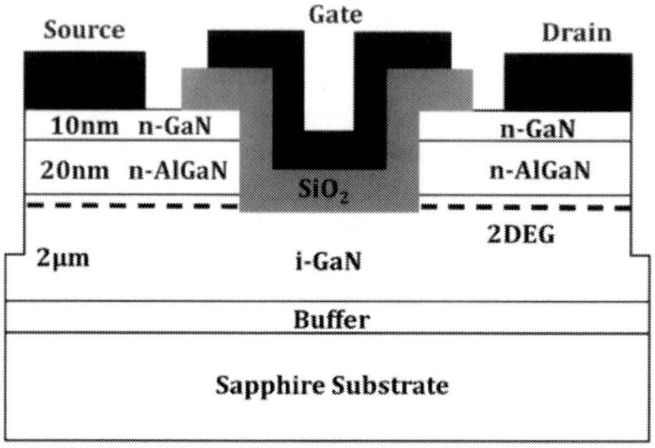

Fig. 1 Schematic cross section of ring-type GaN MOSFET on AlGaN/GaN heterostructure

The device fabrication process was based upon standard photolithography and lift-off technologies. The SiO_2 film of about 500 nm thickness deposited by tetraethylorthosilicate (TEOS) plasma enhanced chemical vapor deposition (PECVD) was used as the etching mask. The gate recess was conducted utilizing ICP system with BCl_3 and $SiCl_4$ gas. The etching conditions of five samples can be divided into three kinds (listed in Table 1).

(1) Sample No. 1 and No. 2 were done with $SiCl_4$ gas, ICP power of 100 W and 50 W, respectively, bias power of 20 W, $SiCl_4$ gas flow rate of 3 sccm and working chamber pressure of 0.25 Pa (termed as: No. 1 $SiCl_4$ 100W/20W and No. 2 $SiCl_4$ 50W/20W, respectively). The etching rates of these two samples were about 1.2 nm/min and 1.0 nm/min, respectively.

(2) Sample No. 3 and No. 4 were etched with mixed BCl_3/Cl_2 gas (20/20 sccm), ICP power of 100 W, bias power of 25 W and at a working chamber pressure of 0.6 Pa. The etching depths were 110 nm and 60 nm respectively (termed as: No. 3 BCl_3/Cl_2, 110 nm and No. 4 BCl_3/Cl_2, 60 nm, respectively), and the etching rates were about 30 nm/min.

(3) Sample No. 5 (termed as: No. 5 $SiCl_4$+Cl_2) was etched by two steps, and the average etching rate was about 20 nm/min. The first step was done by $SiCl_4$ gas with a flow rate of 4 sccm, ICP power of 50 W, bias power of 100 W and at a working chamber pressure of 0.25 Pa. The second step was done by Cl_2 gas with a flow rate of 4 sccm, ICP power of 50 W, bias power of 50 W and at a working chamber pressure of 0.25 Pa.

During the etching processes, the sample stage was cooled by helium gas to keep at the room temperature.

After these dry etching processes, the samples were immersed in HNO_3: BHF = 1:1 solution to remove the possible contamination of Si on the etched surface. As the gate insulator, a 103.5-nm-thick SiO_2 layer was then deposited by PECVD (SAMCO PD-220LC) with silane-based source, followed by annealing at 1000℃ for 10 min in N_2 ambient. After the gate insulator patterning, the source and drain ohmic contacts were formed using Ti/Al/Ti/Au (50/200/40/40 nm) annealed at 850°C for 1 min in N_2 ambient. Transmission line model (TLM) measurement showed that the contact resistance was approximately 0.15-0.36 Ωmm. A Ni/Au (70/30 nm) bi-layer was deposited as the gate metal.

Table 1 Dry etching process conditions of all the samples

No.	Term	ICP Power (W)	Bias Power (W)	Gas	Flow Rate (sccm)	Chamber Pressure (Pa)	Etching Rate (nm/min)	Recess Depth (nm)
1	$SiCl_4$ 100W/20W	100	20	$SiCl_4$	3	0.25	1.2	60
2	$SiCl_4$ 50W/20W	50	20	$SiCl_4$	3	0.25	1.0	60
3	BCl_3/Cl_2, 60 nm	100	25	BCl_3/Cl_2	20/20	0.6	30	60
4	BCl_3/Cl_2, 110 nm	100	25	BCl_3/Cl_2	20/20	0.6	30	110
5	$SiCl_4$+Cl_2	50	100	$SiCl_4$	4	0.25	20	100
		50	50	Cl_2	4	0.25		

3. Results and Discussion

The device used for evaluation is long channel ring-type MOSFET. It has a channel length L of 94 μm and an effective channel width W of 819 μm which was calculated from

$$W = \frac{2\pi L}{\ln r_{out} - \ln r_{in}},$$

where r_{out} and r_{in} is the outer and inner radius. For all the samples, the gate leakage currents were below 10^{-9} A with gate voltage from -10 V to 10 V and drain voltage of 0.1 V, as shown in Fig. 2. In particular, gate leakage is restrained even at a positive gate bias, which is beneficial to achieve the E-mode operation. To extract the capacitance of gate insulator, measurement of C-V characteristics was performed under frequency of 100 kHz and with gate voltage from -10 to 15 V. From Fig. 3, we can observe the hysteresis and the negative threshold voltage from -5.5 to -3.5 V. The difference of threshold voltage by different etching gas is not clear. One reason is the possible positive charges existing in the gate insulator layer leading to a negative shift on the threshold voltage [8], and the other reason may be from the GaN channel layer of this work. Through the evaluation of device isolation, we confirmed that the isolation region had a large leakage current indicating that the GaN buffer layer is slight n-type rather than semi-insulating type.

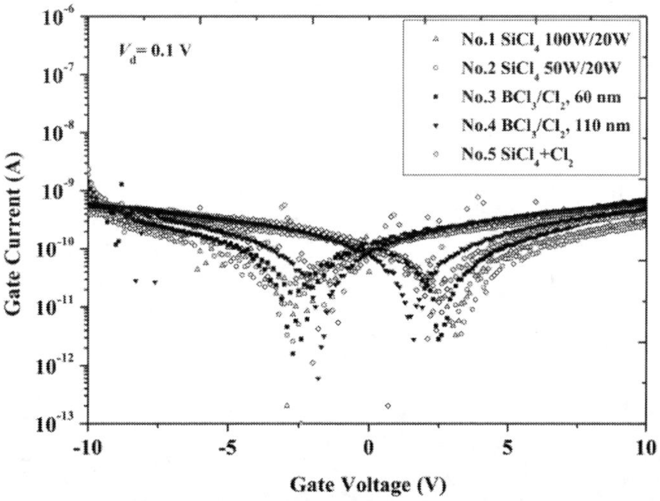

Fig. 2 Gate leakage current of GaN MOSFETs of all the samples with gate voltage from -10 to 10 V.

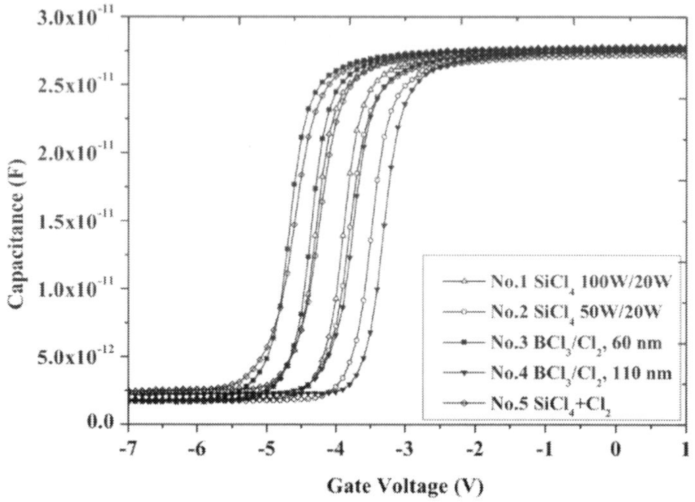

Fig. 3 The segment of *C*-*V* characteristics curve of GaN MOSFETs of all the samples with gate voltage from -7 to 1 V.

Figure 4 shows the I_d-V_d characteristics of the device etched with BCl_3 gas and low etching rate. Drain voltages range from 0 to 20 V and gate voltage range from 10 to -3 V. Device operation up to gate voltage of 10 V was confirmed. A gate capacitance-transconductance method was adopted to determine the electron field-effect mobility. According to the basis of the gradual channel approximation of MOSFETs, field-effect mobility could be calculated as

$$\mu_{FE} = \frac{G_M L}{C_{OX} W V_d},$$

where L is the gate length, C_{OX} is the oxide capacitance per unit area, and G_M is the transconductance. The field-effect mobility of all the samples is showed in Fig. 5. The electron mobility of two devices etched with BCl_3 gas are 141.5 and 136.0 $cm^2V^{-1}s^{-1}$, respectively, as listed in Table 2. The field-effect mobility of the device with the recess depth of 110 nm is only slightly lower than that of the device with the recess depth of 60 nm. Although the etching condition of these two samples was same, but a larger depth meant longer etching time and produced more surface contamination, which would degrade the mobility. However, for the device No. 5 etched with $SiCl_4$ and Cl_2 gas, the field-effect mobility decreased sharply, probably because of much etching surface damage caused by high bias power. The mobility is almost the same for the devices with 20 W bias power and different ICP power, implying that the mobility is more sensitive to the bias power instead of the ICP power. In reference [7], we have reported the influence of etching conditions on the recess profiles, mobility and interface state density for devices with ICP power from 50 to 100 W, bias power from 20 to 60 W and the etching mask of SiO_2. It showed that higher bias power would bring more interface states and lower mobility. It was considered to be due to the much serious surface damage and silicon contamination occurring at higher bias power.

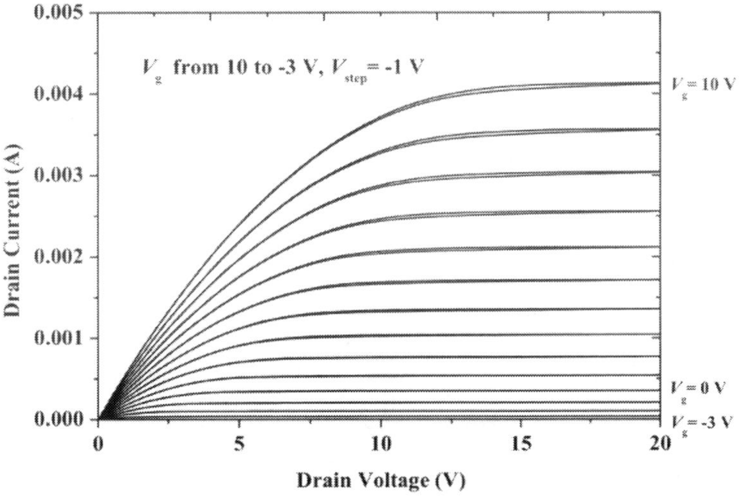

Fig. 4 I_d-V_d characteristics of the device etched with BCl_3 gas and the recess depth of 60 nm.

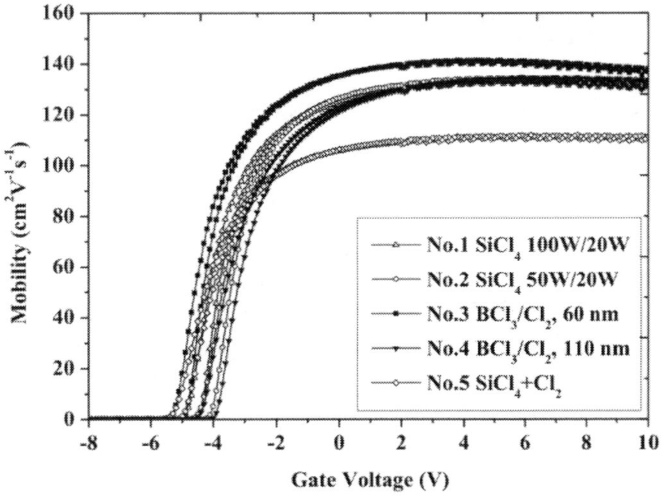

Fig. 5 The field-effect mobility of GaN MOSFETs of all the samples.

Table 2 The measured maximum field-effect mobility and interface state density of all the samples

No	Term	Average of C_{ox} (10^{-8} Fcm^{-2})	Field-effect Mobility μ_{FE} (cm^{2}V^{-1}s^{-1})		Subthreshold Swing S (mV/dec)		Interface State Density D_{it} (10^{11} cm^{-2}eV^{-1})	
			Forward	Reverse	Forward	Reverse	Forward	Reverse
1	$SiCl_4$ 100W/20W	3.45	133.3	133.9	141.5	137.6	2.96	2.82
2	$SiCl_4$ 50W/20W	3.41	133.9	134.5	177.1	154.9	4.21	3.41
3	BCl_3/Cl_2, 60 nm	3.47	141.1	141.5	112.1	139.7	1.91	2.91
4	BCl_3/Cl_2, 110 nm	3.44	133.7	136.0	169.4	149.4	3.97	3.25
5	$SiCl_4+Cl_2$	3.48	111.6	112.0	138.3	176.0	2.87	4.25

The equivalent circuit of the MOS structure can be presented as the oxide capacitance C_{OX} connected in series with a parallel connection of the depletion capacitance C_D and the interface-related capacitance C_{it}. In this case, based on the definition, the subthreshold swing S could be expressed as

$$S = \frac{dV_g}{d\log I_{ds}} = (\ln 10)\left(\frac{kT}{q}\right)\left(1 + \frac{C_D}{C_{OX}} + \frac{C_{it}}{C_{OX}}\right),$$

where k is the Boltzmann constant, T is the absolute temperature, and q is the electron charge. Figure 6 shows the I_d-V_g characteristics of the devices in the subthreshold region. Considering the depletion capacitance is zero, C_{it} can be calculated from the extracted subthreshold swing S. The interface states density D_{it} can then be calculated from C_{it}. The device etched with BCl_3 gas and the recess depth of 60 nm obtained a low interface state density, as listed in Table 2.

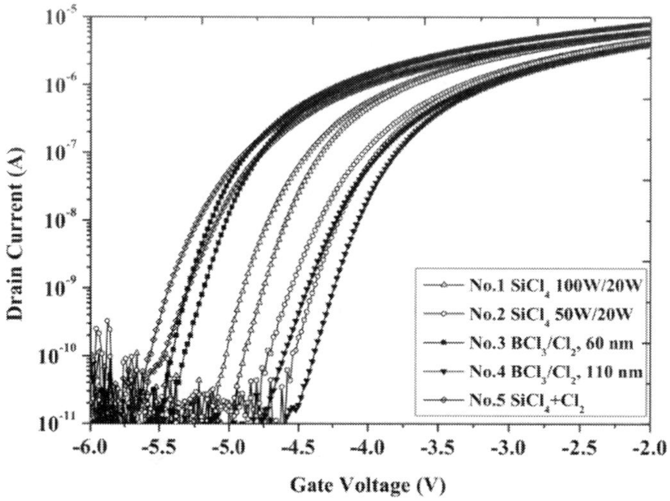

Fig. 6 I_d-V_g characteristics of GaN MOSFETs of all the samples in the subthreshold region

4. Conclusion

The dry recessed-gate GaN MOSFETs on AlGaN/GaN heterostructure etched with BCl_3 gas were fabricated. A high maximum electron mobility of 141.5 $cm^2V^{-1}s^{-1}$ and a low interface state density were obtained. This indicates that BCl_3 is possibly a suitable candidate as an etching gas to fabricate MOSFETs on AlGaN/GaN heterostructure. Further work is necessary to optimize the recess process condition of BCl_3 gas.

References

[1] Mistele D, Rotter T, Röver K S, Paprotta S, Seyboth M, Schwegler V, Fedler F, Klausing H, Semchinova O K, Stemmer J, Aderhold J, and Graul J 2002 *Mater. Sci. Eng.* B **93** 107.
[2] Huang W, Khan T, Chow T P 2006 *IEEE Electron Device Lett.* **27** 796.
[3] Tsai C Y, Wu T L, Chin A 2012 *IEEE Electron Device Lett.* **33** 35.
[4] Ao J P, Nakatani K, Ohmuro K, Sugimo M, Hu C Y, Sogawa Y, Ohno Y 2010 *Jpn. J. Appl. Phys.* **49** 04DF09.
[5] Ao J P, Nakatani K, Y, Sogawa Y, Akamatsu S, Kim Y H, Miyashita T, Motoyama S I, Ohno Y 2011 *Phys. Status Solidi* C **8** 457.
[6] Matsuura K, Kikuta D, Ao J P, Ogiya H, Hiramoto M, Kawai H, Ohno Y 2007 *Jpn. J. Appl. Phys.* **46** 2320.
[7] Wang Q P, Tamai K, Miyashita T, Motoyama S, Wang D J, Ao J P, Ohno Y 2013 *Jpn. J. Appl. Phys.* **52** 01AG02.
[8] Gaind A K, Kasprzak L A 1979 *Solid-State Electron* **22** 303.

Journal of Physics: Conference Series **441** (2013) 012026 doi:10.1088/1742-6596/441/1/012026

Control of electron temperature in SiH_4/H_2 plasma for obtaining high photovoltaic performance in microcrystalline silicon solar cells

Y Sobajima[1,2], H Muto[1], C Sada[1,2], A Matsuda[1,2] and H Okamoto[1,2]

[1]Graduate school of Engineering, Osaka University, Toyonaka, 560-8531, Japan
[2]JST-CREST, Toyonaka, 560-8531, Japan

E-mail: sobajima@ee.es.osaka-u.ac.jp

Abstract. We have proposed two novel processes for the formation of fine n/i interface to improve the photovoltaic performance in substrate-type (n-i-p type) hydrogenated microcrystalline-silicon (μc-Si:H) solar cells whose i layer is deposited at high growth rate of > 2.0 nm/sec; (1) gradual monosilane-(SiH_4)-molecule-introduction method and (2) amorphous silicon (a-Si:H) thin-layer-insertion method. When applying these two methods to the formation process of n/i interface in the solar cells, drastic improvement of the production reproducibility has been achieved in the fabrication process of high efficiency (> 9%) substrate-type μc-Si:H solar cells.

1. Introduction

Plasma-enhanced chemical-vapor-deposition (PECVD) method is popularly used to prepare high quality (low dangling-bond-defect density) hydrogenated microcrystalline silicon (μc-Si:H) films as the same process as hydrogenated amorphous silicon (a-Si:H) films. Since μc-Si:H shows high stability upon light soaking in contrast to a-Si:H, μc-Si:H is recognized as promising materials for application to solar cells [1][2]. However, photoactive-intrinsic-layer (i-layer) thickness in μc-Si:H solar cells is around 2 μm to absorb sufficient sun light. Therefore, high rate growth (> 2.0 nm/sec) of i layer is necessary condition to reduce the production cost of relevant solar cells. To increase the growth rate, being determined by the generation rate of film precursor (silyl radical; SiH_3) in the plasma, high SiH_4-gas-flow rate (high SiH_4 density) and high electron density in the plasma are required [3]. For these purposes, very high frequency (VHF) is used as the power-source frequency to obtain high electron-density plasma with low electron temperature [4][5]. Assuming that the electron-energy-distribution function (EEDF) obeys Maxwell-Boltzmann distribution, the generation-rate ratio of short lifetime species (Si, SiH, and SiH_2 whose generation-threshold energy is more than 9 eV) to film-growth precursors (SiH_3 whose threshold is almost 8 eV) is kept low under low electron-temperature conditions [6][7], resulting in a possible growth of high quality (showing low dangling-bond-defect density) μc-Si:H films. On the other hand, high quality μc-Si:H films are obtained under high working-pressure conditions by making a good use of chemical reactions of short lifetime species (Si, SiH, and SiH_2) with H_2 molecule in the plasma [8], despite that these reaction-rate constants are almost two orders of magnitude lower than the reactions of short lifetime species with SiH_4 molecule. Although high working-pressure condition is effective to decrease the defect density in the resulting film, higher-order-silane molecules (powder-like species) tend to be formed under the coexisting condition of high density-SiH_4 molecules and high density electrons in the plasma. Coexisting condition appears especially at the initial stage of film growth, since SiH_4 molecules are not depleted as yet. Once the powder formation occurs, electron-attachment event to powder gives rise to an elevation of electron temperature in the plasma due to an increase in the electron-loss rate, which leads to an increase of the defect density in the resulting film through an increase of generation-rate ratio of short lifetime species to SiH_3 in the plasma. Therefore, the control of powder-formation process at the initial stage of film growth is a key issue to obtain high quality μc-Si:H-i layer not only at the n/i interface part but also in the bulk-i region in the n-i-p type solar-cell structure.

In this paper, we describe a method to control the production of higher-order-silane molecules (initiation step of powder-formation process) at the initial stage of film growth by applying gradual SiH_4-introduction method to the growth of intrinsic μc-Si:H films in order to realize high-SiH_4-depletion condition from the start. We also propose a novel method to avoid so-called "auto doping" phenomenon of phosphorus (P) atoms into i layer from underlying n layer to form well-controlled fine n/i-interface in the n-i-p-type μc-Si:H solar cells whose i layer is prepared at a high growth rate more than 2 nm/sec.

2. Experimental methods

2.1. Control of starting procedure for film growth and plasma diagnosis

To control the powder-formation process and obtain high quality μc-Si:H, we have adopted a gradual SiH_4-introduction method to adjust the SiH_4-partial pressure from the start of film growth. The intrinsic μc-Si:H film was prepared using the capacitively-coupled VHF(100 MHz)-PECVD system with a conventional shower-head cathode at the deposition rate of 2.2 nm/sec. H_2-flow rate, power density, electrode distance, and substrate temperature were fixed at 168 sccm, 0.17 W/cm^2, 6.0 mm, and 213 °C, respectively. The SiH_4-gas-introduction scheme is shown in Fig. 1. After turning on the H_2 plasma with working pressure of 9.0 Torr, SiH_4 was fed into the plasma with two different introduction times (15 sec and 120 sec from 0 to 3 sccm) as shown in the figure. The waiting times of 5 sec and 30 sec were set before starting the SiH_4 introduction in order to stabilize the H_2 plasma and to know the effect of the waiting time on the properties of n/i interface in the solar cell structure.

Optical emission spectroscopy (OES) is used as the monitoring system of plasma parameters. To know the change of electron temperature in the plasma, H_α-emission ($\lambda = 656.3$ nm) intensity is utilized in this study.

In the steady state plasma, density of H_α in the plasma is given by

$$\frac{d[H_\alpha]}{dt} = N_{eh}\sigma_h v_e [H_2] - \frac{[H_\alpha]}{\tau_{H_\alpha}} = 0, \quad \frac{[H_\alpha]}{\tau_{H_\alpha}} = I_{H_\alpha} \tag{1}$$

where N_{eh}, σ_h, and v_e are electron density for H_α formation, dissociative excitation-cross section of H_2 to H_α, and electron thermal velocity, respectively. The H_α-emission intensity ($I_{H\alpha}$) is then given by

$$I_{H_\alpha} = N_{eh}\sigma_h v_e [H_2] \propto f(T_e). \tag{2}$$

Under the high H_2-dilution-ratio (sufficient H_2 density [H_2]) condition, H_α-emission intensity is sensitive only to the high energy electron density for H_α formation, being the function of electron temperature in the plasma. Therefore, when monitoring the H_α-emission intensity, we know the change of electron temperature in the highly H_2-diluted SiH_4/H_2 plasmas.

2.2. Fabrication of n-i-p type μc-Si:H solar cells

Structure of μc-Si:H solar cell is Asahi type-U glass/ Ag/ ZnO:Al/ n-i-p μc-Si:H/ top-TCO/ Ag-grid. Intrinsic μc-Si:H layer was adjusted at 2 μm in thickness. Asahi type-U glass was used as the substrate to unify the surface roughness of solar cells, in which the textured surface structure scatters the incident sun light for effective light absorption by the extension of light pathway [9]. The root-mean-square (RMS) roughness of Asahi type-U glass is approximately 40 nm. The Ag layer was prepared by

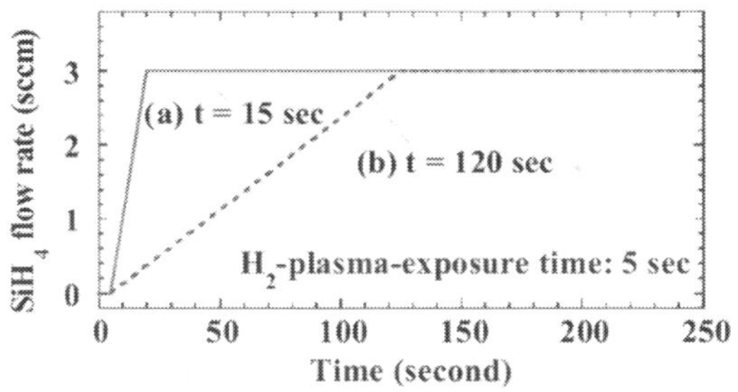

Figure 1. SiH_4-introduction schemes for two different SiH_4-introduction times (t).

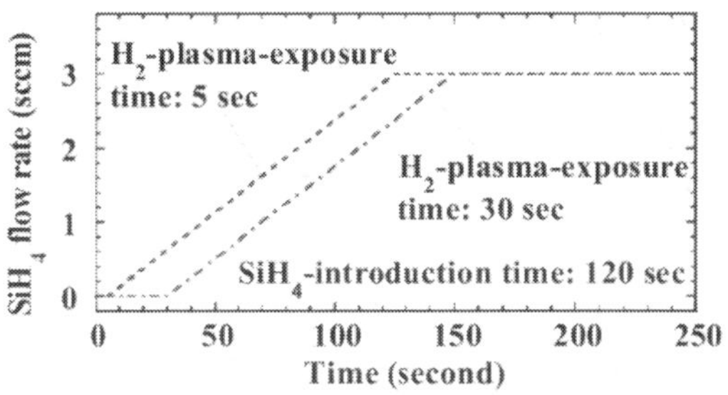

Figure 2. SiH_4-introduction schemes with two different H_2-plasma-exposure times.

radio frequency (RF: 13.56 MHz) magnetron-sputtering method as the back electrode. Top transparent conductive oxide (TCO) was composed of titanium-doped indium oxide (InTiO) and aluminum-doped zinc oxide (ZnO:Al) stacked layers for reducing the incident-light reflection with refractive-index engineering [10]. Each TCO was prepared by RF-magnetron-sputtering method. Phosphorus-doped n-type μc-Si:H and boron-doped p-type μc-Si:H was prepared by RF-PECVD.

To control the powder-formation process especially at the initial stage of i-μc-Si:H growth, gradual SiH_4-introduction method was adopted with the same time-evolution scheme as shown in Fig. 1. Thinking of the process reproducibility, another time-evolution scheme for SiH_4-gas introduction was adopted as shown in Fig. 2, where the waiting time was varied to stabilize H_2 plasma before the introduction of SiH_4. To understand the effect of waiting time, i. e, the exposure time of n layer in the n-i-p type solar-cell structure to high flux density of atomic hydrogen from the plasma, two different waiting times of 5 sec and 30 sec were tested.

3. Results and discussion

3.1. Control of electron temperature in the plasma

Figure 3 shows the time evolution of H_α-emission intensity (electron temperature in this case) for two different SiH_4-introduction times. Both for (a) and (b), the electron temperature is once decreased after introduction of SiH_4 molecules into H_2 plasma, being explained as follows. H_2 ions (H_2^+ and H_3^+) are the predominant positive ions in the H_2 plasma. When increasing the ratio of heavier positive ions such as SiH_3^+ by the SiH_4 introduction into the H_2 plasma, the electron temperature is decreased owing to the reduction of am-bipolar diffusion-loss rate of electrons. The electron temperature then starts to increase by the further increase of SiH_4-flow rate, which is due to an increase of electron-loss rate owing to the electron-attachment events to the powder-like higher-order-silane molecules (Si_nH_{2n+2}, n > 5) produced in the plasma. In addition, at the turning point of electron-temperature after showing the minimum point, i. e, from (A) to (B) and (A') to (B') in Fig. 3, the structure of plasma is changed from

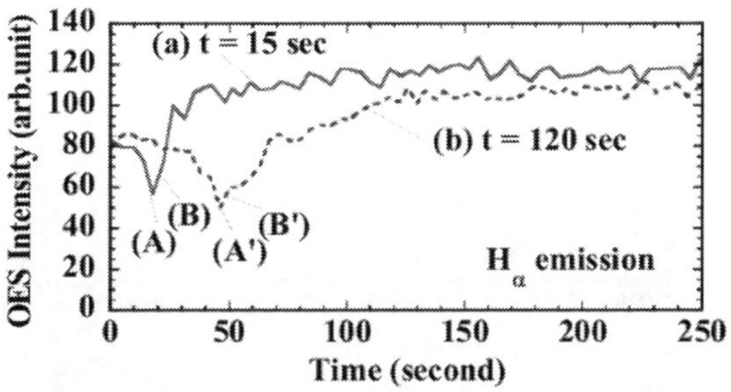

Figure 3. Time evolution of H_α-emission intensity for two different SiH_4-introduction times.

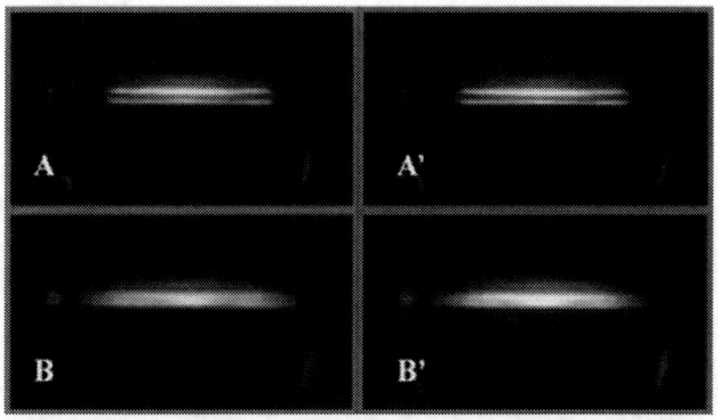

Figure 4. Picture of the plasma at the time (A), (B), (A'), and (B') in Fig.3 showing the structure transition from double (A and A') to single emission (B and B').

double (two lines are seen near the sheath regions between plasma and electrodes) to single (only one bright region is seen between both electrodes) structure as shown in Fig. 4. Drastic change in the structure of plasma as shown in Fig. 4 indicates also a possible evidence of the increase of electron temperature. At the steady-state plasma (during the growth of bulk μc-Si:H-i layer), the electron temperature shows an obvious difference by the difference of the SiH_4-introduction time. Namely, when the SiH_4-introduction time is short (t = 15 sec in this case), powder formation severely occurs at the initial stage of film growth, causing higher electron temperature in the steady state plasma in contrast to the case of long SiH_4-introduction time (t = 120 sec).

It is emphasized here that the change in the dangling-bond-defect density in the bulk region of μc-Si:H deposited under the steady state plasma could be predicted when monitoring the optical emission intensity of specific line (H_α in the case of μc-Si:H growth at high rate).

3.2. External quantum efficiency spectra for μc-Si:H solar cells

External quantum-efficiency spectrum (QE) was measured to evaluate the difference of photovoltaic (PV) performance in the n-i-p type μc-Si:H solar cells whose i layer was prepared under two different SiH_4-introduction-time conditions. Results are shown in Fig. 5. QE of the solar cell, whose i layer was fabricated with short SiH_4-introduction time (t = 15 sec), shows a slight deterioration in the wavelength range longer than 550 nm as shown in Fig. 5(a) in contrast to the case shown in Fig. 5(b). This suggests that the bulk-defect density of i-layer in this solar cell fabricated with shorter SiH_4-introduction time exhibits higher value due to higher electron temperature in the plasma during i-layer-film growth as is predicted in Fig. 3. When i layer exhibits higher defect density, photo-generated electrons and holes in i layer are recombined in the entire part of i layer, showing wide range deterioration in the QE spectrum as is seen in Fig. 5(a).

Next, we have studied the effect of H_2-plasma-exposure time before introducing SiH_4 on the resulting n-i-p solar-cell performance. Two H_2-plasma-exposure times were tested for the initial stage of i-layer growth, e. g, H_2-plasma-exposure times were set at 6 sec and 30 sec before introducing SiH_4 with an identical SiH_4-introduction time of 120 sec. Figure 6 shows the QE spectra for two μc-Si:H solar cells whose i layer was fabricated with different H_2-exposure times before introducing SiH_4. As is clearly seen in the figure, QE spectrum for the solar cell (c) shows a remarkable deterioration in the wavelength range at around 700 nm in contrast to that for the solar cell (b) in Fig. 6. We have proposed a model to explain the severe deterioration of QE spectrum for the solar cell (c) in Fig. 6, in which the auto doping of phosphorus occurs from underlying n layer into i layer by the influence of high atomic hydrogen-flux density especially at the starting stage of i-layer growth in the case of solar cell (c) in the figure. This model is plausible and supported by two experimental facts described below.

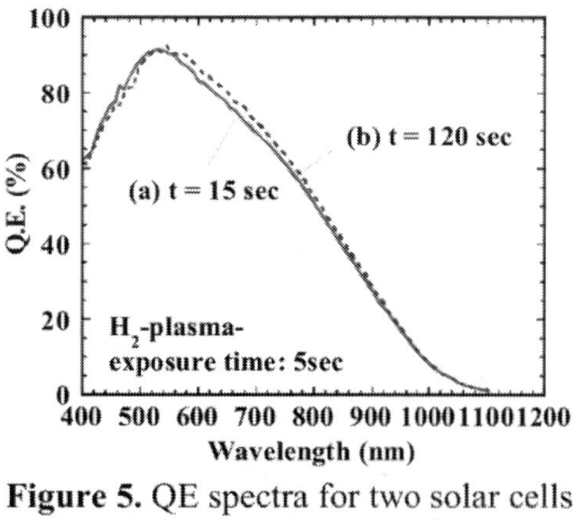

Figure 5. QE spectra for two solar cells whose i-layer growth was performed with different SiH_4-introduction times (t) of 15 sec (a) and 120 sec (b).

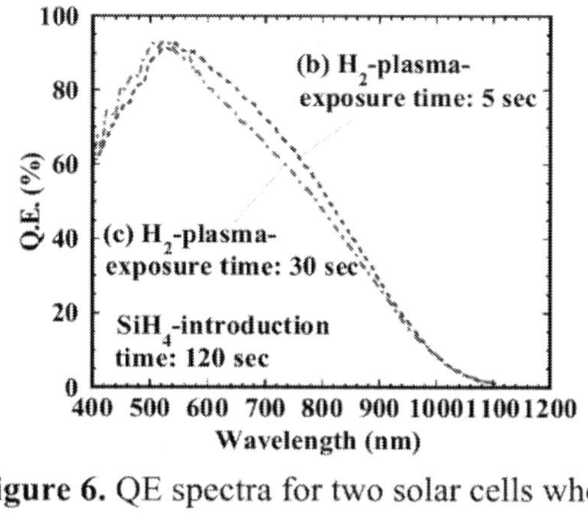

Figure 6. QE spectra for two solar cells whose i layer was prepared with different H_2-plasma-exposure times of 5 sec (b) and 30 sec (c).

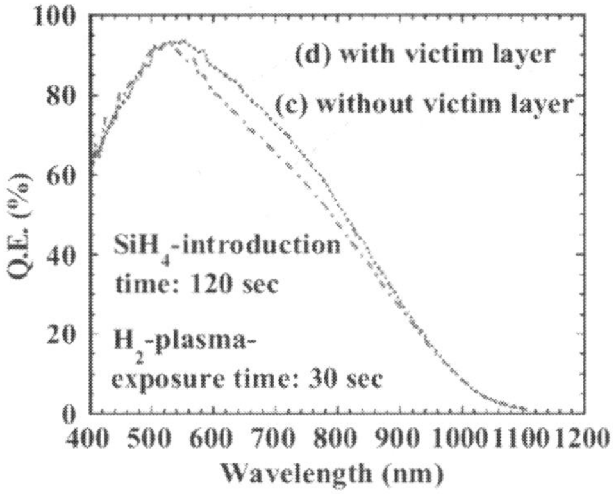

Figure 7. QE spectra for two solar cells fabricated using longer H_2-plasma-exposure time of 30 sec with (d) and without (c) depositing a-Si:H-victim layer on n layer.

(1) when the phosphorus-doped n-type μc-Si:H films are exposed to H_2 plasma produced under the same condition as the initial stage of i-layer growth, films are etched out with quite high rate of 0.3 nm/sec. (2) when the intrinsic a-Si:H film is placed just adjacent to n-type μc-Si:H film and they are exposed to H_2-plasma together for 30 sec, the electric conductivity of intrinsic a-Si:H film near the n-type μc-Si:H film is strictly increased from 1.6×10^{-10} S/cm to 3.5×10^{-10} S/cm, indicating that silicon and phosphorus atoms etched out by the H_2 plasma are deposited on the surface of adjacent a-Si:H film through a chemical transport process (auto doping). Therefore, under the condition of i-layer growth for the solar cell (c), silicon and phosphorus are etched out from the surface of underlying n layer and incorporated into i layer grown simultaneously. Incorporation of phosphorus in i-layer region near n/i interface causes the formation of "flat band" in the solar cell band diagram, resulting in the severe deterioration in QE of the solar cell in the wavelength region at around 700 nm due to the weakened electric field near n/i interface.

To prevent the phosphorus auto doping from underlying n layer into i layer in the n-i-p type solar-cell-fabrication process under the condition for obtaining high PV performance with very high reproducibility, we propose a novel method, in which thin intrinsic a-Si:H layer is deposited (called as "victim" layer) on n layer with proper thickness to be vanished by the H_2-plasma exposure. When 10 nm-thick a-Si:H layer is deposited on n layer as a victim layer, for instance, this layer is vanished by 30-sec H_2-plasma exposure completely before introducing SiH_4. It should be noted here that any auto doping is not observed during the victim-layer growth when using high quality a-Si:H-growth condition with pure SiH_4. Figure 7 shows QE spectra with and without using victim a-Si:H layer under the conditions of 30-sec H_2-plasma-exposure time and 120-sec SiH_4-introduction time. As is seen in the figure, QE spectrum (d) shows a remarkable improvement in the wavelength region at around 700 nm, meaning that the phosphorus auto doping from underlying n layer into growing i layer is successfully suppressed and flat-band formation near the n/i interface is avoided.

4. Conclusion

We have proposed two novel processes to form fine n/i interface in n-i-p type μc-Si:H solar cells, e. g, (1) gradual SiH_4-introduction method to avoid the powder formation at the starting process of i layer growth and (2) victim a-Si:H-layer-insertion method to prevent phosphorus auto doping from underlying n layer into i layer by the influence of H_2-plasma exposure also at the starting stage of i-layer growth.

High SiH_4 depletion is achieved by the gradual SiH_4-introdaction method from the start of film growth, by which the formation of higher order SiH_4 molecules (powder-like species) is well controlled. By the control of powder-formation process, electron temperature in the steady-state plasma during i-layer growth is decreased through the reduction of electron-attachment events to powders in the plasma, giving rise to the growth of high quality intrinsic μc-Si:H layers with low dangling-bond-defect density.

However, high atomic hydrogen-flux density from the plasma at the initial stage of gradual SiH_4-introduction process gives rise to the auto doping of phosphorus into growing i layer from underlying n layer in the fabrication process of n-i-p type μc-Si:H solar cells. To avoid phosphorus auto doping, we have proposed a novel method, in which thin a-Si:H layer is deposited on n layer surface as a victim layer with proper thickness to be vanished by H_2-plasma exposure. Utilization of this victim layer, phosphorus auto doping from underlying n layer is successfully suppressed. When applying these two methods to the fabrication process of n-i-p type μc-Si:H solar cells whose i layer is grown at high rate more than 2 nm/sec, high solar cell performance (> 9% in efficiency) is achieved with fairy good reproducibility.

Acknowledgement
This work was partly supported by Grants-in-Aid from the New Energy and Industrial technology Development organization (NEDO), and by a Global COE Program (Core Research and Engineering of Advanced Materials-Interdisciplinary Education Center for Materials Science) MEXT, Japan.

References

[1] Yamamoto K Yoshimi M Suzuki T Tawada Y Okamoto Y and Nakajima A 1998 *Proc. Mater. Res. Soc. Symp.* **507** p 131
[2] Meier J Dubail S Fluckinger R Fischer D Keppner H and Shah A 1994 *Proc. 1st World Conf. Photovolt. Energy Conf. (Hawai)* vol 1 (IEEE) p 409
[3] Kondo M, Fukawa M, Guo L and Matsuda A 2000 *J. Non-Cryst. Solids* **266-269** 84
[4] Fukawa M Suzuki S Guo L Kondo M and Matsuda A 2001 *Sol. Energy Mater. Sol. Cells* **66** 217
[5] Chantana J Tsutsui Y Sobajima Y Sada C Matsuda A and Okamoto H 2011 *Jpn. J. Appl. Phys.* **50** 045806
[6] Matsuda A, Takai M, Nishimoto T and Kondo M 2003 *Sol. Energy Mater. Sol. Cell.* **78** 3
[7] Matsuda A 2004 *J. Non-Cryst. Solids* **338–340** 1
[8] Sobajima Y, Matsuda A and Okamoto H 2011 *Proc. Mater. Res. Soc. Symp.***1321** p375
[9] Müller J, Rech B, Springer J, and Vanecek M 2004 *Solar Energy* **77** 917
[10] Sobajima Y Muto H Shinohara Y Sada C Matsuda A and Okamoto H 2011 *Tech. Digest 21st Int. Photovoltaic Science and Engineering Conf. (Fukuoka)* 4A-2O-06

Development of electrically conductive DLC coated stainless steel separators for polymer electrolyte membrane fuel cell

Yasuo Suzuki, Masanori Watanabe, Tadao Toda, Toshiaki Fujii

Plasma Ion Assist Co., Ltd Nakagawara 34-1, Kamitoba, Minami-ku, Kyoto-city, Kyoto, 601-8142 JAPAN

E-mail:y.suzuki@plasma-ion.co.jp

Abstract. Polymer electrolyte fuel cell (PEFC) as one of generation devices of electrical power is rapidly expanding the market as clean energy instead of petroleum and atomic energy. Residential fuel cell goes into quantity production and introduction of fuel cell for use in automobiles starts in the year 2015 in Japan. Critical subject for making fuel cell expand is how to reduce cost of fuel cell. In this paper we describe about separator plate which domains large ratio of cost in fuel cell stack. In present time, carbon is used in material of residential fuel cell separator. Metal separators are developed in fuel cell for use in automobiles because of need of mechanical strength at first. In order to make fuel cell expand in market, further cost reduction is required. But the metal separator has problem that by using metal separator contact resistance occurred by metal corrosion increases and catalyst layer and membrane degrade. In recent time we found out to protect from corrosion and dissolution of metals by coating the film of porous free conductive DLC with plasma ion implantation and deposition technology that we have developed. Film of electrically conductive DLC was formed with high speed of 13μm/hr by ICP plasma, and coating cost breakout was performed.

1. Introduction

Residential fuel cell goes into quantity production in the year 2015 and fuel cell for use in automobiles starts to be produced in the year 2015 and they are mass-produced in the year 2020 in Japan. Ratio of separator cost in stack of fuel cell is 10~20% and reduction of separator cost is important in the spread of fuel cell. In present time carbon separators are used but their cost reduction is difficult. Stainless steel is expected in stead of carbon separator from a viewpoint of cost. It is required that separator cost of residential fuel cell is 150yen in the year 2015 and separator cost of fuel cell for use in automobiles is 50yen in A4 size in the year 2020. The electrically conductive DLC coated stainless steel separator by method of developed plasma ion based ion implantation and deposition is superior to contact resistance and corrosion resistance, and it became possible that separator cost was remarkably reduced by high speed deposition equipment.

2. Characteristics required to separators

2.1. Construction of separator

Fuel cell is constructed from anode electrode, electrolyte membrane, cathode electrode and separators which hold these from both sides. Fuel cell is constructed from single cell which is assembled in series. Anode electrode is constructed from gas diffusion layer (GDL) which consists of carbon paper for diffusing H_2 gas/Air, and catalyst layer.

2.2. Characteristics required to separators
Characteristics required to separators is electrically conductive, little in contact resistance between separator and GDL which is anode electrode or cathode electrode and superior to corrosion resistance of H_2SO_4 solution which occurs by reacting with electrolyte membrane and water. And it is superior to cost performance to be specially required to fuel cell for use in automobiles.

3. Conductivities of DLC

3.1. What is DLC
DLC is Diamond Like Carbon. It is a-C:H which consists of diamond structure sp^3 and graphite structure sp^2. DLC has insulation properties similar to diamond, high hardness and corrosion resistance.

3.2. Method of DLC film formation
Generally DLC film is formed by method of arc-ion plating and sputtering. Formed film is porous because DLC film is formed by sputtering and arc-evaporating solid carbon target. Therefore it can not be avoided to be inferior to corrosion resistance. In method of developed PBIID [1,2,3,4,5] technology (Plasma Based Ion Implantation and Deposition), methane and acetylene gas are ionized by simultaneously applying RF and high pulsed voltage, carbon ion is implanted and deposited to substrate, and electrically conductive DLC film is formed. This film is porous free and superior to corrosion resistance. PBIID has two method, CCP (Capacitively Coupled Plasma) and ICP (Inductively Coupled Plasma). Figure1 shows comparison of two methods. Feature of ICP method is that plasma density is very high and deposition rate is very fast.

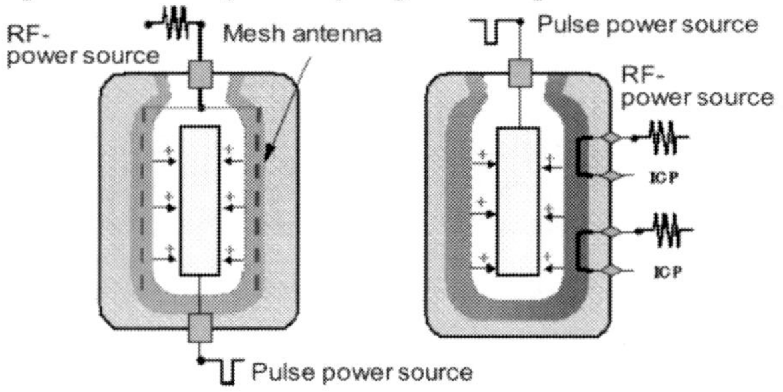

No.	1	2
Plasma	Antenna method(CCP)	ICP method
RF-Power source structure	Flat plate /Cylindrical mesh	One turn Coil
Plasma density	$\sim 3 \times 10^{10}$ cm^{-3}	$\sim 2 \times 10^{11}$ cm^{-3}
Plasma uniformity	±20%	±20%
Deposition speed	1μm/hr	5μm/hr (※)

※The distance between substrate and ICP source is 150mm.

Figure 1. DLC film formation by the PBIID method.

3.3. Method to change from insulation properties to electric conduction
DLC has insulation properties essentially to have diamond construction sp^3. When DLC film is formed under 350℃ and B-ion or N-ion is simultaneously implanted to DLC film, insulating DLC film changes to electrically conductive DLC(a-C:H:B, a-C:H:N) film. Electrically conductive DLC film is nano size (10～80nm) crystallized film to grow with pillar-shaped column. Conductivity of DLC film is less than 5mΩ・cm.

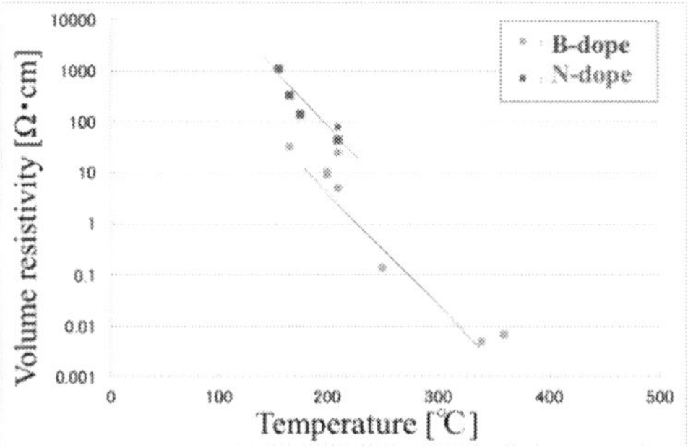

Figure 2. Volume resistivity.

4. Reduction of contact resistance

4.1. Contact resistance (CR) and measurement method

Figure 3 shows measurement method of contact resistance. CR_{sample} is contact resistance between DLC coated sample and GDL. R_{total} is total resistance on measurement. $R_{baseline}$ is resistance except sample.

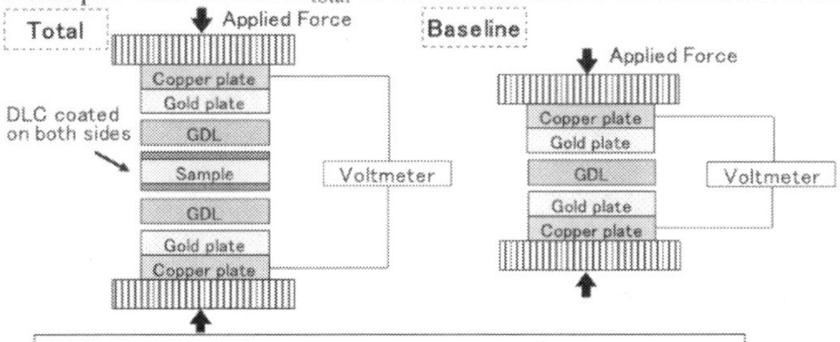

$$CR_{sample} = (R_{total} - R_{baseline})/2 * Area$$

・CR measurement method was four-terminal method.
・The measurement current was 100mA by programmable current supply.
・The GDL was Toray carbon paper, 190μ m thick.

Figure 3. CR Measurement method for coated plates.

4.2. Contact resistance

Electrically conductive DLC film has many contact points which consist of nano size (10~80nm) crystallization to grow with pillar-shaped column. In that result it was able to lower contact resistance by strongly contacting between GDL and electrically conductive DLC film. Figure 4 shows contact resistance on compaction force. It is less than 5mΩ・cm^2.

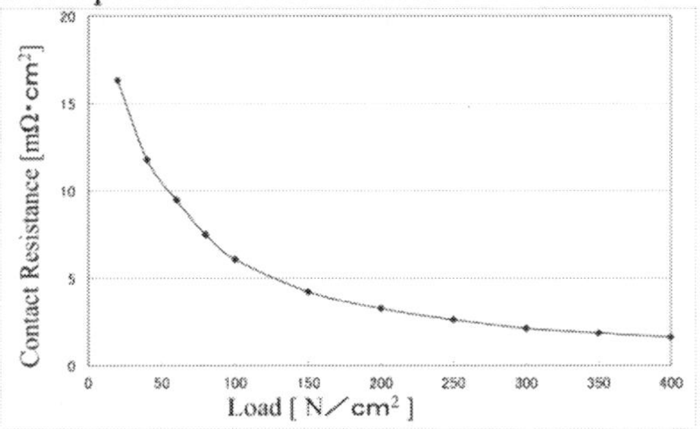

Figure 4. Interfacial resistance dependency on compaction force.

5. Corrosion resistance

5.1. Electrical conductive DLC formation

As shown in Figure6, electrically conductive bi-layer DLC has conductive properties which consists of two kind of DLC film. Top layer DLC film specializes conductive properties. Bottom layer DLC film specializes corrosion resistance. Electrically conductive DLC film is formed at 350℃~. Corrosion resistance DLC film is formed at ~200℃.

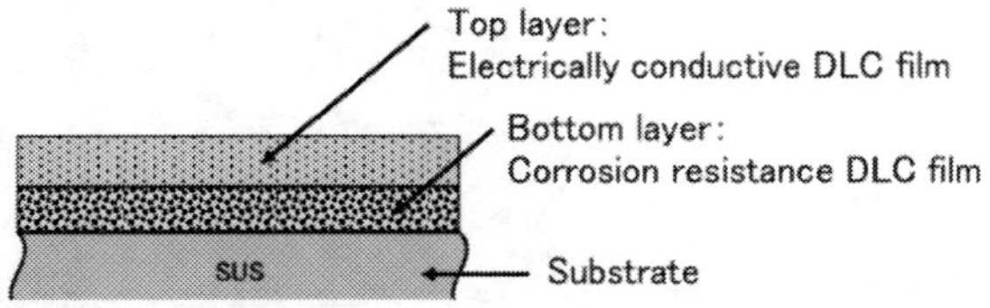

Figure 5. Bi-Layer film.

5.2. Corrosion resistance of electrically conductive DLC film

5.2.1. Anodic polarization curve Figure 6 shows anodic polarization curve with electrically conductive DLC and corrosion resistance DLC. Electrically conductive DLC is nano size film, it has grain boundary. and it is sp^2-rich. Therefore electrically conductive DLC film dissolves by applied voltage and has inferior corrosion resistance. Corrosion resistance DLC is sp^3-rich and has good corrosion resistance, current density is 10^{-6} A/cm^2.

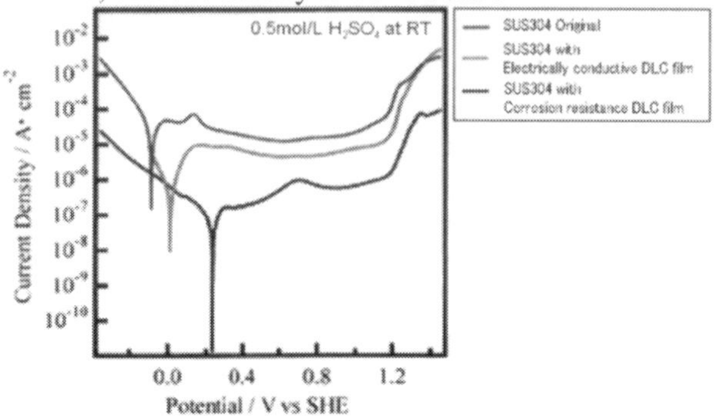

Figure 6. Anodic polarization curve with electrically conductive DLC and corrosion resistance DLC.

5.2.2. Electrochemical corrosion test Figure 7 shows potentionstatic test with H_2SO_4 pH2, 1Vvs SHE, 80℃ and metal dissolution. In potentionstatic test, current density is 10^{-6}A/cm^2. Fe dissolution is 0.02 ppm and very little. This film is superior to corrosion resistance.

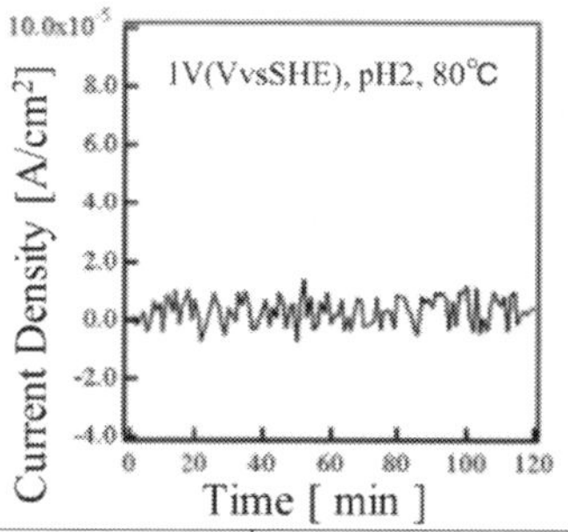

Test condition	Metal contents [ppm]			
	Fe	Cr	Ni	Mo
1V, pH2, 80℃	0.02	0	0.002	0

Figure 7. Metal contents in ppm after immersion test in distilled water in test condition by ICP analyzer for test specimen.

6. Low Cost

6.1. high deposition rate equipment

We manufactured one unit to produce separators continuously in an in-line system consisted of many units in the future. In figure 8 shows conceptual diagram of PIAD(Plasma Ion Assisted Deposition) and figure 9 shows one unit of PIAD system. ICP source and mating box are attached in the front and the backside of equipment. Separator which is set up on 50cm square carrier is going and returning in right and left between ICP sources (space 10cm).Electrically conductive DLC is formed by heating, Ar-cleaning, N-ion implantation, and CH_4/C_2H_2 deposition in serial order.

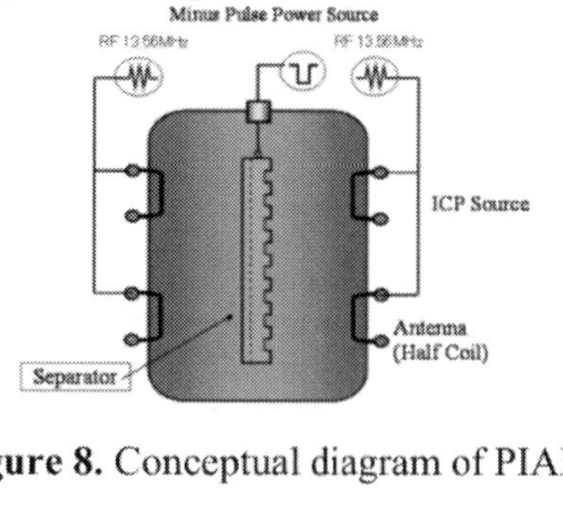

Figure 8. Conceptual diagram of PIAD.

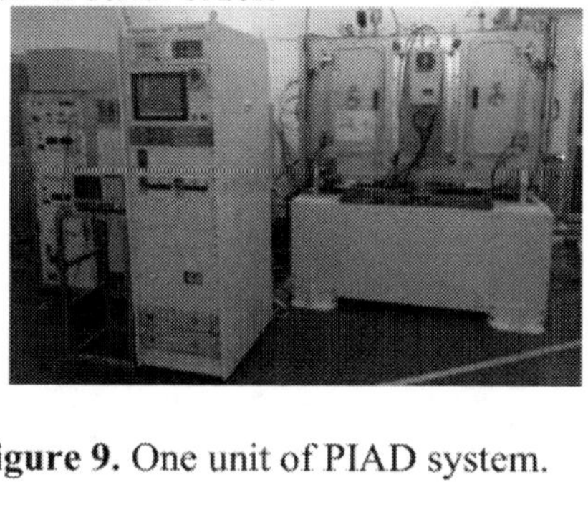

Figure 9. One unit of PIAD system.

6.2. Deposition rate

Deposition rate is 5μm/hr by using ICP plasma with RF power 3kW, distance 15cm from ICP. Figure 10 shows deposition rate in case of distance 5cm from ICP. Deposition rate is proportional to the applied pulse voltage on substrate.

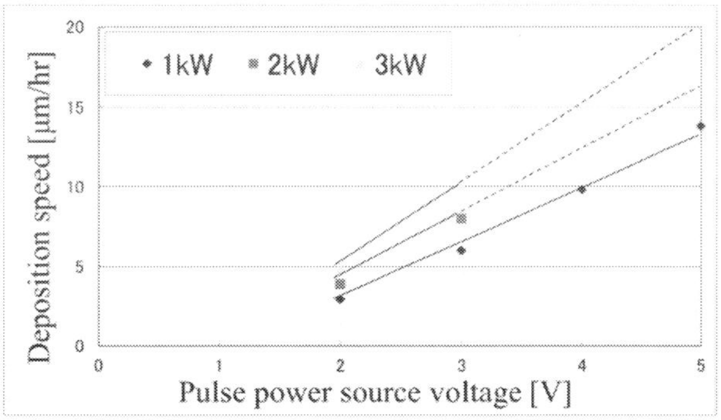

Figure 10. Deposition rate.

6.3. Reducing the coating cost

The coating cost depends on deposition rate. The faster deposition rate, the cheaper coating cost. In case of deposition rate 20μm/hr, film thickness 150nm, and substrate size A4, the coating cost is estimated to be 150yen.

7. Cell performance of electrical generation

7.1. Single Cell Performance

By using electrically conductive DLC coated stainless (SUS316L) steel separator, Electrical generation test of single cell for 1000 hours was carried out. The active electrode area was $25cm^2$. Table 1 shows detail test condition. Figure 11 shows single cell. Figure 12 shows cell performance of electrical generation.Figure13 shows change of cell resistance. These results showed same good characteristics in comparison with carbon separator and electrically conductive DLC coated stainless steel separator.

Table 1. Single cell test condition.

Temperature	80℃
Humidification	Saturated humidification
Gas condition	Hydrogen 136ml/min(70%) Oxygen 568ml/min(40%)
Current Density	0.5 A/cm^2

Figure 11. Single cell

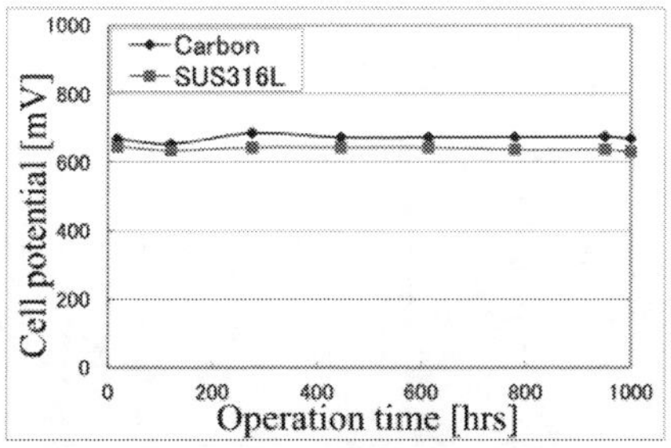

Figure12. Cell performance of electrical generation.

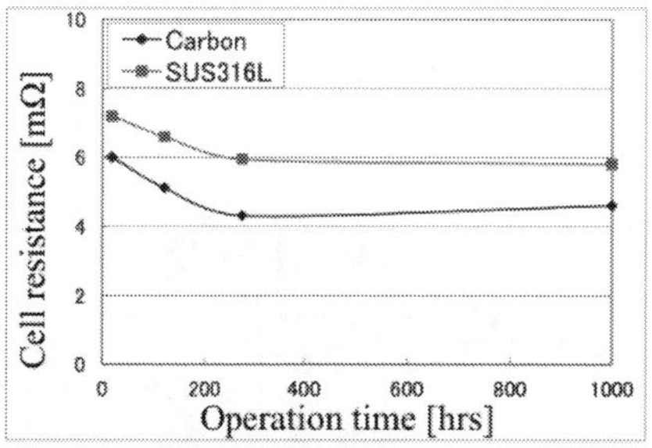

Figure13. Change of cell resistance.

7.2. Stack Cell performance

By using electrically conductive DLC coated stainless (SUS316L) steel separator stack of 5 cells. Electrical generation test for 1000 hours was carried out. The active electrode area was 167cm^2. Table 2 shows detail test condition. Figure 14 shows each cell performance of electrical generation. Figure 15 shows I-V performance before and after electrical generation. Each cell potential for a period of 1000 hours were very stable. And in I-V performance, same good polarization characteristics as carbon separator was gotten.

Table 2. Stack test condition.

Temperature	80℃
Humidification	Saturated humidification
Operation condition	WSS(Weekly Stop & Start)
Current Density	0.3 A/cm^2

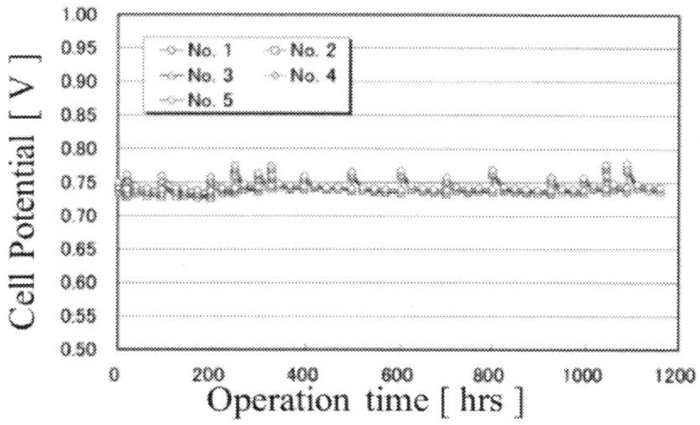

Figure14. Each cell potential for periods when cells were under load.

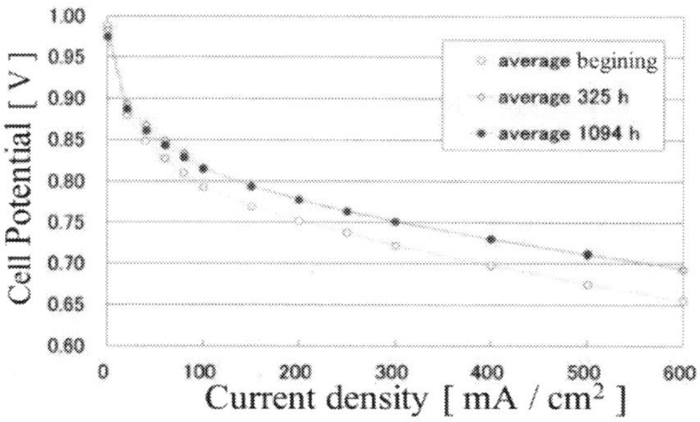

Figure15. Polarization characteristics.

8. Summary

Contact resistance is a few mΩ・cm^2 between GDL constructed with fuel electrode or air electrode, and electrically conductive DLC developed by technology of plasma based ion implantation and deposition .Also current density in corrosion test is less than 10^{-6}A/cm^2 and Fe metal contents after immersion test in distilled water test was 0.02 ppm under condition of 1V, pH2 H_2SO_4, 80℃. Deposition speed of over than 13μm/hr was gotten by using ICP plasma source. In that result possibility was found out that cost of electrically conductive DLC coated stainless steel separator was 150yen at 150nm DLC thin film in A4 size. It is expected that in future if electrically conductive 50nm DLC coated aluminum separator is developed, cost of separator is 50yen in A4 size.

9. References

[1] Conrad J R, Radtke J L, Dodd R A, Worzala F J, and Ngoc C Tran 1987 *J.Appl.Phys.* **62 No.11** 4591-4596
[2] Y.Suzuki 1998 *OYO BUTURI*, **67 No.6** 663-667
[3] Y.Suzuki 2002 *JIEED Japan* **45 No.2** 30-35
[4] Y.Suzuki 2003 *IEEJ Trans.EIS.* **123 No.1**
[5] Y.Suzuki 2006 *JIEED Japan* **49 No.2** 41-46

Hard Coat Layers by PE-CVD Process for the Top Surface of Touch Panel

T Okunishi, N Sato and K Yazawa
R&D Div., Frontier Center, OIKE & Co., Ltd., Japan,

E-mail:t-okunishi@oike-jp.com

Abstract. In order to protect surface from damages, the high pencil hardness and the high abrasion resistance are required for the hard coat layers on polyethylene telephthalate (PET) films for the application of touch panel surface. We have already found that the UV-curing-hard-coat-polymer (UHP) coated PET films show the poor abrasion resistance, while they have the high pencil hardness. It reveals that the abrasion resistance of hard coat layers of the UHP is not simply dependent on the pencil hardness. In this work, we have studied to improve the abrasion resistance of SiOC films as hard coat layers, which were formed by PE-CVD process on UHP coated PET. The abrasion resistance was evaluated by Taber abrasion test. PE-CVD hard coat layers which formed on UHP coater PET films have showed the better abrasion resistance and have the possibility of substitution to the thin glass sheets for touch panel application.

1. Introduction

Recently, a mobile application, such as the smart phone, the tablet and E-paper, is in fashion. The touch panel is used for input-device. A glass sheet is used in the surface of the screen to protect it against the scratch by fingers and stylus pens. However, the glass sheet is poor physical damage, so that it needs the substitution to the glass sheet. The substitution to the glass sheet for the top cover of a touch panel is required in the mobile application. One of candidate of the substitution material is the transparent plastic films because of their flexibility. The plastic film requires the hard coat layer to obtain high abrasion resistance, because the surface of it has the poor abrasion resistance. Therefore, the protective layer on the surface of the plastic film is required to protect any physical damage. One of the protective layers is the UV-curig-hard-coat-polymer (UHP), which satisfies some of the physical damage, but they have not enough abrasion resistance. The hard coat layers formed by CVD process have various kinds of layers, for example, SiO_2, SiOC [1-3], TiO_2 [4, 5], DLC (Diamond like carbon) [6, 7, 8] are known as protective layers. It is well known that the SiOC layers coated with polymer hard coat layers are already used for the top coat of the transparent polycarbonate (PC) windows for actual automobile application, because they have the high abrasion resistance. The specification of the abrasion resistance for automotive windows is defined as the delta haze of the surface after Taber abrasion tests should keep less than 10% in delta haze in ECE R43[9]. The SiOC layers on PC (polycarbonate) sheet meet this criterion. The SiOC layers by CVD process is also expected for the hard coat layers of plastic films for the surface of touch panels. In this work, we have studied the abrasion resistance of hard coat layers by CVD process on UHP coated PET films.

Journal of Physics: Conference Series **441** (2013) 012028 doi:10.1088/1742-6596/441/1/012028

2. Experimental

The structure of hard coat films is shown in Figure 1. The SiOC layers were formed by PE-CVD process on the UHP coated PET films using Roll-to-Roll PE-CVD equipment illustrated in Figure 2. Hexametyldisiloxane (HMDSO) and O_2 were used as the precursor and reactive gases, respectively. The flows of each gas were controlled using mass flow controllers. After the initial evacuation less than 10^{-3} Pa, the SiOC layers were deposited under conditions summarized in table 1. The thickness of the SiOC layers was measured by a surface profiler (P-6, KLA-Tencor). The UHP was used silicon-containing acrylic polymer. PET film (Teijin DuPont Film, Thickness: 188μm) is used as the substrate. The silicon containing acrylic polymers were dried at 80 deg.C in air, and then cured by UV irradiation unit (UC-1501, HOYA-SCHOTT) at 600mJ/cm^2. The thickness of the UHP was about 3μm. The abrasion resistance properties of the hard coat PET films were investigated by delta haze. The delta haze is increment of the haze after the Taber abrasion test using Taber abrasion tester (YASUDA SEIKI SEISAKUSHO Ltd.). The wheels of the Taber tester were used CF-10F, and the load of the wheels was 4.9N [10]. The haze was measured by haze mater (NDH4000, NIPPON DENSHOKU). Compositions of the SiOC layers ware evaluated by X-ray photoelectron spectroscopy (ESCA 5800, ULVAC PHI). The surface of SiOC hard coat layer after the test was observed by a field emission scanning electron microscope (FE-SEM, S-4700, Hitachi).

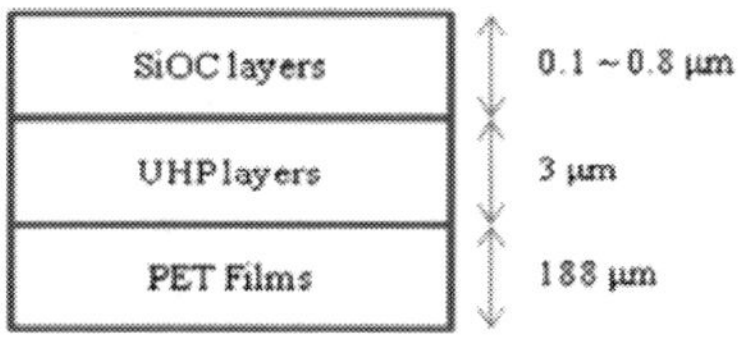

Figure 1. Structure of hard coat films.

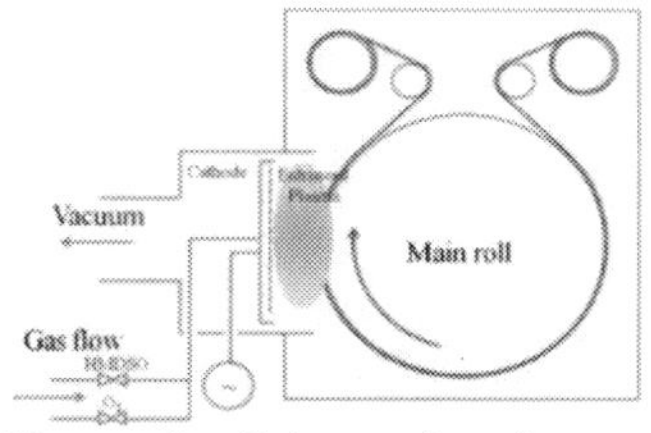

Figure 2. Schematic view of Roll-to-Roll PE-CVD coater.

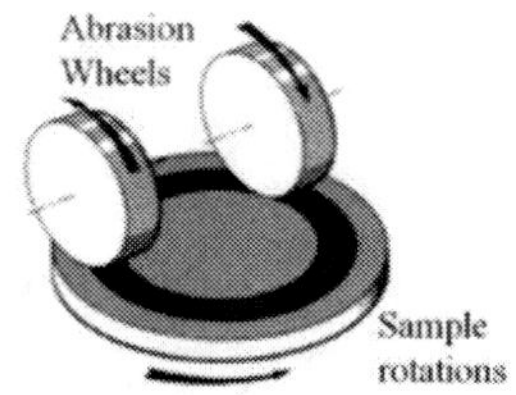

Figure 3. Schematic view of Taber abrasion tester.

3. Results and discussion

Figure 4 shows the appearance of the UHP coated PET film (a) without and (b) with SiOC hard coat layer after Taber abrasion test at 500 cycles. The surface of UHP coated PET is markedly damaged by the abrasion wheels, while the surface of SiOC layers on the UHP coated PET is not markedly damaged. Changes of haze value before and after Taber abrasion tests of the SiOC layers with 400nm thick on the UHP coated PET films and UHP coated PET films shown in Figure 5. The haze value of UHP coated PET rises drastically from 0.6% to 45% with increasing number of cycles in abrasion test from 0 to 500 cycles. On the other hand, the haze value of the SiOC hard coat layers on the UHP coated PET rises gradually from 0.76% only to 4.0%. It means that the SiOC coated films with 400nm thick has very high abrasion resistance. The pencil hardness of both hard coat layers showed the same value, 2H pencil hardness. These results indicate that the Taber abrasion resistance of hard coat layers has no explicit correlation with the pencil hardness of the surface.

Journal of Physics: Conference Series **441** (2013) 012028 doi:10.1088/1742-6596/441/1/012028

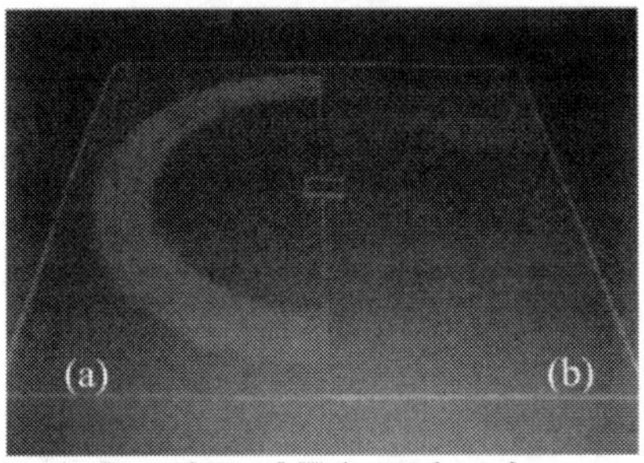

Figure 4. Results of Taber abrasion test. (a): UHP coated PET, (b): SiOC hard coat layers on UHP coated PET.

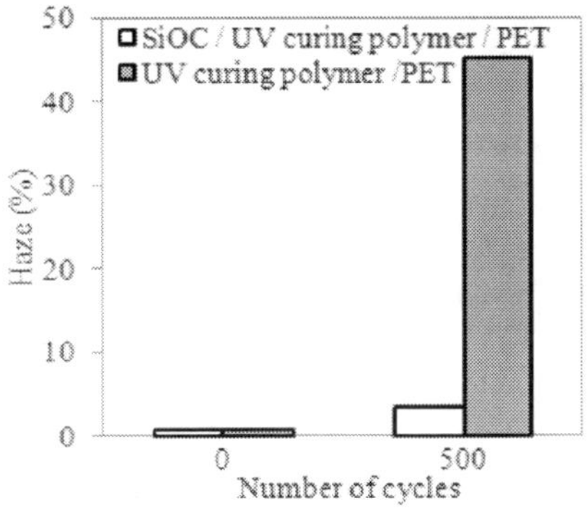

Figure 5. Haze value before after Taber abrasion tests.

The delta haze depends on the thickness of the SiOC layers, as shown in Figure 6. Delta haze decreases a first and shows the minimum value, 4% in 400nm -500nm thick, with increasing the thickness of the SiOC layers.

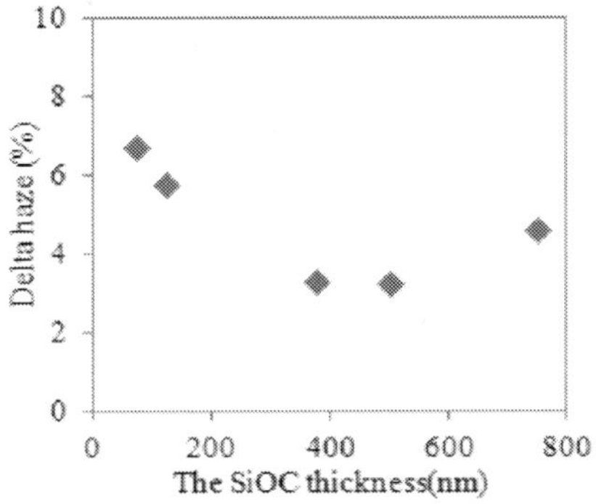

Figure 6. Delta haze vs. thickness of the SiOC hard coat layers.

Delta haze as a function of the amount of HMDSO flow, keeping constant flow of O_2 gas with 100, 350 and 550 sccm are shown in Figure 7. The delta haze of O_2 flow as 350 and 550 sccm in O_2 gas flow exhibit no marked dependence on the HMDSO flow, while delta haze of O_2 flow a 100 sccm keeps stable when HMDSO less than 15 sccm and then it increase, when HMDSO flows exceed 15 sccm. It is expressed in Figure 8, delta haze is re-plotted as a function of flow ratio of HMDSO / (HMDSO + O_2). In Figure 8, the delta haze seems to be almost constant up to 15% in the ratio, and they increases rapidly when the ratio of HMDSO / (HMDSO + O_2) exceed 15%. It seems that the excess HMDSO exist another un-reacted oligomer in SiOC layers, which should show the low abrasion resistance, when ratio of HMDSO / (HMDSO + O_2) exceed 15%.

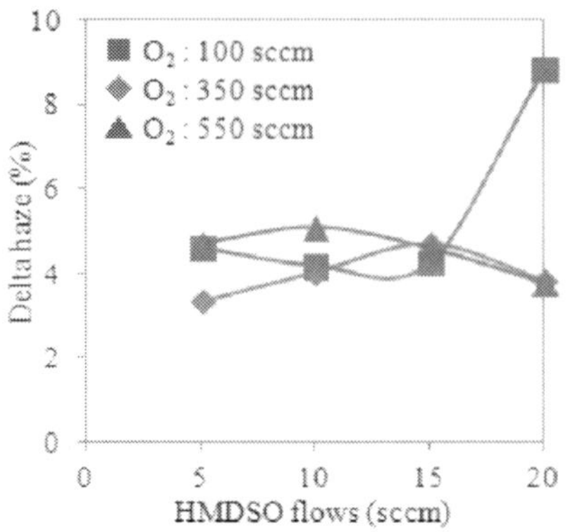

Figure 7. Delta haze vs. HMDSO flow.

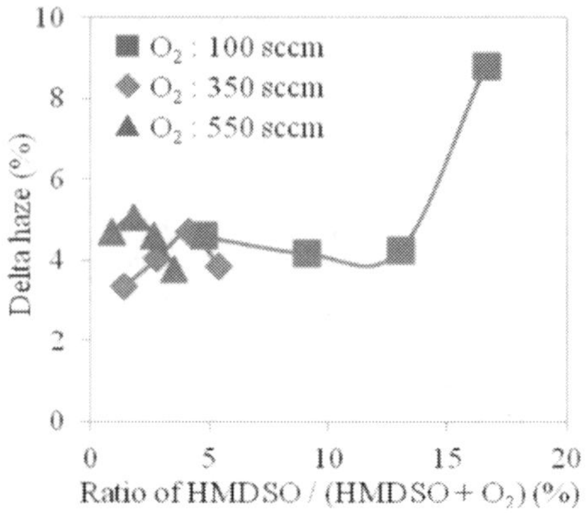

Figure 8. Delta haze vs. ratio of HMDSO / (HMDSO + O_2).

Figure 9 shows the relationship between compositions of the SiOC layers and deposition conditions. The atomic concentrations of carbon in the SiOC layers are increased with increasing the rate of HMDSO / (HMDSO+O_2) during deposition. The atomic concentration of carbon in the SiOC layers depends on the amount of incompletely reacted precursors. It is thought that the SiOC layers which have almost no carbon are formed as inorganic hard materials. Figure 10 shows the relation between the atomic concentration of carbon in the SiOC layers and delta haze. The delta hazes of films at carbon atomic concentration of 10% or lower are almost constant, and they are sharply increased at 20% of carbon atomic concentration.

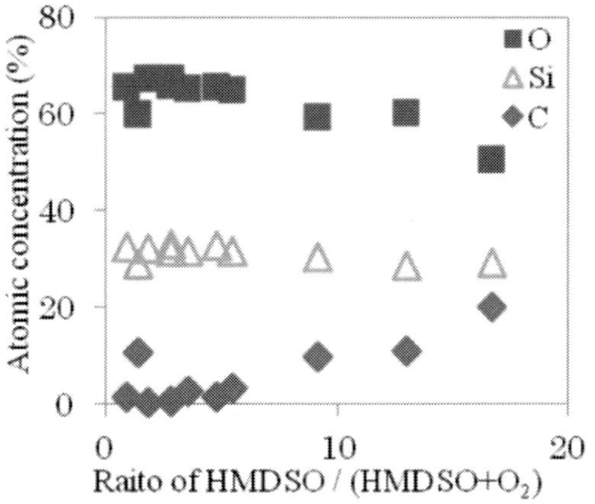

Figure 9. Atomic concentration vs. Ratio of HMDSO / (HMDSO+O_2).

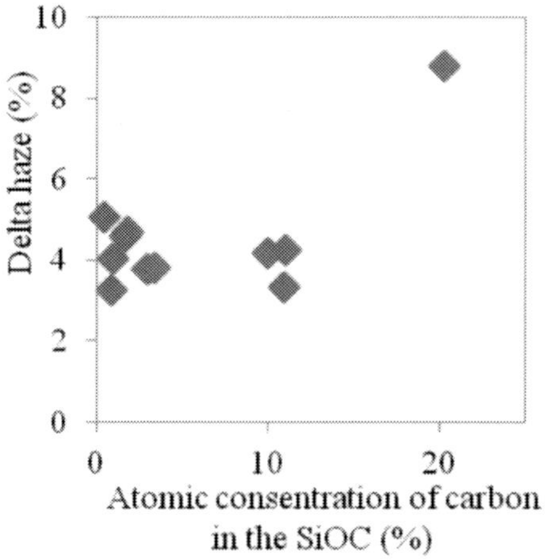

Figure 10. Delta haze vs. Atomic concentration of carbon in the SiOC.

Figure 11 shows the figure plotted the number of cycles to the delta haze. Two different deposition conditions are summarized in table 2. It is thought that the difference of delta haze of the SiOC layers deposited under both conditions in less than 100 cycles in Taber abrasion tests is not important because these results are within the range of variation. It is noted that the SiOC layers deposited under Condition (A), (B) have clear differences in changes of delta haze with increasing the number of cycles. The delta haze of SiOC layers formed under Condition (B) on UHB coated PET slightly depends on the number of cycles and indicates less than 4% even at 500 cycles. On the other hand, in case of Condition (A), the delta haze greatly increases with the number of cycles and reaches to 9% at 500 cycles. It is easily estimated that difference of delta haze of both films will expand wider at 1000 cycles or more. The SEM micrographs of the SiOC and UHP surfaces after Taber abrasion tests at 500 cycles is shown in figure 12. Notation (a), (b) indicate the surface of the SiOC layers deposited under Condition (A), (B), respectively. Notation (c) indicates the surface of the UHP without the SiOC layers. The surface of the SiOC layers deposited under Condition (A) tested after Taber abrasion tests at 500cycles has two clear scratches added to shallow traces, which show 9% in delta haze. On the other hand, it is found that the surface of the SiOC layers under Condition (B) has only shallow traces, so that the delta haze shows 4%. The surface of the UHP has many clear scratches more than these of the SiOC layers deposited under Condition (A), which show 45% in delta haze. These results indicate that the values of delta haze are related to the number of clear scratches on the surface. It is thought that the surface of the SiOC layers, such as inorganic hard material, deposited under Condition (B) has shallow traces by rubbing the surface with sliding taber wheels. On the other hand, the surface of the SiOC layers deposited under Condition (A) which contains some incompletely reacted precursor as organic parts have clear scratches by abrasion by taber wheels. In addition, UHP without SiOC layers, which has more organic behavior compared the SiOC layers deposited under Condition (A) in abrasion, have many clear scratches.

Table 2. Deposition conditions at the SiOC layers formed by PE-CVD process.

Condition	HMDSO flow (sccm)	O_2 gas flow (sccm)	Ratio of HMDSO / (HMDSO + O_2)
A	20	100	16.7
B	10	350	2.7

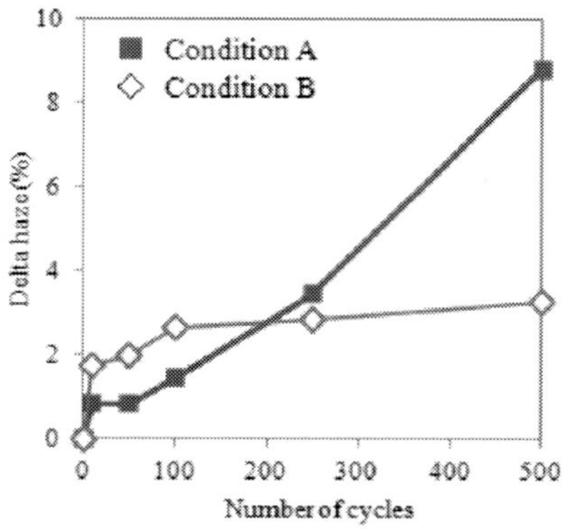

Figure 11. Changes of delta haze vs. Taber abrasion test cycles.

Figure 12. SEM micrographs of the SiOC and the UHP surface after performed Taber abrasion test at 500cycles. (a): The surface of the SiOC layers deposited under Condition (A), (b): The surface of the SiOC layers deposited under Condition (B), (c): The surface of the UHP

4. Conclusion

The SiOC hard coat layers formed by PE-CVD process on UHP coated PET using HMDSO and O_2 gases, in order to add the high abrasion resistance to PET films. The abrasion resistance of SiOC layers on UHP coated PET has been investigated by varying deposition conditions and SiOC thickness. Abrasion resistance has marked dependence on thickness of the SiOC layers and also strongly influenced by depositing conditions. It is clear that the SiOC hard coat layers formed under ratio HMDSO / (HMDSO + O_2) = 2.7% with 400 nm thick should show the excellent abrasion resistance and the delta haze is 4%. The SiOC hard coat layers are expected to have the possibility as a candidate for the substitution to cover glasses for touch panel application.

5. References

[1] Madocks J, Marcus P and Morse P 2006 *Proc. Tec. Conf. on Society of Vacuum Coaters* 569
[2] Madocks J, Morse P and George M 2006 *Proc. Tec. Conf. on Society of Vacuum Coaters* 233
[3] Schmauder T, Nauenburg K D, Kruse K and Ickes G 2008 *Thin Solid Films* **516** 3904-3909
[4] Mathur S and Kuhn P 2006 *Surf. Coat. Technol* **201** 807-814
[5] Mills A, Hill G, Bhopal S, Parkin I P and O'Neil S A 2003 *J. Photochem. Photobiol. A* **160** 185-194
[6] Yehoda J E, Taylor J and Kimock F M, 2011 *Proc. Tec. Conf. on Society of Vacuum Coaters* 261
[7] Hnghes T J, Kimock F M 1993 *Proc. Tec. Conf. on Society of Vacuum Coaters* 139
[8] Daniels B K 1992 *Proc. Tec. Conf. on Society of Vacuum Coaters* 260
[9] ECE R43
[10] ISO 3537

Growth control of ZnO nano-rod with various seeds and photovoltaic application

Hyunwoong Seo[1,2*], **Yuting Wang**[1], **Giichiro Uchida**[1], **Kunihiro Kamataki**[1], **Naho Itagaki**[1], **Kazunori Koga**[1], and **Masaharu Shiratani**[1,2]
[1]Graduate School of Information Science and Electrical Engineering, Kyushu University, 744 Motooka, Nishi-ku, Fukuoka 819-0395, Japan
[2]Center of Plasma Nano-interface Engineering, Kyushu University, 744 Motooka, Nishi-ku, Fukuoka 819-0395, Japan

E-mail: woong_1980@hotmail.com

Abstract. ZnO has attracted much interesting as one of unique materials. Especially, it is suitable for the easy fabrication of nano-structures such as rod, wire and tube as well as particles. ZnO nano-rod is one of good sensitized electrodes because it has good electron transfer and is easily fabricated. In the chemical bath deposition process, seed layer plays an important role in the growth of nano-rod. This work investigated and analyzed the effect of seed layer on the growth of ZnO nano-rod. Fabricated nano-rods were applied to dye-sensitized solar cell. For better performance, ZnO was surface-modified by TiO_2. Surface-modified ZnO had improved electron transfer and wider surface area. Consequently, the current and fill factor were much improved and overall performance was also enhanced with them.

1. Introduction

ZnO has attracted in a wide range of research fields with its multiple properties such as non-toxicity, good electrical conductivity, stability in a hydrogen plasma atmosphere, and low cost. It has been widely studied in the photovoltaics with its band-gap between 3.1 and 3.4 eV. Especially, ZnO has been expected to be comparable to TiO_2 because of its higher electronic mobility and similar energy level of the conduction band [1,2]. However, overall conversion efficiency of ZnO-based solar cell has been lower than that of TiO_2-based solar cell, hence the enhancement on its performance is still a challenging study [3-7]. ZnO is more advantageous than TiO_2 in the structural aspect because it is suitable for the easy fabrication of nano-structures such as rod, wire, and tube. ZnO nano-rods (NRs) were used for the photovoltaic application in this work [8-12]. Among a range of various fabrication methods for ZnO NRs, the chemical bath deposition (CBD) is widely used [13-17]. In the CBD, the seed layer plays an important role in the formation and growth of NR. However, most of related researches have mainly focused on only growth materials as ZnO source and the growth methods. The seed layer was often not specified although the seed layer affects the growth characteristics of NRs. In this work, the effect of the seed layer on the growth of ZnO NR was investigated. The characteristics of ZnO NR such as the growth speed, NR density, and the cristallinity were analyzed according to materials and fabrication of seed layer. After that, fabricated NRs were applied to the photochemical cell as the sensitized electrode. For better performance, ZnO NRs were surface-modified by TiO_2. In order to verify the influence of seed layer and surface modification, the photovoltaic performance, the electrochemical impedance, and structural properties were examined.

2. Experimental

4 kinds of seed layers including a reference were applied for the fabrication of ZnO NR. One ZnO seed was deposited on Fluorine-doped tin oxide (FTO) substrates (15 Ω per sq., Nippon Sheet Glass Co., Ltd.) by RF magnetron sputtering with ZnO target. Working pressure was 0.3 Pa of Ar and 100 W RF power with 13.56 MHz was applied. Substrate temperature was fixed at 500 °C during the deposition and deposition time was 270 s. Other ZnO seeds were deposited on FTO substrates by spin-coating with 50 mM zinc acetate dihydrate ($Zn(CH_3COO)_2{\cdot}H_2O$) and zinc nitrate hexahydrate ($Zn(NO_3)_2{\cdot}6H_2O$) aqueous solutions [18-23]. For uniform layer, they were spin-coated at 3000 rpm for 60 s and at 4000 rpm for 60 s again. Spin-coated layers were sintered at 200 °C for 10 min finally. Their reactions were showed in equation 1 and 2. As a reference, bare FTO substrate was also used without any seed layer.

$$Zn(CH_3COO)_2H_2O \rightarrow ZnO + 2CH_3COOH \tag{1}$$

$$\begin{aligned} &Zn(NO_3)_2 6H_2O + H_2O \rightarrow Zn^{2+} + 2NO_3^- + 7H_2O, \\ &Zn^{2+} + NO_3^- + 2e^- \rightarrow ZnO + NO_2^- \end{aligned} \tag{2}$$

For the growth of ZnO NRs, FTO substrates with and without seed layers were dipped in 50mM $Zn(NO_3)_2{\cdot}6H_2O$ aqueous solution. Hexamethylenetetramine (HMT) was added to solution because it is difficult to grow NR based on only $Zn(NO_3)_2{\cdot}6H_2O$. For the vertical growth, the substrates were reversely dipped and NRs were grown from top to down. In this case, the growth speed was relatively slower than that of normal growth but it was possible to make more uniform and vertical NRs. Temperature was fixed at 90 °C during the growth. Grown NRs were sintered at 500 °C for 30 min. They were characterized by their morphological and compositional properties. The scanning electron microscopy (SEM, S-4200, Hitachi) operated at 15 kV was used to characterize their nano-structure. And their formation was ascertained by measuring X-ray diffraction (XRD, X'pert PRO MRD, Philips) under optimized operating conditions of 30 mA and 40 kV. TiO_2 solution was prepared for the surface modification of ZnO NRs. 2 ml titanium isopropoxide was slowly added in 20 ml ethanol. After adding 1 ml distilled water and 1ml hydrochloric acid as a catalyst, the solution was turned to be milky under stirring and aged for 24 h. And then, ZnO NR electrodes were dipped in TiO_2 solution for 1 min. Taken electrodes out of solution were dried and sintered at 450 °C for 30 min.

Fabricated ZnO NRs were applied to dye-sensitized solar cell (DSC). DSCs were fabricated as follows. FTO/ZnO NR electrodes were soaked in a 0.2 mM N719 dye (cis-bis (isothiocyanato) bis (2, 2'-bipyridyl-4,4'-discarboxylato)-ruthenium(II)-bis-tertabutylammonium, $RuC_{58}H_{86}N_8O_8S_2$) solution for 24 h. Dye-sensitized NR electrodes were rinsed with ethanol and dried using a stream of nitrogen. The platinized counter electrode with a thickness of about 0.1 μm was sputtered at a power of 150 W. After that, both electrodes were sealed using a thermoplastic hot-melt sealant (SX 1170-25, Solaronix) with a thickness of 25 μm. The sealed DSCs were completed by injecting a redox electrolyte through a pre-drilled hole into the counter electrode. The redox electrolyte consisted of 0.5 M LiI, 0.05 M I_2 and 0.5 M 4-tertbutylpyridine in acetonitrile.

Before their characterization, the completed cells were stored in the dark under open-circuit conditions for 24 h to allow the electrolyte to penetrate into nano-rods. The photovoltaic performance was measured under 1 sun (air mass 1.5, 100 mW/cm^2) by a source meter (Model 2400, Keithley Instrument, Inc.). During their irradiance and characterization, the cells were covered with a black mask fitting the active area of the cell. The irradiated cell area was 0.20 cm^2. The I-V characteristic curve and equations 3 were used to calculate the short-circuit current (I_{SC}) and density (J_{SC}), open-circuit voltage (V_{OC}), fill factor (FF) and the overall efficiency (η). The internal impedance of DSCs was measured by electrochemical impedance spectroscopy (EIS, SP-150, Biologic SAS). The EIS spectra were measured over the frequency range from 10 mHz to 1 MHz at room temperature. The applied bias voltage and AC amplitude were set at V_{OC} of the DSC and 10 mV, respectively. The electrical impedances were characterized using the Nyquist diagram.

$$\eta = \frac{P\max}{Pin} \times 100 = \frac{FF \cdot Voc \cdot Jsc}{Pin} \times 100(\%) \tag{3}$$

3. Results and discussion

Figure 1 shows SEM images of ZnO NRs on (a) bare FTO substrate, (b) sputtered ZnO seed, (c) $Zn(CH_3COO)_2 \cdot H_2O)$ seed, and (d) $Zn(NO_3)_2 \cdot 6H_2O$ seed. ZnO NRs were not grown well. NRs were randomly grown and re-grown based on the generated rods because there is no seed as the basis of growth. Sputtered seeds were formed as rod type from the first and they were so uniform and vertical. However, they were different from rods grown by the CBD. ZnO NRs were grown independently of sputtered seed layer and they were not grown uniformly and vertically. In case of $Zn(CH_3COO)_2 \cdot H_2O)$ seed, NRs were well grown but some rods were partially changed to film by the aggregation. Aggregated rods made the surface area reduced and many pores were blocked. On the other hand, ZnO NRs were well grown on $Zn(NO_3)_2 \cdot 6H_2O$ seed. The growth was also uniform and vertical without the aggregation.

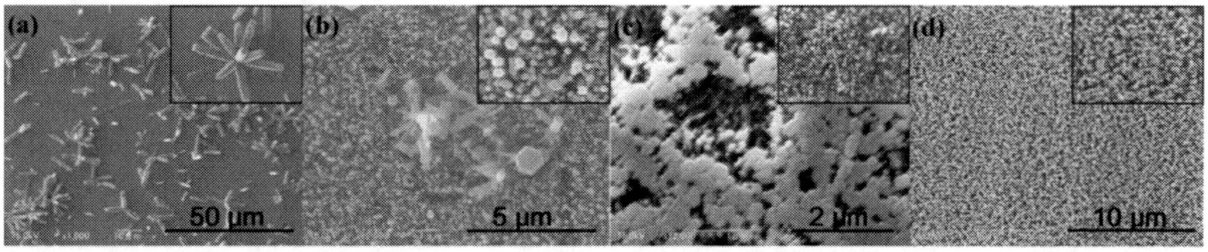

Figure 1. SEM images of ZnO nano-rods on (a) bare FTO substrate, (b) sputtered ZnO seed, (c) $Zn(CH_3COO)_2 \cdot H_2O$ seed, and (d) $Zn(NO_3)_2 \cdot 6H_2O$ seed.

These results were confirmed by XRD measurement. XRD patterns show their composition in figure 2. ZnO NRs were mainly grown to (100) and (101) by the CBD. The growth was the best with $Zn(NO_3)_2 \cdot 6H_2O$ seed like SEM results. The growth was the worst without seed and CBD growth was different from sputtered seed. Therefore, it was concluded that $Zn(NO_3)_2 \cdot 6H_2O$ was used for the seed layer.

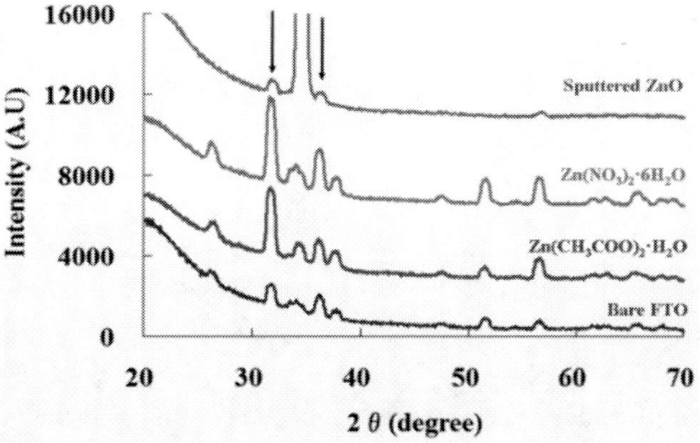

Figure 2. XRD patterns ZnO nano-rods on according to seed layers.

Journal of Physics: Conference Series **441** (2013) 012029 doi:10.1088/1742-6596/441/1/012029

Figure 3 shows (a) rod diameter and (b) normalized efficiency according to HMT ratio. HMT concentration was varied from 10 to 50 mM. $Zn(NO_3)_2{\cdot}6H_2O$ and aqueous solution provided Zn^{2+} and O^{2-} ions as the source of growth. HMT was hydrolyzed in the solution and produced OH^- during the growth. Heterogeneous nucleation of ZnO crystals was formed on the surface with Zn^{2+} and OH^-. After that, the crystals grew as NR. The increase of HMT ratio led to increase the diameter. The performance was not matched with the increase of diameter. It was increased up to 40 mM HMT but decreased since then. Therefore, next step was conducted with fixed concentration of HMT as 40 mM.

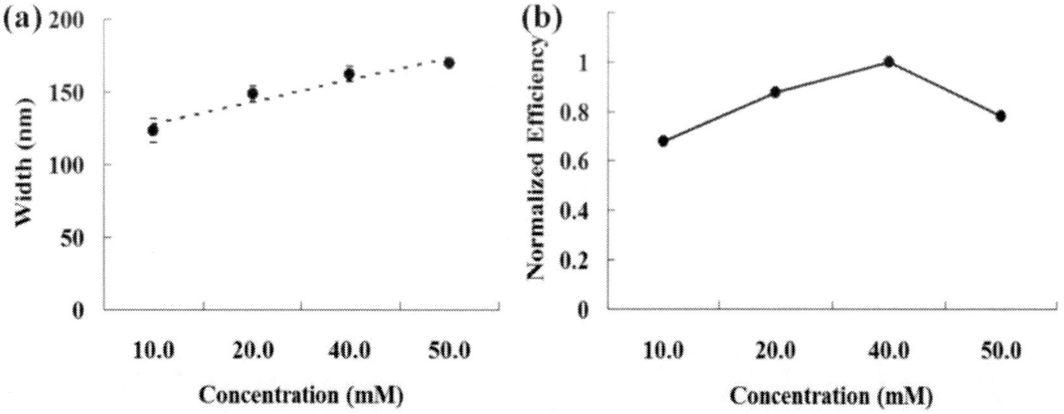

Figure 3. (a) Rod diameter and (b) normalized efficiency of DSCs according to HMT ratio.

Figure 4 shows (a) the NR length and (b) performance change according to the growing time from 1 to 8 h. NRs were reasonably grown. It was almost proportional to the growing time although they were not grown within 1 h. Longer rods led to increase of the photocurrent because more dye molecules were adsorbed on wider surface of longer rod. However, the performance was not proportional to the length of rod although the current was continuously increased. It came from the worse electron transfer. Figure 5 shows I-V characteristic curves of ZnO NR DSCs according to the growing time. As mentioned above, the current was the largest with the longest growing time but FF was decreased. It meant that longer rod had much charge recombination from the contact with redox electrolyte. Then, overall performance was deteriorated with bad electron transfer.

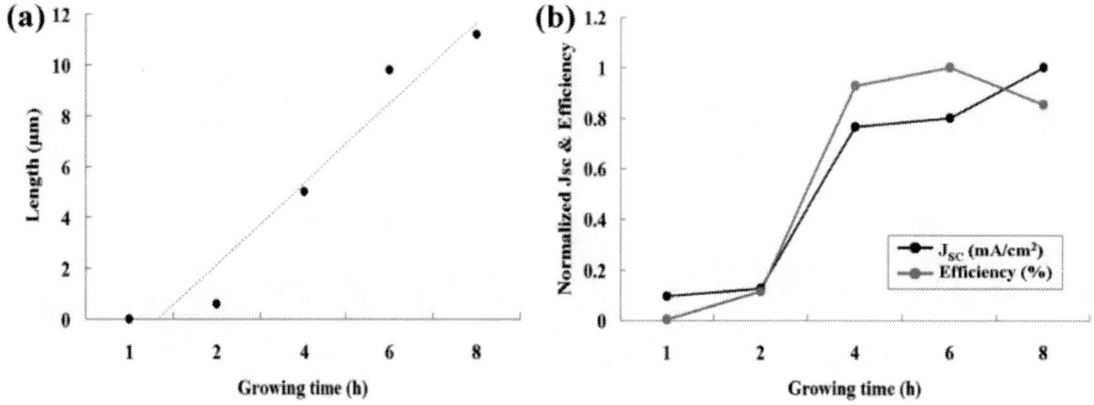

Figure 4. (a) Rod length and (b) performance change according to growing time.

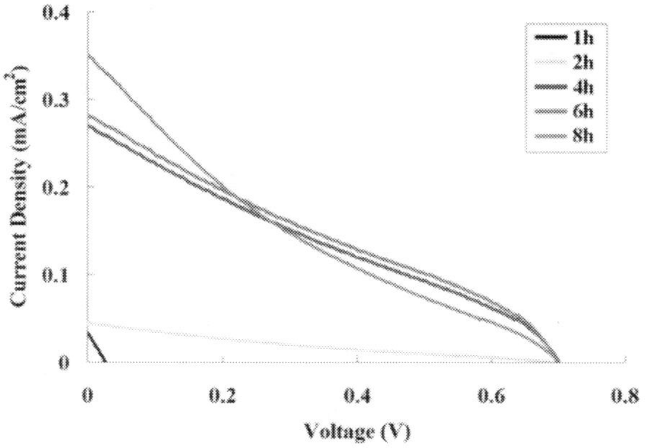

Figure 5. I-V characteristic curves of ZnO nano-rod DSCs according to growing time.

For better performance, ZnO NRs were surface-modified by TiO_2 because overall performance including current and FF was still low in spite of the optimization of growth condition. Figure 6 shows (a) I-V characteristic curves and (b) EIS diagram of ZnO NR DSCs with and without the surface modification. After the surface modification, FF and the photocurrent were much increased while V_{OC} was slightly decreased. The conduction band of TiO_2 was more positive than that of ZnO. Then, the difference between the conduction band and redox potential was reduced, hence V_{OC} was decreased. On the other hand, the surface area was increased and charge recombination was much decreased by the surface modification. More dye molecules were adsorbed and more electrons were photo-generated. Also, the dark current was reduced and it confirmed the reduction of charge recombination. Accordingly, J_{SC} was increased. Improved electron transfer made FF much enhanced from 0.28 to 0.43. In the EIS analysis, internal impedance was much deceased. Decreased impedance proved the improvement of electron transfer at the interfaces of ZnO NR/dye/electrolyte. Consequently, improved current and FF led to enhance the overall performance after the surface modification.

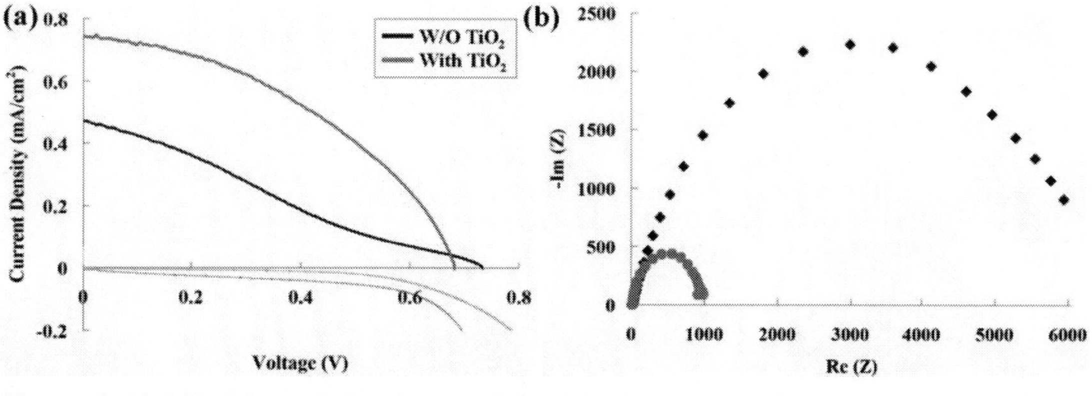

Figure 6. (a) I-V characteristic curves and (b) EIS diagram of ZnO nano-rod DSCs with and without the surface modification.

4. Conclusions

In this work, ZnO NRs were fabricated from various seed layers. $Zn(NO_3)_2 \cdot 6H_2O$ was suitable as the seed and NRs were also grown uniformly and vertically. In the growth process, HMT ratio affected the diameter of NR and the length of NR was longer with the increase of growing time. ZnO NRs fabricated under the optimized condition were applied to DSC but its performance was bad because of low FF, especially. For better performance, ZnO surface was modified by TiO_2. The surface modification improved electron transfer and reduced the charge recombination and increased the surface area. Consequently, the current and FF were much improved, resulting in the enhancement of overall performance.

5. Acknowledgements

This work was partially supported by TEMCO Memorial Foundation and New Energy and Industrial Technology Development Organization (NEDO).

6. References

[1] Hsu C C and Wu N L 2005 *J. Photoch. Photobio. A* **172** 269
[2] Marci G, Augugliaro V, Lopez-Munoz M J, Martin C, Palmisano L, Rives V, Schiavello M, Tilley R J D and Venezia A M 2001 *J. Phys. Chem. B* **105** 1033
[3] Law M, Greene L E, Radenovic A, Kuykendall T, Liphardt J and Yang P D 2006 *J. Phys. Chem. B* **110** 22652
[4] Galoppini E, Rochford J, Chen H, Saraf G, Lu Y, Hagfeldt A and Boschloo G 2006 *J. Phys. Chem. B* **110** 16159
[5] Law M, Greene L E, Johnson J C, Saykally R and Yang P D, 2005 *Nat. Mater.* **4** 455
[6] Xu C K, Shin P H, Cao L L, Wu J M and Gao D 2010 *Chem. Mater.* **22** 143
[7] Lupan O, Guerin V M, Tiginyanu I M, Ursaki V V, Chow L, Heinrich H and Pauporte T 2010 *J. Photochem. Photobiol. A* **211** 65
[8] Xu Y, Fang X and Zhang Z 2009 *Appl. Surf. Sci.* **255** 8743
[9] Masuda Y and Kato K 2009 *Cryst. Growth Des.* **9** 3083
[10] Masuda Y and Kato K 2008 *Cryst. Growth Des.* **8** 275
[11] Masuda Y and Kato K 2008 *Cryst. Growth Des.* **8** 2633
[12] Masuda Y, Kinoshita N, Sato F, Koumoto K 2006 *Cryst. Growth Des.* **6** 75
[13] Izaki M and Omi T 1997 *J. Electrochem. Soc.* **144** L3
[14] Yoshida M, Collins-Martinez V, Amezaga-Madrid P and Aguilar-Elguezabal A 2002 *Thin Solid Films* **419** 60
[15] Peulon S and Lincot D 1998 *J. Electrochem. Soc.* **145** 864
[16] Sang B, Yamada A and Konagai M 1998 *Jpn. J. Appl. Phys.* **37** L206
[17] Minegishi K, Koiwai Y, Kikuchi Y, Yano K, Kasuga M and Shimizu A 1997 *Jpn. J. Appl. Phys.* **36** L1453
[18] Akhtar M S, Khan M A, Jeon M S and Yang O B 2008 *Electrochim. Acta* **53** 7869
[19] Seo H, Son M K, Kim J K, Shin I, Prabakar K and Kim H J 2010 *Phys. Scr. T* **139** 014029
[20] Rani S, Suri P, Shishodia P K and Mehra R M 2008 *Sol. Energ. Mat. Sol. C.* **92** 1639
[21] Shinde V R, Gujar T P and Lokhande C D 2007 *Sol. Energ. Mat. Sol. C.* **91** 1055
[22] Kakiuchi K, Saito M and Fujihara S 2008 *Thin Solid Films* **516** 2026
[23] Masuda Y and Kato K 2008 *Thin Solid Films* **516** 2474

Synthesis of AlB_{12} and YB_{66} Nanoparticles by RF Thermal Plasmas

Sooseok Choi, Jiro Matsuo and Takayuki Watanabe

Dept. Environmental Chemistry and Engineering, Tokyo Institute of Technology, 4259-G1-22, Nagatsuta-cho, Midori-ku, Yokohama 226-8502, Japan

E-mail: watanabe@chemenv.titech.ac.jp

Abstract. Boron-rich compounds of AlB_{12} and YB_{66} nanoparticles were synthesized in radio frequency (RF) thermal plasmas. Yttrium tetraboride or aluminium powders with 10 μm in diameter and boron powder with 45 μm in diameter were evaporated in the high temperature region of the thermal plasma, and then metal boride nanoparticles were formed in the tail region of the plasma with rapid quenching. Boron-rich compounds were identified by X-ray diffractometry in the product. Polyhedral shaped nanoparticles about 20 nm in size were measured in Al-B system by transmission electron microscopy, while YB_{66} particles about 50 nm in size were cubic in morphology. The composition of raw powder and the input power of RF thermal plasma were controlled to enhance the content of boron-rich metal borides in as-prepared nanoparticles. Since boron has very high melting point and evaporation temperature, high boron content in the raw powder and high input plasma power were preferable to synthesize AlB_{12} and YB_{66} nanoparticles.

1. Introduction

Boron-rich metal borides composed of polyhedral boron subunits and metal atoms have excellent material properties of high melting point, hardness, corrosion resistance, thermionic electron emission, and neutron absorption ability with low density [1]. Since they have various applications including mechanical tools, chemical catalysts, electronic devices and nuclear materials, typical production methods such as chemical vapour deposition and a long-time solid reaction in high temperature have been introduced [2, 3]. Although boride nanoparticles are strongly required in many applications due to advantages of nanoparticles in optical, chemical, mechanical, electromagnetic characteristics, they are difficult to be obtained from solution which is typically used for the preparation of nanoparticles. In this work, boron-rich compounds of AlB_{12} and YB_{66} were prepared by radio frequency (RF) induction thermal plasmas. The RF thermal plasma is receiving much attention to synthesize nanoparticles, because it has sufficiently high temperature for the evaporation of any kinds of precursor and rapid quenching rate for nucleation and condensation [4]. In order to increase the content of boron-rich compounds in final products, effects of boron composition ratio in raw feeding powders and the input power of RF thermal plasma on the synthesis of AlB_{12} and YB_{66} were examined in the present work.

2. Experimental setup

Experimental setup and operating conditions are presented in figure 1 and table 1, respectively. Raw materials composed of boron (size: 45 μm, purity: 99%, Kojundo Chemical Laboratory Co. Ltd.)

powder and aluminum (size: 10 μm, purity: 99%, Kojundo Chemical Laboratory Co. Ltd.) or yttrium tetraboride (size: 10 μm, purity: 99%, Japan New Metals Co. Ltd.) powder are injected into the RF thermal plasma by a powder feeder. The composition ratio of metal to boron was 1:2, 1:12, and 1:15 for Al-B system, while it was controlled at 1:12, 1:40, 1:66, and 1:100 for Y-B system. The feed rate of raw material was fixed at 100 mg/min. The RF thermal plasma was operated at atmospheric pressure and the input power was controlled from 24 kW to 33 kW. Argon was used for the carrier gas of the raw powder, and argon and helium were used for plasma supporting gases. Evaporated boron and metal were nucleated and condensed in the RF thermal plasma producing nanoparticles that were found on a water cooling inner chamber and a filter. Collected particles were analyzed by transmission electron microscopy (TEM, JEM-2010F, JEOL Ltd.) and X-ray diffractometry (XRD, Miniflex, Rigaku Corp.).

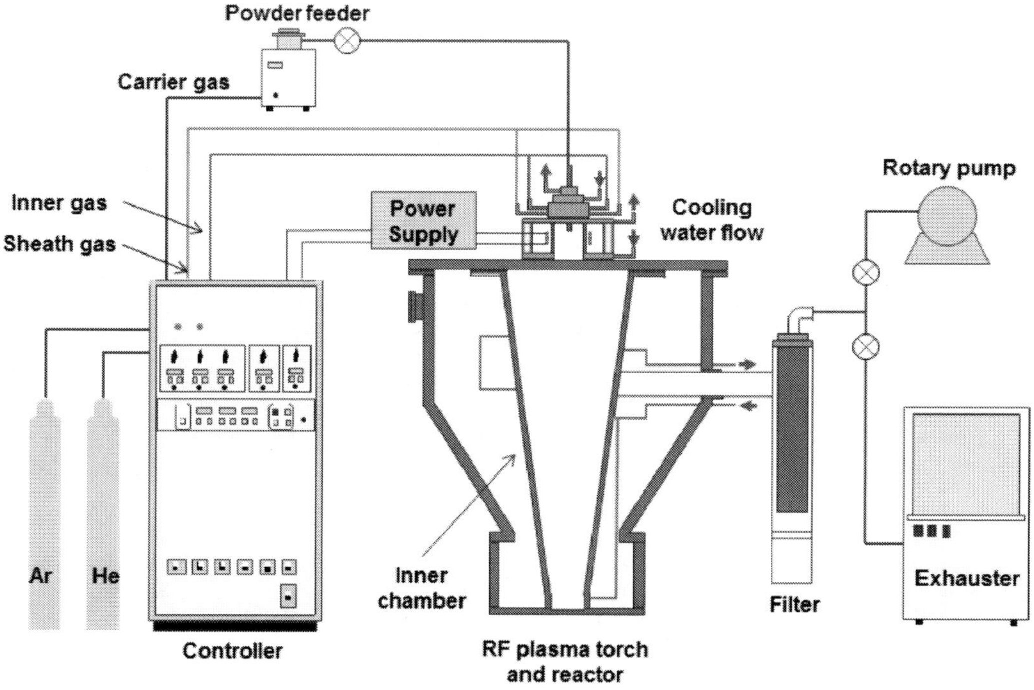

Figure 1. Experimental setup for the synthesis of boron-rich metal boride nanoparticles.

Table 1. Experimental conditions.

Raw powder size	Al and YB_4 = 10 μm, B = 45 μm
Metal to boron ratio (***R***)	Al:B = 1:2, 1:12, 1:15 Y:B = 1:12, 1:40, 1:66, 1:100
Powder feed rate	100 mg/min
Operating pressure	101.3 kPa
Radio frequency	4 MHz
Input power (***P***)	24, 27, 30, 33 kW
Sheath Gas	Ar 60 L/min and He 5 L/min
Inner Gas	He 5 L/min
Powder Carrier Gas	Ar 3 L/min

3. Results and discussion

TEM images of synthesized nanoparticles in Al-B and Y-B systems are presented in figures 2 and 3, respectively. These images were obtained at the highest boron composition ratio in the raw powder among used conditions in the present work at the fixed input plasma power of 30 kW. In a RF thermal plasma process for the synthesis of nanoparticles, spherical shaped products are typically produced. In the present work, however, polyhedral nanoparticles which were about 20 nm in average diameter were synthesized in the case of Al-B system as seen in figure 2. Such polyhedral morphology seems to be caused by the structure of icosahedral B_{12} and B_{20} units which compose AlB_{12} [5]. In figure 3, rectangular shaped nanoparticles about 50 nm in the length of each side were found in the product when YB_4 and B powders were used as raw materials. The cubic shape is unique morphology for YB_{66} nanoparticles [6]. Therefore, successful synthesis of boron-rich metal borides nanoparticles of AlB_{12} and YB_{66} were confirmed from the observation of TEM images.

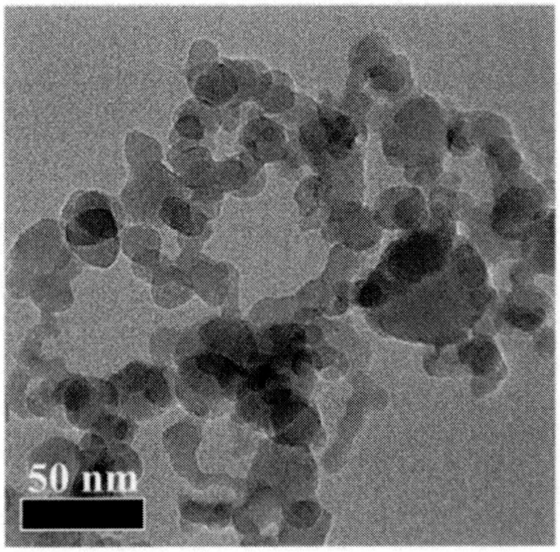

Figure 2. TEM image of nanoparticles prepared by the RF thermal plasma in Al-B system at operating conditions of 1:15 for Al to B ratio in the raw powder and 30 kW for the input plasma power.

Figure 3. TEM image of nanoparticles prepared by the RF thermal plasma in Y-B system at operating conditions of 1:100 for Y to B ratio in the raw powder and 30 kW for the input plasma power.

In order to increase the quantity of boron-rich compound in the product, the composition ratio of boron in the raw powder and the input plasma power were changed. Figures 4 and 5 present XRD data for the final product from the RF thermal plasma treatment of Al and B raw powders. First, the intensity of AlB_{12} is increased with increasing boron composition ratio in the raw powder as seen in figure 4. The nucleation temperature of boron is higher than that of aluminum according to the synthesis mechanism of metal borides based on the homogeneous nucleation model [7, 8]. As a result, boride nanoparticles are produced from the condensation of boron and aluminum monomers on boron nuclei. Therefore, enough quantity of boron nuclei and monomers should be secured by increasing the composition ratio of boron in the raw powder for the synthesis of boron-rich metal boride nanoparticles. In addition, the peak for AlB_{10} is also produced together with AlB_{12} as shown in figure 4, because both AlB_{12} and AlB_{10} nanoparticles are composed of icosahedral B_{12} units. Therefore, boron clustering from boron nuclei and monomers takes place at first, followed by chemical reactions between boron clusters and Al monomers in the high temperature environment of RF thermal plasma forming AlB_{12} and AlB_{10} nanoparticles.

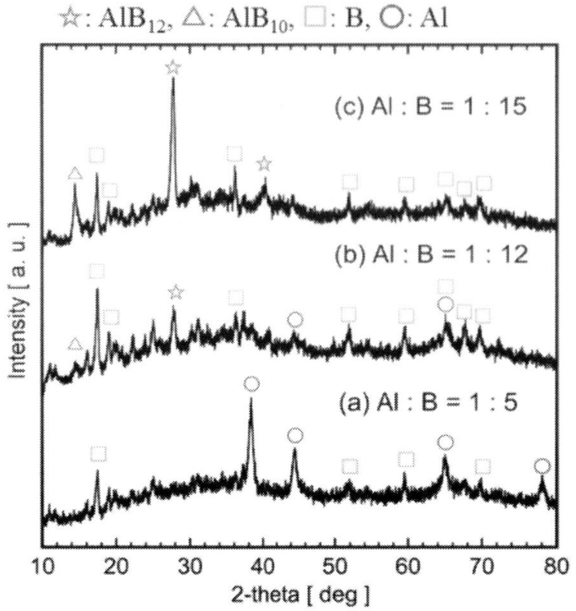

Figure 4. XRD data for products in Al-B system in different composition ratio of aluminium to boron in raw material at the fixed input power of 30 kW: (a) 1:5, (b) 1:12, and (c) 1:15.

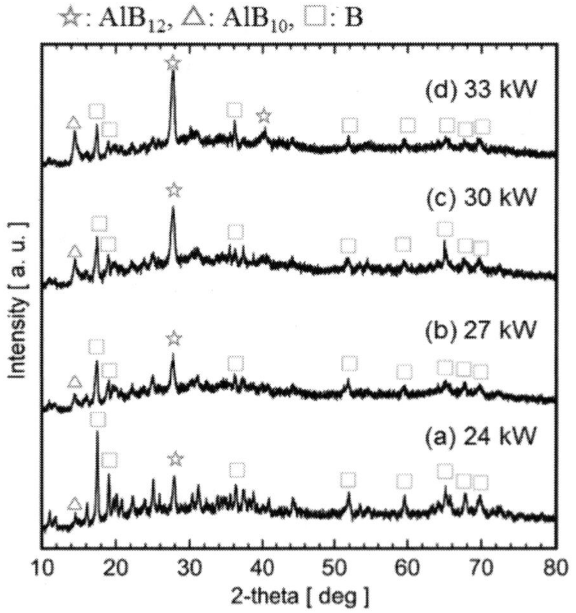

Figure 5. XRD data for products in Al-B system in different input plasma power at the fixed raw material composition of Al:B = 1:15 : (a) 24 kW, (b) 27 kW, (c) 30 kW, and (d) 33 kW.

In the synthesis of boride nanoparticles in the RF thermal plasma, boron vapour should be reach at the saturation vapour pressure for nucleation. Since boron has very high melting point and evaporation temperature of 2 349 K and 4 200 K, respectively, high temperature and enthalpy of the RF thermal plasma are required to produce large amount of boron nuclei and monomers. Therefore, the content of AlB_{12} in the product is increased with increasing the input plasma power as shown in figure 5.

Experimental results on the synthesis of YB_{66} from evaporated YB_4 and B powders are presented in figures 6 and 7. The composition ratio of boron in the raw powder was controlled at the fixed input plasma power of 30 kW. As shown in figure 6, peak intensity for YB_{66} is increased with increasing the

Journal of Physics: Conference Series **441** (2013) 012030 doi:10.1088/1742-6596/441/1/012030

composition ratio of boron in the raw material. Although YB_{66}-related peaks at 22° and 25° for 2-theta values are appeared in the lowest boron contents in the raw material as shown in figure 6 (a), these two peaks also correspond to YB_4. In addition, XRD pattern only for YB_{66} was very weak to be detected in figure 6 (a). Therefore, high boron content in the raw material is essentially required for the synthesis of YB_{66} nanoparticles in the RF thermal plasma. However, the peak intensity of YB_{66} is still weak compared with that of YB_4 even in the highest boron composition ratio in the raw material as depicted in figure 6 (d). In order to synthesize YB_{66} nanoparticles in the RF thermal plasma, large amount of boron nuclei and monomers are required because YB_{66} is composed of 13-icosahedron unit of $(B_{12})_{12}B_{12}$ which is called supericosahedron [9].

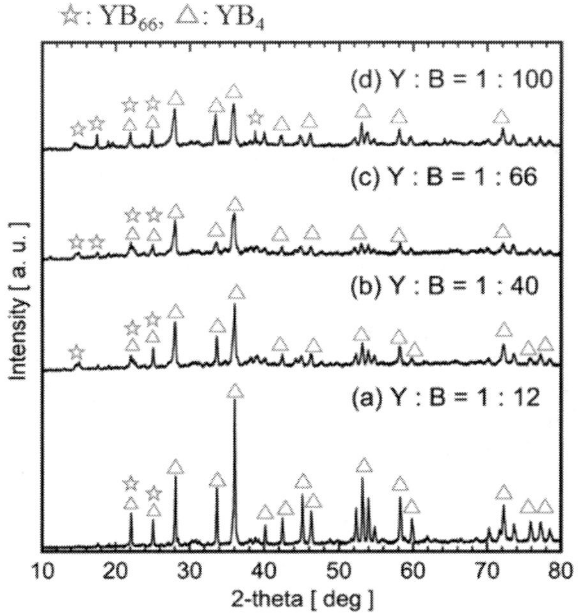

Figure 6. XRD data for products in Y-B system in different composition ratio of yttrium to boron in raw material at the fixed input plasma power of 30 kW: (a) 1:12, (b) 1:40, (c) 1:66, and (d) 1:100.

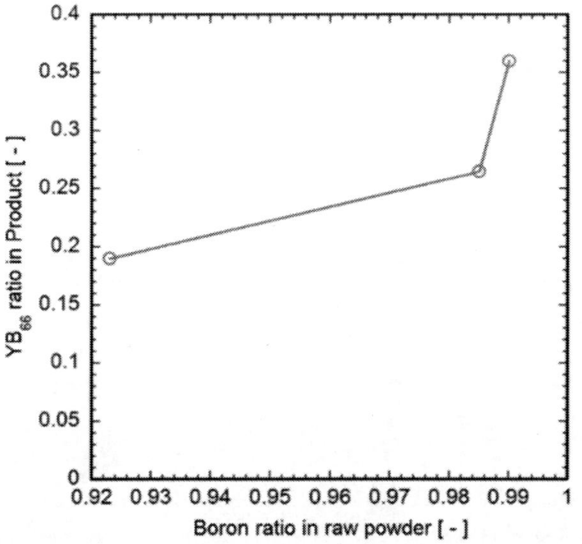

Figure 7. YB_{66} particle ratio in synthesized nanoparticles evaluated by TEM images according to boron composition ratio in raw material.

Since YB_{66} has unique cubic morphology, TEM images for nanoparticles produced in Y-B system were analyzed to evaluate the ratio of YB_{66} nanoparticles in the final product. From the analysis of 200 particles for each condition, the ratio of rectangular nanoparticle in TEM images is presented in figure 7. The number of YB_{66} nanoparticle is increased with increasing the composition ratio of boron in the raw material, and this result is in a good agreement with XRD results in figure 6.

4. Conclusion

Boron-rich metal boride nanoparticles of AlB_{12} and YB_{66} were successfully synthesized from the vapour phase of raw powders in the RF thermal plasma. Boron nuclei and monomers form boron clusters, and they react with metal monomers in the high temperature environment of the RF thermal plasma producing boron-rich metal boride nanoparticles. Since AlB_{12} is composed of icosahedral B_{12} and B_{20} units, polyhedral nanoparticles is observed in the TEM image of final products synthesized from Al and B raw powders. Whereas cubic YB_{66} nanoparticles is found when YB_4 and B powders is used as the raw material. Because boron clusters are required to form boron-rich compound, the content of boride nanoparticles in final products is increased with increasing the composition ratio of boron in the raw material. In addition, boron-rich compound is more easily synthesized in the condition of high input plasma power due to high melting point and evaporation temperature of boron.

References

[1] Albert B and Hillebrecht H 2009 *Angew. Chem.Int. Ed.* **48** 8640
[2] Wang G, Brewer J R, Chan J Y, Diercks D R and Cheung C L 2009 *J Phys. Chem. C* **113** 10446
[3] Liu Y, Lu W J, Qin J N and Zhang D 2007 *J Alloys Compd.* **431** 337
[4] Kruis F E, Fissan H and Peled A 1998 *J Aerosol Sci.* **29** 511
[5] Higashi I 2000 *J Solid State Chem.* **154** 168
[6] Huang J Y, ishigaki T, Tanaka T and Horiuchi S 1998 *J Mater. Sci.* **33** 4141
[7] Girshick S L, Chiu C P and McMurry P H 1990 *Aerosol Sci. Technol.* **13** 456
[8] Cheng Y, Shigeta M, Choi S and Watanabe T 2012 *Chem. Eng. J* **183** 483
[9] Richards S M and Kasper J S 1969 *Acta Cryst.* **B25** 237

Synthesis of Niobium Boride Nanoparticle by RF Thermal Plasma

Yingying Cheng, Sooseok Choi, and Takayuki Watanabe

Dept. Environmental Chemistry and Engineering, Tokyo Institute of Technology, Yokohama, Japan

E-mail: watanabe@chemenv.titech.ac.jp

Abstract. Experimental study has been conducted to investigate the formation mechanism of transition metal boride nanoparticles in radio frequency thermal plasmas. The effect of nucleation temperature on the synthesis of niobium boride nanoparticles was investigated. The operating condition of powder feed rate plays an important role in the synthesis of boride nanoparticles. The mass fraction of niobium boride in the product was decreased with increasing powder feed rate. The main reason is that the evaporation time of niobium is long and the existence area of evaporated niobium is extended along with the thermal plasma flame at high powder feed rate. The diameter of the product increases along with the increase of powder feed rate due to increased monomer density in the plasma. In addition, high melting and boiling temperatures of niobium leads to high mass fraction over 96% of niobium boride in the collecting filter.

1. Introduction

Radio frequency (RF) thermal plasmas has many unique advantages including high enthalpy, high chemical reactivity, large plasma volume, long residence time, and selective oxidation or reduction atmosphere according to the required chemical reactions. The high temperature of thermal plasma enables the evaporation of a large amount of raw materials even with high melting and boiling temperatures [1-3]. The formation of nanoparticles in supersaturated state by homogeneous nucleation and heterogeneous condensation can be accomplished due to the rapid quenching rate in the tail flame [4-7]. Furthermore, it is available to synthesize nanoparticles with high purity by RF thermal plasma, because the thermal plasma is generated in the torch without an internal electrode.

Niobium borides are recognized as potential candidates for the application of high-temperature structural materials due to their excellent properties. Attractive properties of niobium borides are high melting temperature, high mechanical strength, high thermal and electrical conductivity, and excellent chemical stability [8]. The RF thermal plasma which can provide very high temperature without oxygen is a promising alternative to conventional methods in the synthesis of niobium boride nanoparticles.

In Nb-B system, the niobium with low saturation vapor pressure has high nucleation temperature compared with that of boron with relatively high saturation vapor pressure. As a result, the formation mechanism of niobium boride is different from that in Ti-B system where boron has lower saturation vapor pressure than titanium. In Ti-B system, boron nucleates first at the upstream region of thermal plasma flow, followed by the condensation of titanium monomers on boron nuclei. Since, boron nuclei on the liquid state have relatively wide temperature region, titanium boride are easy to fabricate. On the other hand, niobium vapor with low saturation vapor pressure reach supersaturation state first and starts to nucleate in Nb-B system. The nucleation temperature is a critical factor for the synthesis of boride nanoparticles in the RF thermal plasma [10].

The purpose of this study is to examine the effect of nucleation temperature on the synthesis of niobium boride nanoparticles in the RF thermal plasma. The experiments were carried out with controlling the powder feed rate, because it changes nucleation temperatures of niobium and boron which play an important role in the formation of niobium boride nanoparticles in the RF thermal plasma. The nanoparticle products collected from the inner wall of the reaction chamber and the collection filter were compared in terms of phase composition and crystalline size based on X-ray diffractometry (XRD).

2. Experimental Procedure

The schematic apparatus of RF thermal plasma system for the production of boride nanoparticles is shown in figure 1. The system consists of a plasma torch, a reaction chamber, a particle collection filter, and a power supply. The plasma torch is composed of a water-cooled quartz tube and a water-cooled induction coil (3 turns), coupling its electromagnetic energy to the plasma at a frequency of 4 MHz. The feeding powders were niobium (particle size: 20 μm, purity: 98%, Kojundo Chemical Laboratory Co. Ltd., Japan) and crystalline boron (particle size: 45 μm, purity: 99%, Kojundo Chemical Laboratory Co. Ltd., Japan). Niobium and boron particles were mixed at the fixed initial molar composition before they were injected into the RF thermal plasma through the powder feeder.

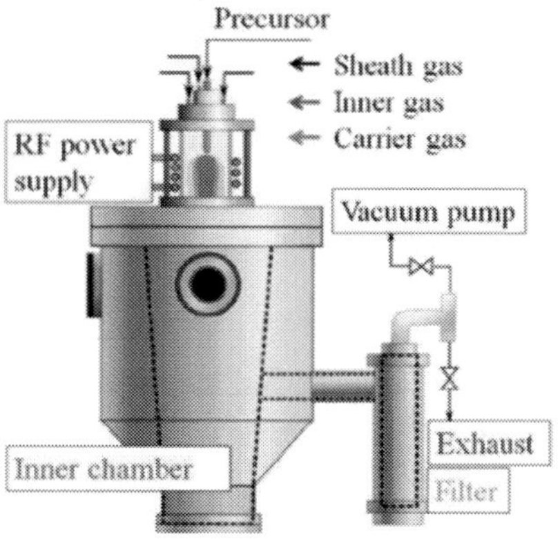

Figure 1. Experimental set-up of RF thermal plasma for the synthesis of metal boride nanoparticles.

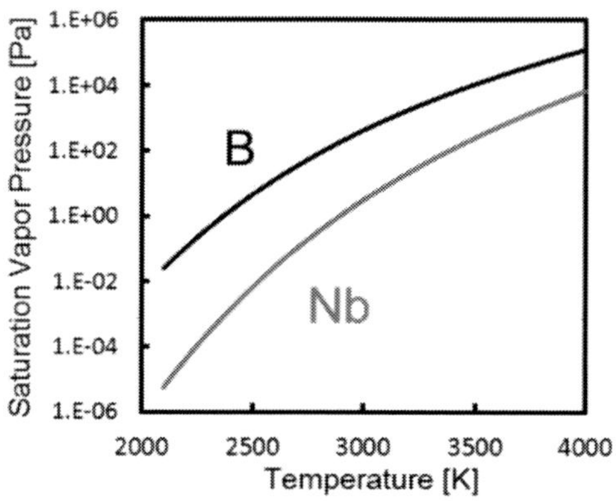

Figure 2. Relationship between saturation vapour pressures and temperature in Nb-B system.

After the precursors were introduced into the plasma region with argon carrier gas, they were instantaneously evaporated due to the very high enthalpy of the RF thermal plasma. Therefore, niobium boride nanoparticles are synthesized from the gas phase.

The operating conditions are summarized in Table 1. Argon was introduced as the carrier gas of 3 L/min, the inner gas of 5 L/min, and the sheath gas of 60 L/min. Helium of 5 L/min was also used as the sheath gas. The sheath gases were injected from the outer slots of the plasma torch located between the injection tube and the quartz tube to protect the inner surface of the quartz tube and to stabilize the plasma. The powder feed rate was 0.2, 0.5, and 1.0 g/min and the molar content of boron in the feeding powder was controlled from 33.3 to 80.0 at.%. The RF plasma was operated at the fixed input power of 30.4 kW under atmospheric pressure.

The synthesized nanoparticles were characterized for phase identification by XRD (MXP3TA, Mac Science). The crystalline size was calculated from the full widths at the half maximum (FWHM) of the most intensive diffractions according to the Scherrer's equation. Quantitative phase analysis by using

XRD data was carried out based on the adiabatic method [9]. This method evaluates the mass fraction of each phase from the relative intensity of the diffraction peak of product. The mass fraction of X phase can be calculated by following equation (1)

$$W_X = \frac{I_X}{K_A^X \sum_{X=A}^{N} \frac{I_X}{K_A^X}} = \frac{I_X}{K_{Ti}^X \cdot \left(\frac{I_{Nb}}{K_{Nb}^{Nb}} + \frac{I_{NbB}}{K_{Nb}^{NbB}} + \frac{I_{NbB_2}}{K_{Nb}^{NbB_2}}\right)} \quad (1)$$

where, X (= Nb, NbB, NbB_2) denotes each phase in the product. I_X presents the intensity of X phase in the product from the XRD spectrum. K_A^X is the ratio of the reference intensity ratio (RIR) values of X phase to that of the reference phase A, *i.e.*, $K_A^X = RIR_X/RIR_A$. In the experimental measurement of Nb-B system, three peaks for Nb, NbB and NbB_2 were identified in XRD spectrum, and Nb was selected as the reference materials. Based on the powder diffraction file cards, RIR values for Nb, NbB and NbB_2 are determined to be 19.34, 5.43, and 8.26, respectively.

Table 1 Operating condition for the synthesis of niobium boride nanoparticles

Process parameters	Value
Sheath gas and flow rate	Ar-He (60:5) 65 L/min
Inner gas and flow rate	Ar 5 L/min
Carrier gas and flow rate	Ar 3 L/min
Plasma power	30.4 kW
Reactor pressure	101.3 kPa
Frequency	4 MHz
Powder Feed rate	0.2-1.0 g/min
Boron molar content in feeding powder	33.3-80.0 at.%

3. Nucleation Temperature and Formation Mechanism

The nucleation temperatures of niobium and boron play an important role in the controlled synthesis of niobium boride nanoparticles in the RF thermal plasma. Materials with high saturation vapour pressure leads to low nucleation temperature, while lower saturation vapour pressure results in relatively higher nucleation temperature. Figure 2 shows the saturation vapour pressure according to the temperature in Nb-B system. The nucleation temperatures of niobium and boron at the critical saturation ratio are presented in figure 3. The nucleation temperature was calculated by equation (2) for the nucleation rate J. The particle formation was observed experimentally at $J = 1.0\ cm^{-3}\ s^{-1}$ [11].

$$J = \frac{\beta_{11} N_S^2 S}{12}\sqrt{\frac{\Theta}{2\pi}}\exp\left[\Theta - \frac{4\Theta^3}{27(\ln S)^2}\right] \quad (2).$$

Here, N_S is the equilibrium saturation monomer concentration at temperature T, and it is estimated by the saturation vapour pressure. β_{11}, S, and Θ are the collision frequency function between monomers, the saturation ratio, and the dimensionless surface tension, respectively.

The formation mechanism of boride nanoparticles is based on the nucleation temperature of the constituent components. After the precursors are injected into the plasma, they are instantaneously evaporated due to the high temperature. The vapours are transported with the plasma flow to the reaction chamber where the plasma temperature decreases rapidly. The saturation vapour pressure of evaporated precursors decreases drastically along with the temperature decrease and will fall below their partial pressures in plasma. At this moment, the vapour of raw materials reaches their supersaturation state. The supersaturation of vapours leads to the production of nuclei by homogeneous nucleation. In Nb-B system, niobium vapour is supersaturated first due to its lower saturation vapour pressure than that of boron. Simultaneously, the nucleation of niobium takes place and niobium nuclei are generated on molten state. The partial pressure of niobium vapour decreases rapidly, which is attributed to the vapour consumption by the nanoparticles growth with homogeneous nucleation and heterogeneous condensation. Then the co-condensation of boron and niobium occurs dominantly on the surface of the niobium nuclei. The partial pressure of boron vapour decreases by heterogeneous condensation on niobium nanoparticles. This combined mechanism makes nanoparticles to grow. In addition, the nanoparticles grow further by coagulation among themselves.

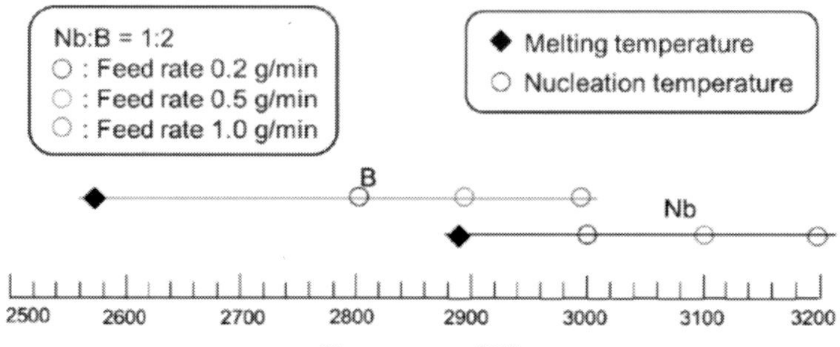

Figure 3. Nucleation temperature at the critical saturation ratio for constituent components of boride.

4. Results and Discussion

Figure 4 shows the XRD spectra of product at the fixed initial composition with different powder feed rates. From the XRD measurement, Nb and NbB_2 were identified. The dominant product was unreacted Nb, while the intensity of NbB_2 peaks was very low. This XRD result indicates that the synthesis of niobium boride nanoparticle is difficult compared with titanium borides [9]. The main reason of the little amount of prepared niobium boride is that the small temperature gap between the nucleation temperature and melting temperature of niobium. The nuclei on the liquid state are important in the formation of metal boride nanoparticles, because chemical reactions between Nb nuclei and B monomers are impossible after the solidification of Nb. Seen from figure 3, niobium nuclei in liquid state have relatively narrow region. The nucleation of niobium takes place only around the niobium nucleation temperature in a short time. Simultaneously, niobium and boron vapours co-condensate on the surface of niobium nuclei. During the high temperature region between the nucleation temperature and melting temperature of niobium, niobium and boron vapours can be mixed well on the liquid state. On the other hand, niobium and boron vapours are not mixed well in the relatively low temperature gap between the melting temperatures of niobium and boron. The relatively narrow temperature gap between the nucleation temperature and melting temperature of niobium means less boron and niobium vapours can be mixed well.

Figure 5 indicates the effect of powder feed rate on the phase composition of product. The mass fraction of niobium boride in the product decreases with increasing powder feed rate from the experimental results. Raw materials are fully evaporated in thermal plasma. With the low powder feed rate, both of boron and niobium are immediately evaporated when they are injected into the thermal plasma. Then the evaporated niobium is nucleated in a limited high temperature area at the upstream of plasma flow, where much niobium and boron co-exist and can react with each other. The amount of raw materials increases at high powder feed rate, though the heat from the plasma keeps constant at the fixed input power of 30.4 kW. The heat transferring from the plasma to each particle requires relatively long time in the case of high powder feed rate rather than the case of low one. In addition, niobium is difficult to be evaporated due to the relatively high melting temperature, evaporation temperature, and latent heat. Therefore, the existence area of evaporated niobium is extended along with plasma flame. The boridation occurs easily in the upstream region of plasma flow due to high temperature and abundant boron vapour, while it is difficult in the downstream region because of relatively low temperature for the boridation and diffuse of boron vapour which is produced in the upstream region. Therefore, less niobium nuclei can react with boron, and the molar ratio of niobium boride nanoparticles in the product decreases at higher powder feed rate.

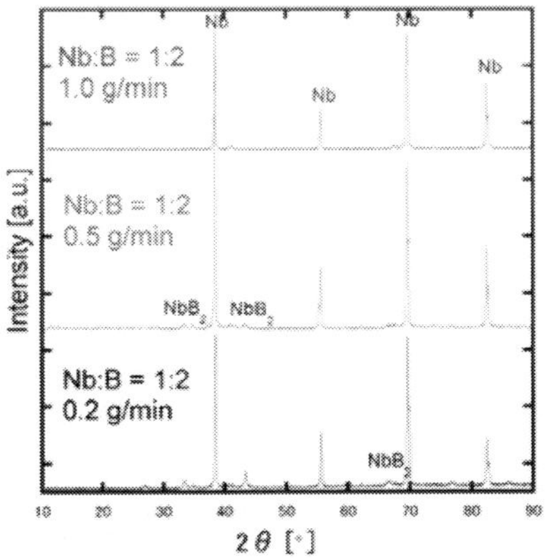

Figure 4. XRD spectra of product collected on the reactor wall at fixed initial composition with different powder feed rates.

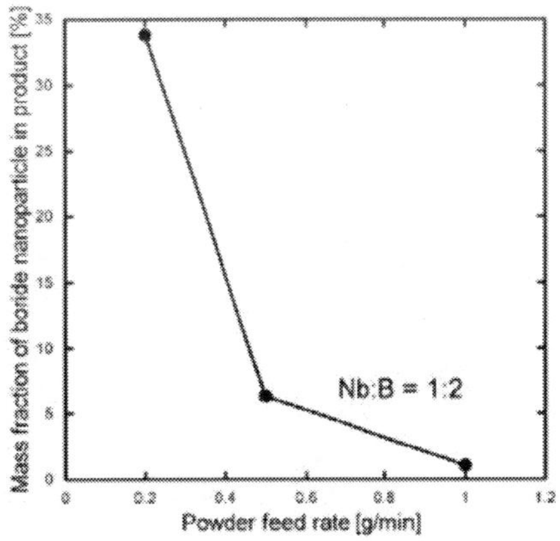

Figure 5. Effect of powder feed rate on the phase composition of product at Nb:B = 1:2.

Effect of powder feed rate on the crystalline diameter of as-prepared niobium boride nanoparticles is shown in figure 6. The crystalline size of the prepared nanoparticles increases along with powder feed rate. In high powder feed rate, the enhanced monomer density in the plasma leads to the high growth rate of niobium boride nanoparticles. In addition, the growth process including the nucleation and co-condensation processes occurs from the nucleation temperature of niobium to the melting temperature of boron. Based on figure 3, the growth time increases with powder feed rate. Furthermore, the nanoparticles are formed at earlier stage at the higher feed rate with relatively large amount of vapour compared with the case of low feed rate. For this reason the nucleation positions shift to the upstream region of plasma. Because the temperature in upstream is high, the nucleation rate of niobium is reduced. As a result, a small number of stable niobium nuclei which share a large amount of vapour are generated. Consequently, large size niobium boride nanoparticles are synthesized in the high feeding rate condition.

Figure 7 demonstrates the XRD spectra of products from the filter and the inner wall of the reaction chamber at the fixed niobium and boron composition of 1 to 3 with powder feed rate of 0.2 g/min. The dominant product from the filter is niobium boride, while the main product from the inner wall is unreacted niobium. Table 2 display the mass fraction of niobium boride in the product from the filter and the inner wall. The filter case has much higher mass fraction of niobium boride in the product than the inner wall case. In the reaction chamber, the high temperature region for the existence of niobium nuclei is limited in the centre region of the plasma flame due to the high melting temperature (2890 K) and boiling temperature (5120 K) of niobium. For this reason, the formation of niobium borides is difficult near the inner wall of the chamber because of the lower temperature. On the other hand, synthesized niobium borides in the centre region are transported to the filter along with plasma flow. Moreover, relatively low niobium boride density of 6.97 g/cm^3 compared with niobium density of 8.57 g/cm^3 promotes enhanced niobium boride fraction in the filter.

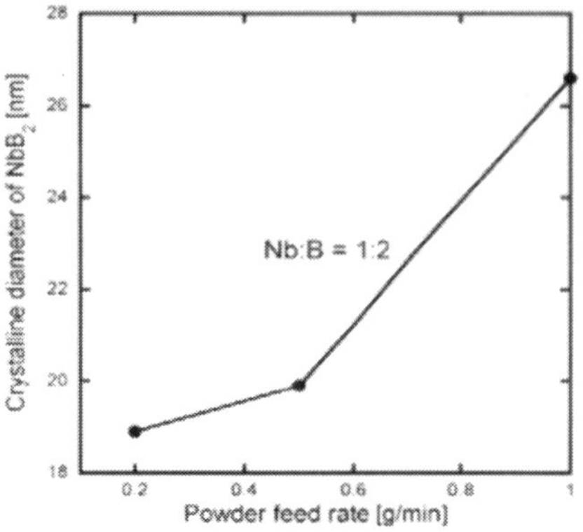

Figure 6. Effect of powder feed rate on the crystalline diameter of product at Nb:B = 1:2.

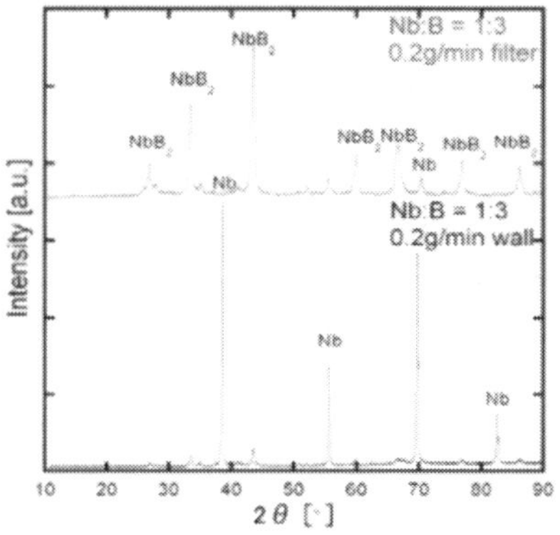

Figure 7. XRD spectrums of product from the filter and inner wall of chamber at the Nb:B = 1:3 with powder feed rate of 0.2 g/min.

Table 2 Mass fraction of niobium boride in the product from the filter and the inner wall of the reactor at the powder feed rate of 0.2 g/min.

Initial Composition	Filter (%)	Wall (%)
Nb:B = 1:2	96.66	33.81
Nb:B = 1:3	96.82	19.34
Nb:B = 1:4	96.75	16.88

5. Conclusions

Experimental study were carried out to prepare niobium boride nanoparticles and to investigate the effect of powder feed rate on the phase composition and crystalline size of product in RF thermal plasma. In this work, niobium boride nanoparticles were successfully synthesized in RF thermal plasma. The nucleation temperature plays an important role in the controlled synthesis of niobium boride nanoparticles in RF thermal plasma. The little production of niobium boride nanoparticles resulted from the relatively narrow temperature gap between the nucleation and melting temperatures of niobium. The mass fraction of niobium boride nanoparticles in the product was decreased with increasing the powder feed rate due to the extended evaporation region of raw materials. The crystalline size of niobium boride nanoparticles increased with the increase of powder feed rate. High mass fraction of the niobium boride nanoparticle in the product was found from the collecting filter due to high melting and boiling temperatures of niobium.

References

[1] Watanabe T, Nezu A, Abe Y, Ishii Y and Adachi K 2003 *Thin Solid Films* **435** 27
[2] Watanabe T and Okumiya H 2004 *Sci. Technol. Adv. Mater.* **5** 639
[3] Szepvolgyi J, Markovic Z, Scheier P and Feil S 2006 *Plasma Chem. Plasma Process.* **26** 597
[4] Shigeta M and Watanabe T 2005 *J. Mater. Res.* **20** 2801
[5] Desilets M, Bilodeau JF and Proulx P 1997 *J. Phys. D: Appl. Phys.* **30** 1951
[6] Gonzalez NYM, Morsli ME and Proulx P 2008 *J. Therm. Spray Technol.* **17** 533
[7] Shigeta M and Watanabe T 2010 *J. Appl. Phys.* **108** 043306
[8] Yeh CL, and Chen WH 2006 *J. Alloy Comp.* **422** 78
[9] Cheng YY and Watanabe T 2011 *J. Chem. Eng. Jpn.* **44** 583
[10] Shigeta M and Watanabe T 2009 *J. Therm. Spray Technol.* **18** 1022
[11] Girshick SL, Chiu CP and McMurry PH 1990 *Aerosol Sci. Technol.* **13** 465

Evaluation of Highly Functional TiO_2 Synthesized by Solution Precursor Spraying in a DC-RF Hybrid Plasma Flow System

J Jang[1], H Takana[2], Y Ando[3], O P Solonenko[4] and H Nishiyama[2]
[1] Department of Mechanical Systems and Design, Graduate School of Engineering, Tohoku University, 2-1-1 Katahira, Aoba-ku, Sendai 980-8577, Japan
[2] Institute of Fluid Science, Tohoku University, 2-1-1 Katahira, Aoba-ku, Sendai 980-8577, Japan
[3] Faculty of Engineering, Ashikaga Institute of Technology, 268-1 Omae-cho, Ashikaga 326-8558, Japan
[4] Khristianovich Institute of Theoretical and Applied Mechanics, SB RAS, Novosibirsk, Russia

E-mail: jang@paris.ifs.tohoku.ac.jp

Abstract. Carbon doped TiO_2 synthesized in a DC-RF hybrid plasma flow system by the solution precursor spraying method was evaluated. The effects of mass fraction of TTB in solution precursor on contents and photocatalytic activity of carbon doped TiO_2 were investigated. Highly functional carbon doped TiO_2 at 11 wt% TTB had higher UV-visible absorption characteristics, which showed higher degradation rate of methylene blue solution under visible light irradiation.

1. Introduction

Titanium dioxide (TiO_2) is an attractive material for environmental purification commercially used because of its environmentally friendly photocatalytic activity. However, the main drawback of typical undoped TiO_2 is low photocatalytic activity activated only under UV light irradiation. Many works have been focused on the preparation of more functional TiO_2 doping with metal and non-metal atoms in order to improve photocatalytic activity in UV and visible light irradiation [1-3]. Among distinct papers published within a decade, D. Chen et al prepared highly functional TiO_2 doping with carbon element using a sol-gel process [1]. Its higher photocatalytic activity was shown by degradation of methylene blue solution only under visible light irradiation, compared with undoped and nitrogen doped TiO_2 [1,4]. However, this sol-gel process requires long processing time and needs further toxic chemicals as catalysts. To overcome these difficulties found in the sol-gel process, a solution or suspension plasma spray method was proposed [5-7].

In previous studies, a DC-RF hybrid plasma flow system was optimized for high performance of the alumina spheroidization process even under a small electric power by thermofluid flow controls to increase plasma enthalpy and heat exchange rate [8-11]. Among these studies, water droplet injection into downstream argon plasma flow in the DC-RF hybrid plasma flow system increased effectively plasma enthalpy. Injected droplets are rapidly vaporized and the vapor is transferred along the chamber wall to upstream argon plasma flow by characteristic backflows, which are occurred from the chamber wall to the vicinity of RF coils. The alumina spheroidization rate is increased without increase in a electric power by higher plasma enthalpy induced by mixing with produced hydrogen and argon plasma flow. Hence, a DC-RF hybrid plasma flow system under a small electric power is feasible to synthesize nano-particles with coupling solution precursor spraying and characteristic backflow transportation.

In this study, highly functional TiO_2 can be synthesized in a DC-RF hybrid plasma flow system by coupling a solution precursor spraying and characteristic backflow transportation. After vaporization of injected droplets of solution precursor in downstream plasma flow, small particles are transferred to upstream plasma flow by the characteristic backflows near the chamber wall. Then, highly functional carbon doped TiO_2 is produced by prolonging the interaction time with plasma flow. Mixing effects of carbon and TiO_2 are discussed by using the X-ray diffraction patterns, Raman spectra, and UV-visible absorption spectra. Furthermore, an effect of carbon doped TiO_2 on the methylene blue degradation under visible light irradiation is investigated.

2. Experimental setup

2.1. DC-RF hybrid plasma flow system

Figure 1 shows a schematic of DC-RF hybrid plasma flow system with a two-phase atomizer in order to inject droplets of solution precursor into downstream plasma flow. The operating pressure of 33 kPa was kept during the operation. DC power of 1.1 kW and RF power of 6.6 kW (4 MHz) were supplied at constant. The working gas was argon. The flow rates of DC plasma forming gas and swirling sheath gas were 5 and 20 S*l*/min, respectively. A gas-liquid atomizer (YS-03, Yaezaki, Japan) was installed at $r = 95$ mm and $z = 195$ mm. Atomizing gas was kept at 6 S*l*/min to achieve sufficient atomization. The average gas temperature was 1038.6 ± 4.8 °C at $r = 0$ mm and $z = 195$ mm without injections of solution precursor and atomizing gas. Further details of this system are obtained in Refs. 8-11

2.2. Preparation of solution precursor and undoped TiO_2 powders

A solution precursor was prepared to be mixed with titanium tetrabutoxide (TTB, Wako Pure Chemicals Industries, Ltd.) and ethanol (Wako Pure Chemicals Industries, Ltd.). TTB is easily dissolved in ethanol, so that the solution precursor is easily atomized as smaller size of droplets. For complete dispersion the solution precursor was stirred for ten minutes by a magnetic stirrer before implement of experiments. The solution precursor was sprayed at a mass flow rate of 1.6 g/min on average by the atomizer. The mass fractions of TTB were 11 and 24 %, respectively. Undoped TiO_2 powders of anatase and rutile (Wako Pure Chemicals Industries, Ltd.) were used for comparison of physicochemical characteristics with carbon doped TiO_2.

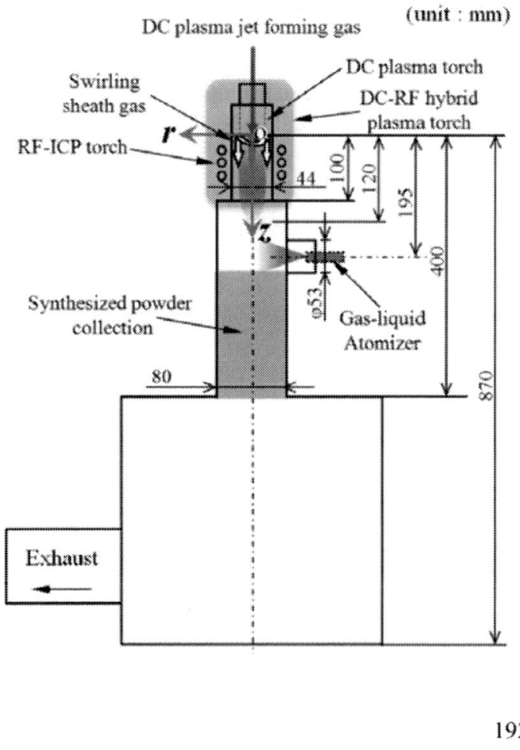

Figure 1. Schematic of DC-RF hybrid plasma flow system with a two-phase atomizer in order to inject droplets of solution precursor into downstream plasma flow.

2.3. Analysis of carbon doped TiO_2

Carbon doped TiO_2 powders were collected at the chamber wall of z = 200 mm to 400 mm. The production rates of carbon doped TiO_2 powders were 0.4 and 0.56 g/h at 11 and 24 wt%, respectively. After annealing at 300 °C to remove all contaminants, analyses of the powders were carried out. However, a part of powders was additionally annealed at 500 °C for the further accurate analysis of X-ray diffraction (XRD; RINT-V, Rigaku Co., Japan) patterns. The morphological analysis was carried out by a transmission electron microscopy (TEM; HF-2000, Hitachi Co., Japan). Successful incorporation of carbon in TiO_2 was analyzed by a Raman spectrometer (NRS-5100, JASCO Co., Japan) at the laser excitation of 532 nm. The UV-visible absorption characteristic was measured by an UV-visible spectrometer (V-7200, JASCO Co., Japan). Additionally, highly functional carbon doped TiO_2 for photocatalytic activity was assessed by degradation of methylene blue solution under visible light irradiation.

3. Experimental results and discussion

3.1. Mechanism of carbon doped TiO_2 synthesis by solution precursor spraying

Figure 2 shows optical spectra with and without solution precursor injection measured at z = 70 mm. A sauter mean size of solution precursor droplets is estimated to be 48 μm from the model proposed by A. Aliseda et al., which shows a good agreement for a liquid mixture [12]. The characteristic time of droplet vaporization is also estimated to be 45 and 16 ms at the temperatures of 1000 and 1500 K. When injecting the solution precursor into downstream plasma flow at z = 195 mm, droplets of the solution precursor are completely vaporized before arriving at the chamber wall, because the maximum power for complete vaporization is estimated to be 0.024 kW. The vapor is transported to upstream plasma flow in the DC-RF hybrid torch by characteristic backflows [11] and successfully dissociated into titanium and carbon in the plasma flow, as shown in Fig. 2. Synthesis of carbon doped TiO_2 is favored under high concentration of carbon and low concentration of oxygen [13]. Excited carbon is clearly observed at wavelengths of 193 and 247 nm. Excited titanium is also observed at wavelengths of 368 and 373-376 nm. Furthermore, the presence of OH band at wavelengths of 306-309 nm indicates the presence of species of oxygen and hydrogen in the plasma flow, because OH band is produced by the chemical reaction with oxygen and hydrogen. However, the amount of

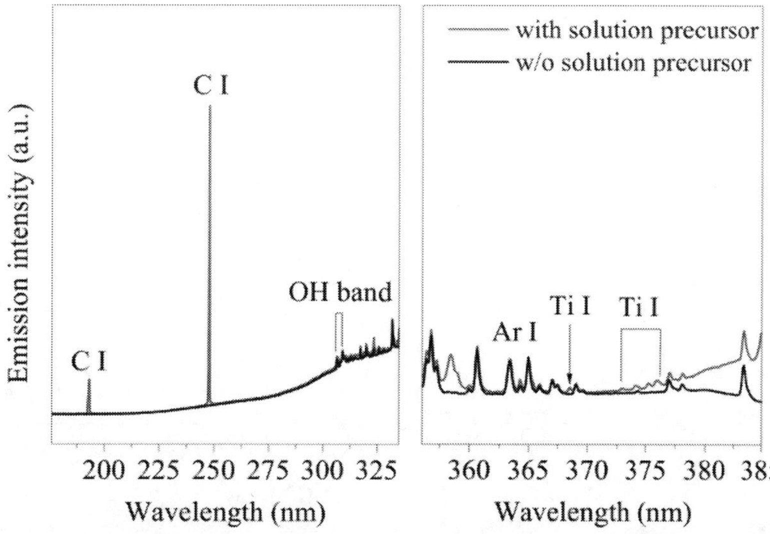

Figure 2. Optical spectra with and without solution precursor injection measured at z = 70 mm.

Table 1. Contents and crystallite size of carbon doped TiO_2 powders

Mass fraction of TTB	Contents (%)			Crystallite size (nm)	
	Anatase	Rutile	Carbon	Anatase (101)	Rutile (110)
11 wt%	25.9	20.3	53.8	11.0	19.3
24 wt%	15.6	40.0	44.4	13.6	15.2

oxygen for the chemical bond with titanium can be decreased to form different chemical species such as OH, H_2O, and CO_x. Then, oxygen vacancy in TiO_2 structure is occurred and carbon is substituted in oxygen vacancy under high concentration of carbon [4,13].

3.2. Analysis of XRD pattern

Figure 3 shows XRD patterns of carbon doped TiO_2 powders. The results of the contents and the average crystallite size are tabulated in Table 1. The XRD patterns indicate that crystal structure of carbon doped TiO_2 is mixed with anatase and rutile, because the phase transformation of quasi-stable anatase phase is preferred in the thermal plasma flow [14]. The contents of anatase and carbon in carbon doped TiO_2 at 11 wt% TTB have 1.66 and 1.21 times larger than that of carbon doped TiO_2 at 24 wt% TTB, respectively. The average crystallite size (L) is estimated by the Scherrer formula [14]:

$$L = (K \cdot \lambda)/(B \cdot cos\theta) \quad (1)$$

where K denotes the shape factor of 0.89, λ is the wavelength of X-ray (Cu Kα), B is the full width half maximum, and θ is the Bragg angle. The crystallite sizes of both phases are similarly estimated in both powders. From the result of the XRD analysis, it is found out that the mass fraction of TTB in solution precursor plays a dominant role for the contents of anatase and carbon in carbon doped TiO_2.

3.3. Morphological characteristics

Figures 4 (a) and (b) show TEM micrographs of carbon doped TiO_2 at 11 and 24 wt% TTB. The crystalline carbon doped TiO_2 and amorphous activated carbon are observed. The crystalline size of carbon doped TiO_2 is mainly several ten nm in both Figs. 4 (a) and (b), which shows a good agreement with the average crystallite sizes in Table 1. However, the average particle sizes of carbon doped TiO_2 are 123.2 ± 40 and 104.2 ± 36 nm at 11 and 24 wt% TTB, respectively, which are measured by a particle size analyzer (UPA EX250, Microtrac Inc.). It can be explained that nano-crystalline TiO_2

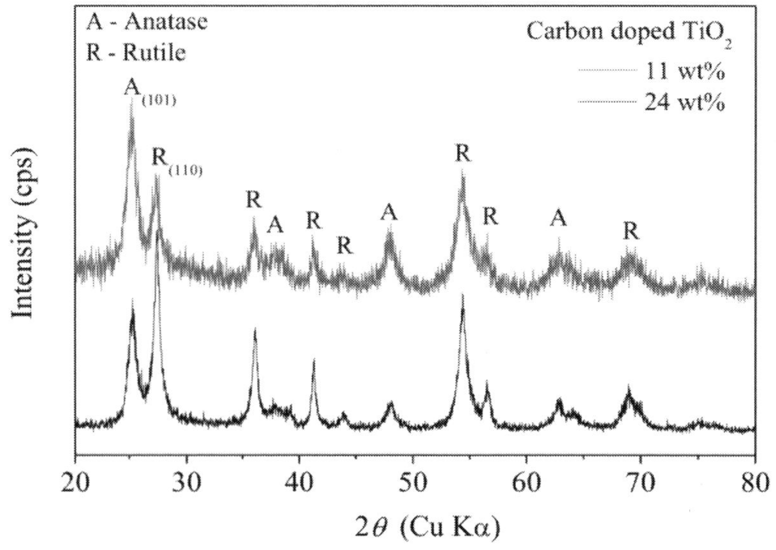

Figure 3. XRD patterns of carbon doped TiO_2 powders.

particles are preferred to be agglomerated in plasma flow, as shown in Fig. 4. In Fig. 4 (a), carbon doped TiO_2 with a lattice spacing of 0.35 nm crystallized in anatase TiO_2 is observed. In Fig. 4 (b), carbon doped TiO_2 with a lattice spacing of 0.32 nm crystallized in rutile TiO_2 is observed.

3.4. Raman spectrum analysis

Figures 5 (a) and (b) show Raman spectra for comparison of carbon doped TiO_2 with undoped TiO_2. In Fig. 5 (a), undoped anatase TiO_2 shows three Raman peaks with high intensities at 379.2, 504.2, and 626.2 cm^{-1}. On the other hand, the Raman spectrum of carbon doped TiO_2 at 11 wt% TTB is observed for shifted three Raman peaks of anatase TiO_2 for 386.5, 506.5, and 630.5 cm^{-1} and three Raman peaks of graphite at 1351.5, 1597.5, and 2668.2 cm^{-1}. From this result, carbon doped TiO_2 at 11 wt% TTB is inferred to mainly form anatase TiO_2 and small shifts of Raman peaks are caused by rutile presence and carbon incorporation. In Fig. 5 (b), undoped rutile TiO_2 shows two Raman peaks with high intensities at 436.2 and 600.2 cm^{-1}. On the other hand, the Raman spectrum of carbon doped TiO_2 at 24 wt% TTB is observed for shifted two Raman peaks of rutile TiO_2 at 398.3 and 586.3 cm^{-1} and three Raman peaks of graphite at 1327.2, 1567.3, and 2672.3 cm^{-1}. Carbon doped TiO_2 at 24 wt% TTB is inferred to form mainly rutile TiO_2 and large shifts of Raman peaks are caused by anatase presence and carbon incorporation [15].

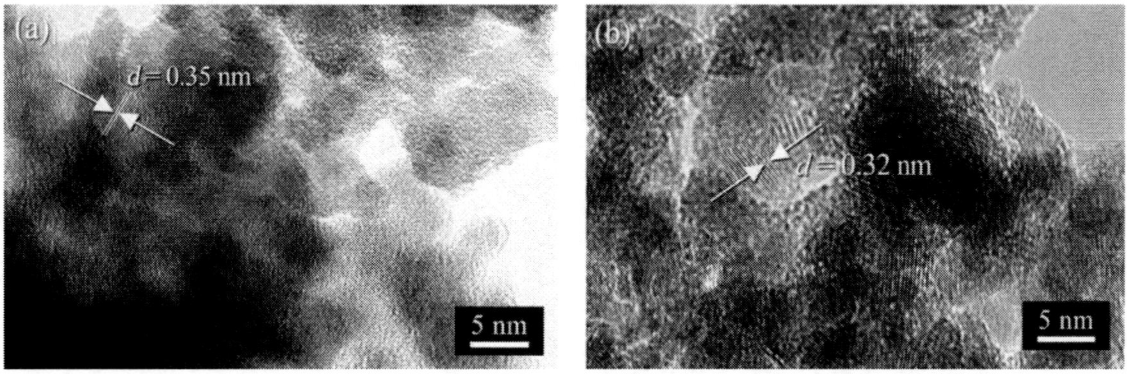

Figure 4. TEM micrographs of carbon doped TiO_2 at (a) 11 and (b) 24 wt% TTB.

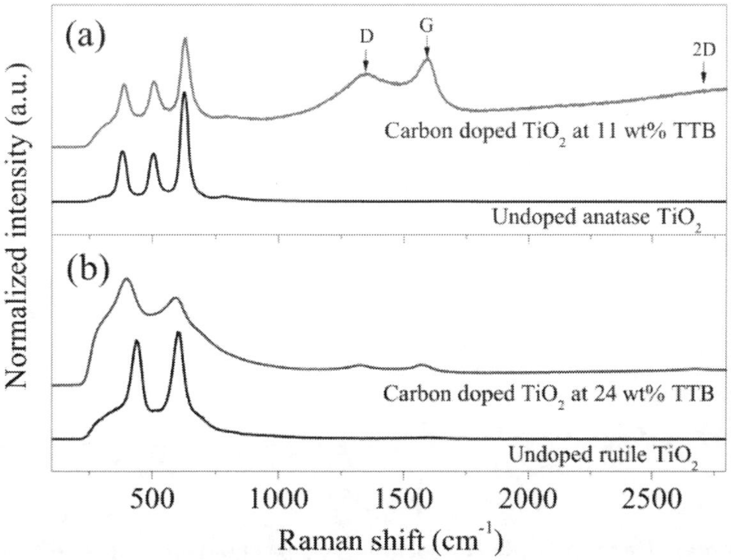

Figure 5. Raman spectra for comparison of carbon doped TiO_2 at (a) 11 and (b) 24 wt% TTB with undoped TiO_2.

3.5. UV-visible absorption characteristics

Figure 6 shows the effect of carbon doped TiO_2 on UV-visible absorption. Compared with undoped anatase TiO_2, carbon doped TiO_2 at 24 wt% TTB, in spite of being carbon and a larger content of rutile, slightly increases in UV-visible absorption by a cut-off wavelength of around 380 nm [16]. On the other hand, carbon doped TiO_2 at 11 wt% TTB shows higher UV-visible absorption by a cut-off wavelength of around 630 nm. It can be explained by that larger contents of carbon and carbon incorporation with TiO_2 increase UV-visible absorption more [17]. Then, UV-visible absorption of carbon doped TiO_2 is sensitively dependent on the contents of carbon doped TiO_2.

3.6. Degradation of methylene blue solution

Figure 7 shows the effect of carbon doped TiO_2 on degradation of methylene blue solution under visible light irradiation using a light-emitting diode (LED) of 0.6 W. Methylene blue solution with 0.3 mg/L is degraded by the reactive oxygen radicals produced by photocatalytic activity of TiO_2 under visible light irradiation. Carbon doped TiO_2 at 11 wt% TTB completely degrades the methylene blue solution even under visible light irradiation due to increase in visible light absorption, as shown in Fig. 6. However, degradation of carbon doped TiO_2 at 24 wt% TTB remains to be 35 %. Photocatalytic

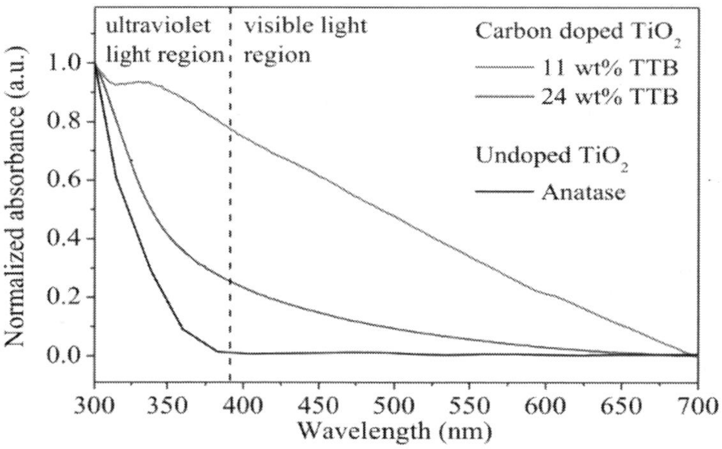

Figure 6. Effect of carbon doped TiO_2 on UV-visible absorption.

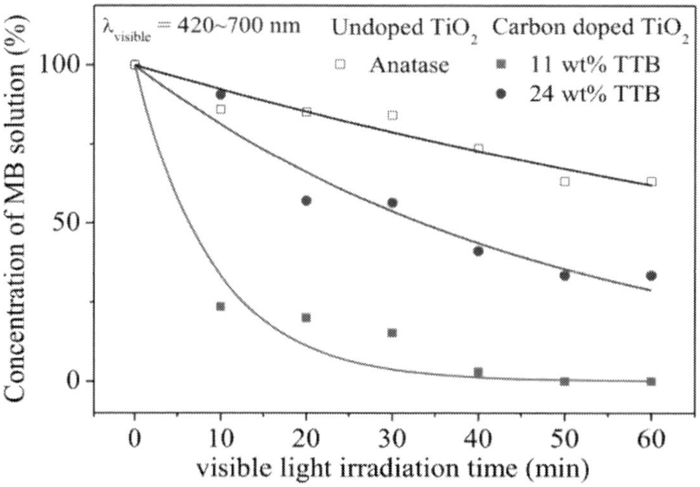

Figure 7. Effect of carbon doped TiO_2 on degradation of methylene blue solution under visible light irradiation.

activity of undoped anatase TiO_2 shows the worst degradation of methylene blue solution under visible light irradiation due to low visible light absorption, as shown in Fig. 6. Therefore, highly functional carbon doped TiO_2 at 11 wt% TTB improves degradation rate of methylene blue solution even under visible light irradiation due to larger contents of anatase TiO_2 and carbon.

4. Conclusions

In the present paper, highly functional carbon doped TiO_2 powders were successfully synthesized by solution precursor spraying in a DC-RF hybrid plasma flow system. The results are summarized as follows;

(1) Solution precursor droplets are vaporized and dissociated sufficiently in the upstream plasma flow by the backflow transportation. This is verified from the presence of titanium and carbon by the optical emission spectroscopy.

(2) The contents of carbon doped TiO_2 are dependent on the mass fraction of TTB in the solution precursor. Larger contents of anatase TiO_2 and carbon are obtained at 11 wt% TTB than at 24 wt% TTB.

(3) Highly functional carbon doped TiO_2 at 11 wt% TTB is validated from higher degradation of methylene blue solution that of carbon doped TiO_2 at 24 wt% TTB even under visible light irradiation.

Acknowledgements

The present study was partly supported by a grant from the Global COE program (2012). The authors would like to give our sincere thanks to Mr. K. Katagiri and Mr. T. Nakajima of the Institute of Fluid Science (IFS), Tohoku University, Japan, for useful technical supports.

References

[1] Chen D, Jiang Z, Geng J, Wang Q, and Yang D 2007 *Ind. Eng. Chem. Res.* **46** 2741-2746
[2] ConPg Y, Zhang J, Chen F, and Anpo M 2007 *J. Phys. Chem. C* **111** 6976-6982
[3] Zaleska A 2008 *Recent Pat. Eng.* **2** 157-164
[4] Leary R and Westwood A 2011 *Carbon* **49** 741-772
[5] Brinley E, Babu K S, and Seal S 2007 *J. Miner. Met. Mater. Soc.* **59** 54-59
[6] Fazilleau J, Delbos C, Rat V, Coudert J F, Fauchais P, and Pateyron B 2006 *Plasma Chem. Plasma. Proc.* **26** 371-391
[7] Delbos C, Fazilleau J, Rat V, Coudert J F, Fauchais P, and Pateyron B 2006 *Plasma Chem. Plasma. Proc.* **26** 393-414
[8] Nishiyama H, Onodera M, Igawa J, and Nakajima T 2009 *J. Therm. Spray Tech.* **18** 593-599
[9] Takana H, Jang J, Igawa J, Nakajima T, Solonenko O P, and Nishiyama H 2011 *J. Therm. Spray Tech.* **20** 432-439
[10] Jang J, Takana H, Solonenko O P, and Nishiyama H 2011 *J. Fluid Sci. and Tech.* **6** 729-739
[11] Jang J, Takana H, Park S, and Nishiyama H 2012 *J. Therm. Spray Tech.* **21** 900-907
[12] Aliseda A, Hopfinger E J, Lasheras J C, Kremer D M, Berchielli A, and Connolly E K 2008 *Int. J. Multi. Flow* **34** 161-175
[13] Valentin C D, Pacchioni G, and Selloni A 2005 *Chem. Mater.* **17** 17-26
[14] Feng L, Xu D, and Lei A 2008 *J. Therm. Spray Tech.* **17** 473-477
[15] Lohse B H, Calka A, and Wexler D 2007 *J. Alloys. Comp.* **434-435** 405-409
[16] Tsukamoto D, Tsukamoto D, Shiraishi Y, Sugano Y, Ichikawa S, Tanaka S, and Hirai T 2012 *J. Am. Chem. Soc.* **134** 6309-6315
[17] Lia Y, Lib X, Lic J, and Yinc J 2006 *Water Res.* **40** 1119-1126

Journal of Physics: Conference Series **441** (2013) 012033 doi:10.1088/1742-6596/441/1/012033

Production and modification of hollow powders in plasma under controlled pressure

Igor P Gulyaev
Institute of Theoretical and Applied Mechanics SB RAS
ul. Institutskaya 4/1, Novosibirsk, 630090, Russia

E-mail: gulyaev@itam.nsc.ru

Abstract. Conducting plasma processes under high or low pressure is an efficient way to affect the heat, mass and momentum exchange in a two-phase flow and this technique is widely used in such well-developed technologies as low-pressure plasma spraying (LPPS) and high pressure plasma-chemical processes. In addition operating pressure is a key parameter in novel plasma process for modification of hollow powders properties. Plasma processing of porous ceramic powders is an effective method of producing hollow spheres (HOSP) with predefined properties. Regardless the method hollow powders were produced their geometric and structural properties can be adjusted by re-melting in plasma of certain pressure: low pressure processing will expand hollow spheres and high pressure – contract it. Regulating the outer diameter of hollow sphere allows adjusting its shell thickness, apparent density, gas pressure in the cavity etc. Preliminary experiments with zirconia hollow powders demonstrated good agreement with theoretical estimations of HOSP properties. The same technique can be used for adjusting properties of ceramic hollow powders produced by different methods, including cost effective fly-ash particles (cenospheres).

1. Introduction

Hollow ceramic spheres are widely as filler in composite materials for weight reduction, improving its heat and sound isolating properties. Applications that do not request strict phase composition and geometric properties of spheres can utilize microspheres produced from coal ash – cenospheres [1]. On the contrary technologies demanding decisive chemical compound of hollow spheres, its size and shell thickness require specially produced powders. Hollow spheres were found to be used in systems for separation of light gases like helium and hydrogen [2]. Plasma sprayed zirconia thermal barrier coatings produced of hollow spherical (HOSP) powder demonstrate decreased heat conductivity compared to coatings produced of compact powders [3] which is essential for gas turbine and jet engine production. The reason for different coating microstructure is likely to be explained with peculiar fluid dynamics during spreading of hollow droplet which produces a counter-jet [4]. Aside from spreading process hollow powders undergo complete melting in thermal jet due to thinner wall which can broaden possibilities of spraying refractory materials using methods with relatively low gas temperature (HVOF, detonation spraying).Plasma processing of porous particles is an effective way of production of hollow spherical powders of various materials including ceramics [5]. Formation of hollow spheres takes place owing to surface melting of material and generation of liquid surface film which entraps the gas contained in pores. Further complete melting and heating of the particle heating doesn't affect the mass (moles) of entrapped gas, however the size of the sphere can change. Final dimensions of the sphere are defined by balance of inner gas pressure P_g, ambient pressure P_{amb} and surface (Laplace) pressure P_{surf} at material solidification temperature [6]. During traditional plasma treatment of powders (spheroidization) at atmospheric pressure outer diameter D_p, shell thickness Δ_p and apparent density ρ_{app} are predefined by the porosity p of initial material. As shown below, the ambient pressure P_{amb1} is a key parameter governing diameter and apparent density of produced hollow spheres. Further manipulation by these properties can be achieved by re-melting obtained HOSP at different pressure P_{amb2}. In the case that the hollow spheres need to be expanded, i.e. lower the shell thickness and apparent density, re-melting should be conducted under lowered pressure $P_{amb2} < P_{amb1}$. In the case that produced powder needs to be densified re-melting should be conducted at raised pressure $P_{amb2} > P_{amb1}$.

2. Experimental setup

Experiments on plasma processing of powders under controlled pressure were conducted using setup shown in fig. 1. DC plasma torch (ITAM SB RAS, Novosibirsk) with channel diameter 8 mm was used for generating plasma. It was connected to pressure-tight water-cooled plasma chamber of 30 L volume, provided with plasma gas line and powder transport line. Typical operating conditions were nitrogen plasma gas flow rate 20-60 slpm, arc current 200-250 A (electric power 30-60 kW), achieved pressure range 0.5 – 2.5 atm.

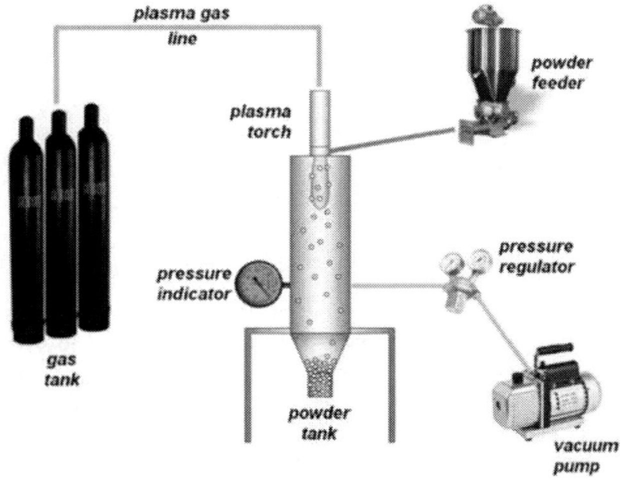

figure 1. Scheme of experimental setup for variable pressure plasma processing.

Working gas flowrate was measured using RMB series rotameters (Dwyer) and PFM series flowmeters (SMC). Pressure in the plasma chamber was measured using MPX5700 pressure sensors (Freescale Semiconductors). During operation of plasma torch there is a constant gas flow into the chamber therefore maintaining preselected pressure requires controlled evacuation of gas. This was implemented by special pressure regulator, which could be connected to a vacuum pump for operation under lowered pressure. The minimum operating pressure was limited by throughput of the vacuum pump, while maximum pressure – mainly by stable operation of plasma torch and durability of applied pneumatic lines.

3. Theoretical model of hollow sphere formation

The necessary condition for hollow powder formation in a plasma jet is presence of sufficient amount of gas inside a particle of initial material. This can be provided by a number of methods: microporosity of the particles, preliminary gas sorption, gas release due to thermal decomposition or chemical reaction. The proven method of ceramic hollow spheres production is plasma treatment of agglomerated powders obtained by spay drying technology [5]. Such particles present agglomerates of diameter 20 – 200 μm, consisting of bound 3 -5 μm grains. The individual agglomerate contains 10^3 – 10^5 grains and has volume porosity about 50%.

Consider the simplest scenario of hollow sphere formation from a porous particle with diameter D_{p0} and open porosity p (all the pores are interconnected), which is being uniformly hated. When the porous agglomerate enters plasma jet its material as well as the inner gas is being heated (fig. 2). While the gas expands it leaves the volume of the particle, however the volume of pores remains the same. In the moment when particle surface reaches the melting temperature the liquid shell is being formed. It captures and fixes the mass of the gas inside the particle. The trapped gas gathers into the single central cavity, the size of which depends on temperature and ambient pressure. The final outer diameter D_p of produced hollow sphere is defined by pressure balance on the droplet surface at material solidification temperature.

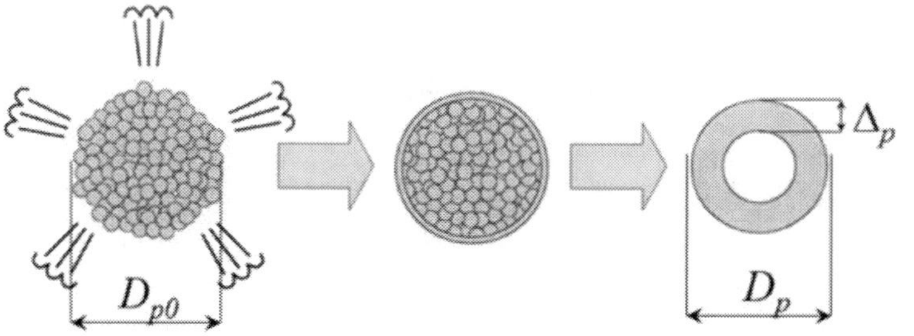

figure 2. The hollow sphere formation scenario from porous agglomerate

Let's define the outer diameter D_p of produced hollow droplet and relative shell thickness $\delta_p = \Delta_p / D_p$ corresponding to the melting temperature of material T_m, i.e. final dimensions of the sphere after cooling and solidification. The total volume of the pores in initial agglomerate is $V_{g0} = \frac{\pi}{6} D_{p0}^3 \cdot p$, and the mass of trapped gas is $m_g = V_{g0}\rho_g$, where ρ_g - gas density at temperature T_m and pressure being equal to ambient pressure P_{amb}. After complete melting and formation of single gas cavity the shrinkage of melted droplet takes place due to the action of additional surface (Laplace) pressure P_{surf}, so the final volume of the gas cavity will reduce to the value. The droplet diameter D_p and relative shell thickness δ_p can be calculated with the pressure balance on the droplet surface at the given temperature T_m:

$$\frac{m_g}{M}\frac{RT_m}{V_g} = P_{amb} + \frac{4\sigma}{D_p}\left(1 + \frac{1}{1-2\delta_p}\right), \tag{1}$$

where M – molar mass of the gas and σ - surface tension of melted material. The left-hand side of the equation is pressure in the gas cavity P_g calculated by the ideal gas law. It is counterbalanced by the sum of the ambient atmosphere pressure P_{amb} and the surface pressure P_{surf} calculated considering both surfaces of the liquid shell. Equation (1) contains two unknown variables, namely D_p and δ_p. However one of them can be easily expressed through another using the mass balance $(1-2\delta_p) = \sqrt[3]{1-6m_p/\pi\rho D_p^3}$, where mass of the particle m_p is defined by diameter and porosity of initial agglomerate $m_p = \frac{\pi}{6}\rho D_{p0}^3(1-p)$. Iterative solution of equation (1) (minutely discussed in [6]) provides equilibrium diameter D_p and relative shell thickness δ_p of the hollow droplet as a function of initial powder porosity p, material melting temperature T_m and ambient pressure P_{amb}. In practical operations it is more convenient to measure apparent density of the spheres $\rho_{app} = \rho_{ZrO2}\left(1-(1-2\delta_p)^3\right)$. Properties of zirconia used for calculations are as follows: density 5.6 g/cm^3, surface tension at melting point 0.43 N/m, melting temperature 3000 K.

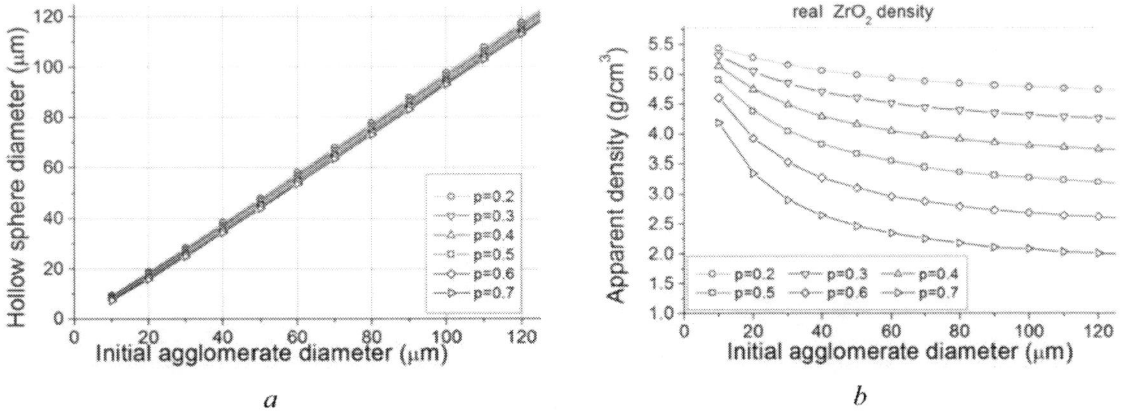

figure 3. Calculated dependence of zirconia hollow sphere diameter (a) and apparent density (b) on initial agglomerate powder diameter for various feedstock porosity.

Calculation results in fig. 3, a demonstrate that the diameter of formed zirconia sphere D_p is only slightly (by 2-10%) less than initial agglomerate diameter D_{p0} and almost independent of porosity in the range $p = 0.2 - 0.7$. This result is typical for ceramic materials like zirconia, alumina and silica. On the other hand the porosity p is the key parameter governing the apparent density (or shell thickness δ_p) of the formed sphere. Figure 3, b shows expected result: the less initial porosity of the feedstock – the higher is apparent density of produced spheres.

The influence of ambient pressure P_{amb} on parameters of hollow zirconia spheres produced from 100 μm agglomerates with porosity p=0.6 is presented in table 1. Gas inside the pores of initial particle, which originally had pressure P_{amb}, is being compressed by additionally applied pressure P_{surf}, which reduces the droplet diameter to the value $D_p<D_{p0}$. The final pressure in the gas cavity is equal to the sum of ambient pressure P_{amb} and surface pressure P_{surf} . As one can see $P_{surf}\sim$ 0.4 atm, therefore formation of hollow spheres under comparable ambient pressure P_{amb}=0.5 atm causes noticeable reduction of particle diameter down to 90 μm and corresponding rise of particle apparent density. However in case of P_{amb}=2.5 atm the surface pressure is minor compared to ambient pressure, so the shrinkage of droplet is small enough and its diameter D_p=97.5 μm is almost equal to initial D_{p0}.

Table 1. Calculated parameters of hollow zirconia spheres produced from initial particles with diameter D_{p0}=100 μm and porosity p=0.6 under various ambient pressures P_{amb}

P_{amb}, atm	D_p, μm	ρ_{app}, g/cm^3	δ_p	P_{surf}, atm	P_g, atm
0.5	90	3.1	0.118	0.44	0.94
1.0	94	2.7	0.098	0.41	1.41
2.5	97.5	2.4	0.086	0.4	2.90

The two-step method of production of hollow powder with specified properties becomes available under controlled pressure processing. Assume the hollow sphere of diameter D_{p1} and shell thickness δ_{p1} was formed under ambient pressure P_{amb1} and then re-melted under different ambient pressure P_{amb2} which caused change of diameter to the value D_{p2}. Estimation of new sphere diameter should be carried out by the pressure balance equation (1), where new value of P_{amb2} should be used

and the mass of trapped gas should be calculated by expression $m_g = (P_{amb1} + P_{surf1})\frac{M}{RT_m}\frac{\pi}{6}D_{p1}^3(1-2\delta_{p1})^3$, where surface pressure $P_{surf1} = \frac{4\sigma}{D_{p1}}(1+1/(1-2\delta_{p1}))$.
Figure 4 shows results of calculations for apparent density of zirconia spheres produced from 100 μm initial particles with porosity p=0.5. The white circles display density of the spheres after their formation at a given pressure.

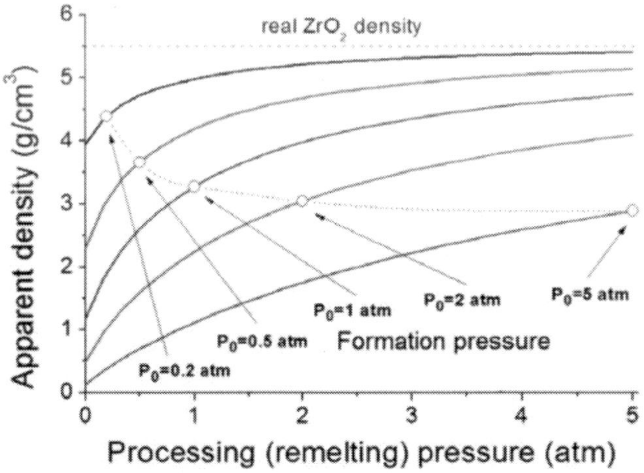

figure 4. Calculated change of apparent density of hollow zirconia spheres produced from 100 μm initial particles with porosity p=0.5

As one can see, regulating formation ambient pressure from 0.2 to 5 atm provides change of powder density in wide range 4.3 - 2.9 g/cm^3. Even more pronounced changes in density can be realized with second stage processing. For example, if the hollow powder produced at P_{amb1} = 5 atm is re-melted at P_{amb2} = 0.2 atm its density will drop down to 0.35 g/cm^3.

4. Results and discussion

Firstly three types of zirconia agglomerated (spray-dried) powders were spheroidized under atmospheric pressure. These feedstock powders had wide-range size distribution 20-120 μm. Average porosity was measured for all 3 types of feedstock powder and was found to be p_{av} = 0.4, 0.5, 0.6. Figure 5 demonstrates apparent density measurements using fluid pyknometry method (accuracy ε=5%) of various size fractions of produced hollow zirconia powders (figures with solid lines). Dash lines are calculated estimations for apparent density of zirconia hollow spheres produced from agglomerates with p = 0.4 and 0.6.

The measured apparent density is systematically higher than calculated one for the small size spheres which can be explained with presence of dense fragments of broken particles shell. Moreover, different fractions of initial agglomerated powders obtained by spray drying can have significantly diverse porosity p [7], and average value p_{av} doesn't account this. This explains very steep curve of measured density for the initial porosity p_{av}=0.5.

In order to eliminate undesired effects of initial powder variable porosity and presence of broken fragments two-step experiments with variable pressure were conducted using narrow powder fraction. First, the hollow zirconia powder of wide size range 20 – 120 μm was produced under raised pressure of P_{amb1} = 2.5 atm. The narrow fraction of 71-80 μm was separated using sieving machine and its apparent density was found to be 4.14 g/cm^3. Second, the separated fraction was re-melted under atmospheric pressure P_{amb2} = 1 atm, and its apparent density was found to be 3.3 g/cm3 (fig. 6).

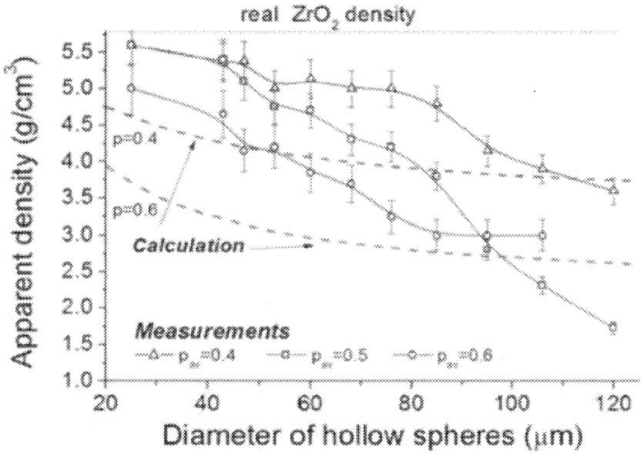

figure 5. Apparent density of hollow zirconia spheres produced from feedstock of various average porosity under atmospheric pressure

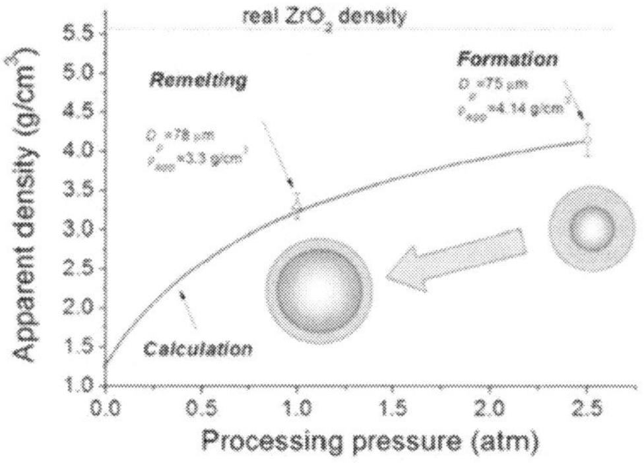

figure 6. Modification of hollow zirconia spheres apparent density. Figures – experiment, solid line - calculation

In order to validate these experimental results calculations using proposed hollow sphere modification model were conducted. The solid line in fig. 6 demonstrates change in zirconia hollow sphere apparent density after processing (re-melting) under controlled pressure. In calculations was used measured value of formation gas pressure P_{ambl}= 2.5 atm, sphere diameter D_p =75 μm (average over the fraction 71-80 μm) and shell thickness $\delta_p = 0.18$ (corresponding to $\rho_{app} = 4.14$ g/cm^3). As one can see model predicts decrease in apparent density (down to 3.23 g/cm^3) of hollow powder almost identical to the measured one. This is very promising result, although it is obvious that more broadened experiments are required to investigate limitations of this technique.

5. Conclusion

Formation of hollow spherical powders in plasma was analyzed and theoretical model described. Novel method of hollow powders modification using plasma under controlled pressure was introduced and first experiments described. The suggested method can be used for of modification of hollow spheres obtained by various techniques, including cost-effective separation of fly-ash particles (cenospheres). Obviously clear understanding of capabilities of introduced method requires broadened experiments under controlled pressure as well as study of adjacent problems such as determination of thermal properties of plasmas and performance of plasma torch (power, efficiency) under non-atmospheric pressures.

Acknowledgements
The reported study was partially supported by RFBR, research project No. 12-08-31150.

References

[1] Drozhzhev V S, Danilin L D et al. 2005 *Proc. World of Coal Ash Conf.*, 11-15 April, Lexington, USA 115-16
[2] Fomenko E V, Anshits N N et al. 2010 *Doklady Physical Chemistry* **435(2)** 202-04
[3] Chi W, Sampath S, Wang H 2006 *J. Thermal Spray Tech.* **15(4)** 773-78
[4] Gulyaev I P, Solonenko O P, Gulyaev P Yu, Smirnov AV 2009 *Tech. Phys. Lett.* **35(10)** 885–888.
[5] Solonenko O P, Gulyaev I P and Smirnov A V 2011 *J. Ther. Sci. Technol*.**6**(2) 219- 34
[6] Gulyaev I P 2010 Ph. D. dissertation ITAM SB RAS, Novosibirsk (In Russian)
[7] Vehring R, Foss W R and Lechuga-Ballesteros D 2007 *Aerosol Sci.* **38** 728-46

Fabrication of iron oxide nanoparticles using laser ablation in liquids

T Iwamoto[1] and T Ishigaki[2]

[1]Institute of Sustainability Research and Education, Hosei University, 2-17-1 Fujimi, Chiyoda-ku, Tokyo 102-8160, Japan

[2]Department of Chemical Science and Technology, Hosei University, 3-7-2 Kajino-cho, Koganei-shi, Tokyo 184-8584, Japan

E-mail: ishigaki@hosei.ac.jp

Abstract. Nanoparticles of iron oxides were synthesized through the ablation of the bulk-targets of Fe, Fe_3O_4, and Fe_2O_3 in water by the irradiation of a Nd:YAG laser. Our samples prepared via the laser ablation method displayed small (ca. 1 nm in diameter) and large (over 5 nm in diameter) particles. The small particles were well-dispersed, whereas large particles were agglomerated. The FeO and Fe_3O_4 phases were formed in the synthesized powders irrespective of the kind of targets. The formation of the other phases, Fe and Fe_2O_3, changed depending on the the kind of targets used in the laser ablation. Poly(*N*-vinyl-2-pyrrolidone) (PVP) as a protective reagent was employed to disperse large particles in our samples, and accordingly, their dispersibility was improved as mole concentration of PVP increased.

1. Introduction

Iron oxide is known broadly as a magnetic material such as a permanent magnet. In iron oxide, three kinds of FeO, Fe_3O_4, and Fe_2O_3 exist and each is applied in different fields. Recently, iron oxide nanoparticles have been synthesized by various routes, and their crystal structure and magnetic property have been investigated in detail [1–7]. At present, new applications like hyperthermia and a drug delivery system using iron oxide nanoparticles are also expected, to say nothing of ordinary applications of pigment, contrast medium, and magnetic devices using iron oxide in a bulk state.

In general, chemical and physical routes have been chosen as syntheses of nanoparticles. The main advantage of chemical routes is to yield a large amount of nanoparticles, while that of physical routes is to fabricate nanoparticles with high-crystallinity. Recently, a laser ablation method in physical routes has been focused on, especially since it is easy and eco-friendly and possesses only a few experimental parameters [8–13]. Moreover, this method can provide free-standing nanoparticles, and hence their intrinsic properties can be examined because nanoparticles prepared by chemical methods are usually protected by organic molecules. However, nanoparticles synthesized by the laser ablation method are often agglomerated, and thus there is an unsolved problem with respect to nanomaterials prepared by this method [8–13].

In the present paper, iron oxide nanoparticles are fabricated via the laser ablation method, and their morphology and crystal structure are investigated. Additionally, to disperse the iron oxide nanoparticles prepared using the laser ablation method, we employ organic molecules as a protective reagent and examine their effect on dispersibility and the crystal structure of the nanoparticles.

2. Experimental procedure

To ablate three kinds of targets (Fe, Fe_2O_3, and Fe_3O_4, purity 99.99%, Kojundo Chemical Laboratory Co., Ltd., Saitama, Japan) and synthesize nanoparticles, a Nd:YAG laser (Quantel Brilliant b, Les Ulis, France) with pulse duration of 6 ns, a wavelength of 1064 nm and a repetition rate of 10 Hz was used [14,15]. Figure 1 schematically illustrates the experimental setup. The target was fixed on the bottom of a glass beaker (20 ml), and distilled water as a solvent was added at a level of 10 mm from the target-surface. The beaker was rotated at 40 rpm and target-surface was irradiated for 1 h by the Nd:YAG laser with a laser spot of approximately 250 μm. Then, colloidal nanoparticles were collected.

Crystal structure of nanoparticles was estimated by using X-ray diffraction (XRD, Rigaku SmartLab, 40 kV, 30 mA, CuKa 0.1541 nm, Tokyo, Japan). Morphology and dispersibility of nanoparticles were investigated by transmission electron microscopy (TEM, JEOL JEM-2100F, Tokyo, Japan).

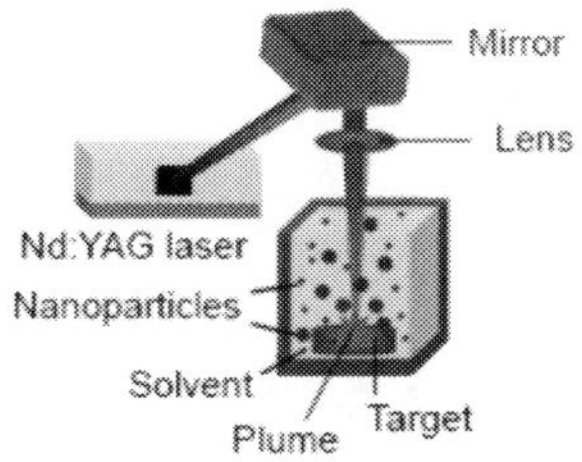

Figure 1. Schematic illustration of experimental setup.

3. Results and discussion

Figures 2(a) and 2(b) show TEM photographs of our sample prepared at laser power of 100 mW for 1 h by using a Fe target. Small particles and large particles co-exist. Figure 2(c) displays a size distribution histogram of small particles seen in Fig. 2(a). The particles exhibit a mean diameter of 1.0 nm and a standard deviation of 0.1 nm. The small particles are highly dispersed, but large particles over 5 nm in diameter are agglomerated. These mono- and highly-dispersed small nanoparticles in the case of the laser ablation method do not appear in the previous literatures on agglomerated iron oxide nanoparticles with 10-50 nm in diameter [8,10,13]. In Fig. 1(d), an XRD pattern of the sample exhibits Fe, FeO, and Fe_3O_4 peaks. Fe_3O_4 is the main product because this pattern corresponds to bulk-Fe_3O_4 profile. Hence, this sample is a mixture of iron and iron oxide.

Figure 3 provides XRD patterns of our samples synthesized at various laser powers for 1 h by using Fe, Fe_2O_3, and Fe_3O_4 targets. In the case of a Fe target, Fe, FeO, and Fe_3O_4 peaks are detected. Fe_2O_3 peaks are also detected as laser power increases. When a Fe_2O_3 target is used, FeO, Fe_2O_3, and Fe_3O_4 peaks are seen and, FeO and Fe_2O_3 peaks increase as laser power increases. When a Fe_3O_4 target is used, FeO and Fe_3O_4 peaks are detected and FeO is decreased by increasing laser power. However, the Fe_2O_3 peak at 33° is detected as laser power increases like in the case of the Fe target. This probably suggests that FeO is oxidized to Fe_2O_3 because temperature and pressure in the plume seen in Fig. 1 increase as laser power increases. Moreover, the kind of targets may influence initial particles evaporated from these targets. In all targets, FeO and Fe_3O_4 peaks are detected in our samples. S.-Warkocka et al. claimed that Fe_3O_4 nanoparticles were reduced to form FeO phase [12].

Fe_3O_4 is the main product in our samples, and Fe_3O_4 diffractions are seen together with other iron oxide ones similar to those in the case of our samples also in Ref. [9,11,12]. Here, we consider the relationship between temperature and O_2 partial pressure in the iron oxide phase, which is determined by the H_2/H_2O equilibrium in the plume at high temperature. In the light of an Ellingham diagram [16], which represents many oxides in thermodynamic states and determines their O_2 partial pressure, our experimental environment probably keeps the most proper balance between temperature and O_2 partial pressure. When target and laser power are altered, a balance of the O_2 partial pressure in the plume is broken by ablating the target and generating new O atoms, the O_2 partial pressure in the plume changes, and then other iron oxides (FeO or Fe_2O_3) are prepared.

IOP Publishing
doi:10.1088/1742-6596/441/1/012034

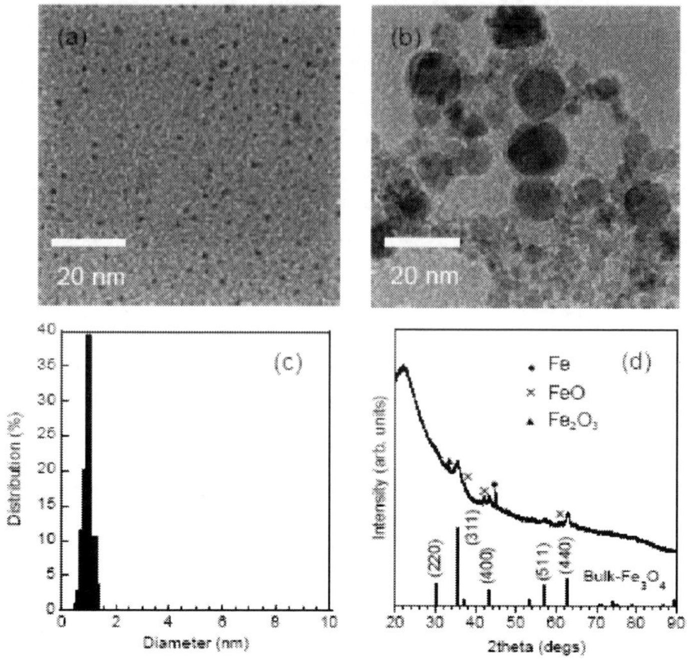

Figure 2. TEM photographs of (a) small and (b) large nanoparticles in our sample prepared at laser power of 100 mW for 1 h using a Fe target. (c) A size distribution histogram of (a). A mean diameter is 1.0 nm and standard deviation is 0.1 nm. (d) An XRD pattern of the sample. The bulk-Fe_3O_4 profile is taken from ICDD, No. 01-089-2355. The values in the graph represent the index of plane for Fe_3O_4.

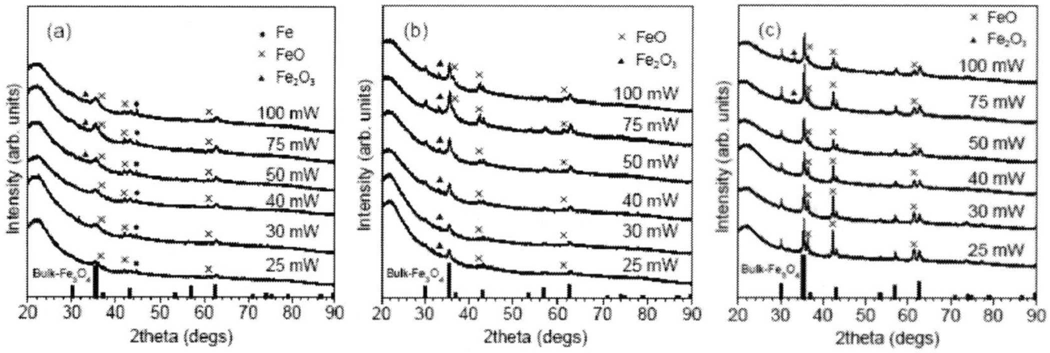

Figure 3. XRD patterns of samples synthesized at various laser powers for 1 h by using (a) Fe, (b) Fe_2O_3, and (c) Fe_3O_4 targets. The bulk-Fe_3O_4 profile is taken from ICDD, No. 01-089-2355.

As shown in Figs. 2(a) and 2(b), small particles are dispersed, whereas large particles are agglomerated. Yang explained the evolution of the laser-induced plasma in liquid and reported that four kinds of chemical reactions occurred inside the plasma and liquid, and the interface between the plasma and liquid [17]. Since the first chemical reaction occurs inside the laser-induced plasma, the small initial particles are synthesized as shown in Fig. 2(a). After that, the second chemical reaction also takes place inside the plasma. However, some initial particles escape from the plume and enter the liquid area, where cooling takes place and further grain growth stops. In the second chemical reaction, the other initial particles remaining in the plasma grow into large particles at high temperature. Therefore, small and large particles exist in our samples.

To avoid the formation of large particles, organic molecules were used as a protective reagent. We tried to synthesize iron oxide nanoparticles by using various organic molecules such as several kinds of amino acids and sodium dodecyl sulfate, and consequently, it was found that poly(*N*-vinyl-2-pyrrolidone) (PVP, an average molecular weight of 55,000, Sigma-Aldrich), which was very useful in chemical syntheses of metallic alloy nanoparticles [18–20], was the best protective reagent in our experiments. Herein, we prepare iron oxide nanoparticles via the laser ablation in water containing PVP by using a Fe_3O_4 target. Figure 4 presents TEM photographs of samples prepared at laser power of 50 mW for 1 h under various mole concentrations of PVP. As PVP concentration increases, large particles are dispersed. Small particles around 1.0 nm in diameter show high-dispersibility as seen in Fig. 2(a) irrespective of PVP concentration (not shown).

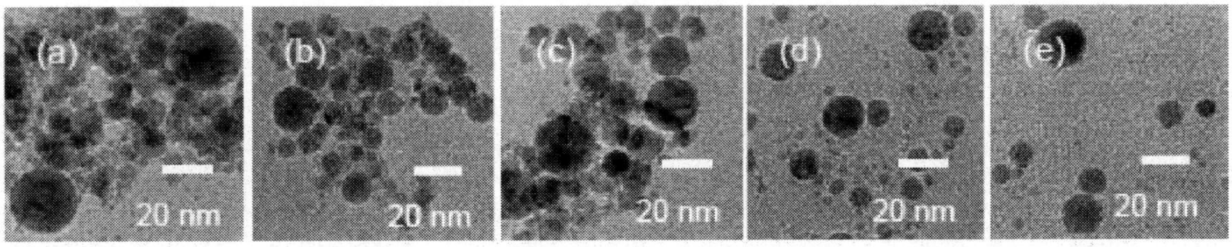

Figure 4. TEM photographs of large particles in samples prepared at laser power of 50 mW for 1 h under various mole concentrations of PVP by using a Fe_3O_4 target.: (a) without PVP and (b) 0.01, (c) 5, (d) 20, and (e) 50 mM of PVP.

Figure 5 exhibits XRD patterns of samples fabricated at laser power of 50 mW for 1 h and various mole concentrations of PVP by using a Fe_3O_4 target. When PVP concentration increases, these diffractions are broad, and hence, crystalline sizes of samples coated by PVP become small. This result has been reported in the previous literature with respect to PVP-coated FePt nanoparticles prepared by chemical routes [20]. By increasing the amount of PVP, the growth of nanoparticles may be retarded due to coating them with a large amount of PVP. PVP as a protective reagent affects dispersibility and crystalline size of nanoparticles irrespective of the kind of syntheses (chemical or physical methods).

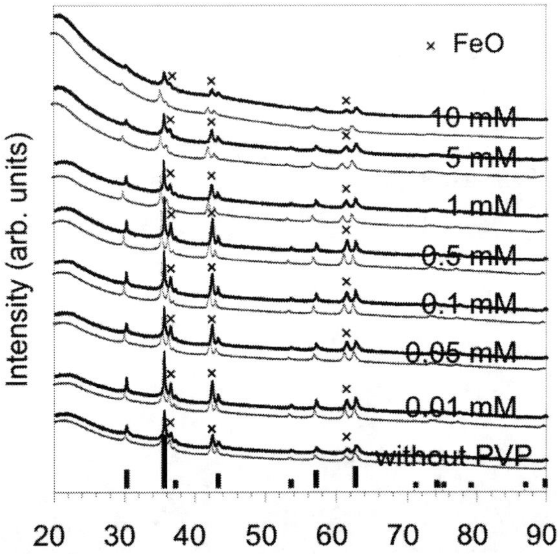

Figure 5. XRD patterns of samples prepared at laser power of 50 mW for 1 h under various concentrations of PVP by using a Fe_3O_4 target. The bulk-Fe_3O_4 profile is taken from ICDD, No. 01-089-2355.

4. Conclusion

In conclusion, the present paper provided some new information. Our samples prepared via the laser ablation method showed small (ca. 1 nm in diameter) and large (over 5 nm in diameter) particles. Targets used in the laser ablation affected crystal structure of our samples. Irrespective of targets, FeO and Fe_3O_4 peaks were detected. When PVP was used as a protective reagent to disperse large particles, their dispersibility was improved as mole concentration of PVP increased. By adding PVP, crystalline sizes became small.

Acknowledgment

We thank Mr. M. Ichihara at Precision Analysis Laboratory, Hosei University, for his assistance in TEM observations.

References

[1] Pérez J A L, Quintela M A L, Mira J, Rivas J and Charles S W 1997 *J. Phys. Chem. B* **101** 8045
[2] Morales M P, Veintemillas-Verdaguer S, Montero M I, Serna C, Roig A, Casas L, Martínez B and Sandiumenge F 1999 *Chem. Mater.* **11** 2058
[3] Sun S and Zeng H 2002 *J. Am. Chem. Soc.* **124** 8204
[4] Teng X and Yang H 2004 *J. Mater. Chem.* **14** 774
[5] Pinna N, Grancharov S, Beato P, Bonville P, Antonietti M and Niederberger M 2005 *Chem. Mater.* **17** 3044
[6] Cai W and Wan J 2007 *J. Colloid Interface Sci.* **305** 366
[7] Laurent S, Forge D, Port M, Roch A, Robic C, Elst L V and Muller R N 2008 *Chem. Rev.* **108** 2064
[8] Zeng X, Wang Z, Liu Y and Ji M 2005 *Appl. Phys. A* **80** 581
[9] Kurland H-D, Grabow J, Staupendahl G, Andrä W, Dutz S and Bellemann M E 2007 *J. Magn. Magn. Mater.* **311** 73
[10] Liu P, Cai W and Zeng H 2008 *J. Phys. Chem. C* **112** 3261
[11] Amendola V, Riello P and Meneghetti M 2011 *J. Phys. Chem. C* **115** 5140
[12] Swiatkowska-Warkocha Z, Kawaguchi K, Wang H, Katou Y and Koshizaki N 2011 *Nanoscale Res. Lett.* **6** 226
[13] Osipov V V, Platonov V V, Uimin M A and Podkin A V 2012 *Tech. Phys.* **57** 543
[14] Al-Mamun S A, Nakajima R and Ishigaki T 2012 *Thin Solid Films* **523** 46
[15] Al-Mamun S A, Nakajima R and Ishigaki T 2013 *J. Colloid Interface Sci.* **392** 172
[16] Ellingham H J T 1944 *J. Soc. Chem. Ind., London* **63** 125
[17] Yang G W 2007 *Prog. Mater. Sci.* **52** 648
[18] Iwamoto T, Matsumoto K, Kitamoto Y and Toshima N 2007 *J. Colloid Interface Sci.* **308** 564
[19] Iwamoto T, Kitamoto Y and Toshima N 2009 *Physica B* **404** 2080
[20] Iwamoto T, Matsumoto K, Matsushita T, Inokuchi M and Toshima N 2009 *J. Colloid Interface Sci.* **336** 879

Effects of Laser Energy Density on Silicon Nanoparticles Produced Using Laser Ablation in Liquid

Hiroki Kobayashi, Pattarin Chewchinda, Hiroyuki Ohtani, Osamu Odawara, Hiroyuki Wada
Tokyo Institute of Technology, 4259 Nagatsuta, Midori-ku, Yokohama 226-8503 Japan

E-mail: kobayashi.h.aw@m.titech.ac.jp

Abstract. We investigated the morphology of silicon nanoparticles prepared using laser ablation in liquid through varying the energy density and laser irradiation time. Silicon nanoparticles were prepared using laser ablation in liquid. A silicon wafer was irradiated in ethanol using a laser beam (Nd: YAG/second harmonic generation, 532 nm). Crystalline silicon nanoparticles approximately 6 nm in size were observed by TEM observation. The quantity of silicon nanoparticles proportionally increased with an increase in energy density greater than the laser ablation threshold. This quantity also increased with an increase in laser irradiation time without saturation due to absorption of the nanoparticles in liquid in the light path.

1. Introduction

Silicon nanoparticles are important for applications such as optoelectronics [1], [2] and bioimaging [3], [4]. Silicon nanoparticles have been researched to synthesize or to apply to applications. Silicon is used as the raw material for a semiconductor chip and is easily acquired. Small nanoparticles less than the de Broglie wave of an electron and the hole or Bohr radius of exciton produce a quantum effect. Such particles are "quantum-dots". A quantum-dot has discrete bands instead of continuous bands and notably produces impact ionization [5]. Impact ionization is also "multiple exciton generation" (MEG). In general, one photon generates an exciton (a pair of electron and hole). However, this phenomenon generates more than 2 exciton pairs from one photon. Silicon can produce 2.6 exciton pairs from one photon (3.4 eV) [6]. The maximum size of a silicon quantum-dot can be estimated using the Bohr radius of an exciton in silicon. The diameter is 8.6 nm [7]. Therefore, silicon nanoparticles with a diameter 8.6 nm or lower are necessary for certain applications, especially MEG-based applications.

Laser ablation in liquid is one method for synthesizing nanoparticles. This method produces nanoparticles by irradiating a target in solution using a laser with high energy density. Laser ablation in the gas phase or under vacuum has been studied since the 1960s [8], while the study of laser ablation in liquid was started recently [9] ~ [11]. In the case of laser ablation in liquid, the target material was metal usually. Later, organic material [12] and ceramics [13] ~ [17] were used as a target. Laser ablation in liquid is advantageous for producing silicon nanoparticles because silicon nanoparticles can be stably maintained on a particle surface by using ethanol as solvent [18]. Moreover, laser ablation in liquid has other advantages: the simplicity of the process and the possible creation of small nanoparticles with relatively small spatial dispersion [19].

Silicon nanoparticles have attracted attention, but it is difficult to produce high quantities by laser ablation in liquid. This study explores fabrication of silicon nanoparticles using laser ablation in liquid for applications such as a quantum-dot-sensitized solar cell. Therefore, production of silicon nanoparticles with a diameter less than 8.6 nm at a certain quantity is the objective for the study herein. Thus, the laser energy density and irradiation time were investigated for their dependence on nanoparticle size and quantity.

2. Experimental

A <100> p-type silicon wafer 600 μm thick with a 11.5 ~ 15.5 Ω cm resistance was used as the raw material. The silicon wafer was cut 2 cm by 2 cm. Before laser ablation, the silicon wafer was cleaned in ethanol using ultrasonication twice for 10 minutes and then dried. The silicon wafer was irradiated with a laser beam, which was not focused with a lens. The unpolished side of the silicon wafer was placed 7 mm beneath the top surface of the ethanol. Ethanol was used as the solvent for laser ablation to prevent oxidation of the silicon during laser ablation. The ethanol was stirred using a magnetic stirrer during laser ablation. The laser beam was a second harmonic generation of a Nd:YAG laser (wavelength: 532 nm; pulse duration: 13 ns; and repetition rate: 10 Hz). The diameter of the irradiated laser beam on the silicon wafer was 14 mm (1.54 cm^2). The energy density varied from 0.17 to 0.63 J/cm^2 (laser energy: 2.6 - 9.7 J/pulse). Irradiation time varied from 30 minutes to 3 hours.

The prepared nanoparticles were identified using Raman spectroscopy. The solution containing silicon nanoparticles was dropped in an Al pan under N_2 atmosphere and then dried. This solution drying procedure was repeated 4 times. After this procedure, black powder was confirmed on the Al pan. To measure the Raman spectrum, the black powder was irradiated with a laser under the atmosphere to measure Raman spectrum. The particle size of the prepared silicon nanoparticles was measured using a transmission electron microscope (TEM). After laser ablation, the solutions were filtered with a Teflon filter (pore size: 100 nm). The solutions were then dropped onto a Cu grid, and the grid was then dried overnight at 70 °C in a vacuum. The quantity of silicon nanoparticles was measured using inductively-coupled plasma mass spectrometry (ICP-MS). Absorption spectra were measured by spectrometer with an integrating sphere.

3. Results and discussion

Figure 1 shows a Raman spectrum of the silicon nanoparticles produced by ablating a silicon wafer in ethanol under an Ar atmosphere. The energy density was 0.17 J/cm^2, and the ablation was performed for 2 hours. Two peaks were observed at 494 cm^{-1} and 518 cm^{-1}. The peak wavelength of the silicon crystal is at 520 cm^{-1}. The particle size of silicon estimated by the value of Raman shift is 4 nm [20]. This peak was slightly shifted as an effect of nanosize silicon, whereas the sample peak at 494 cm^{-1} indicates silicon dioxide [21]. The Raman peak for silicon dioxide is attributed to the silicon wafer native oxide layer or oxidation of the prepared silicon nanoparticles during Raman spectra

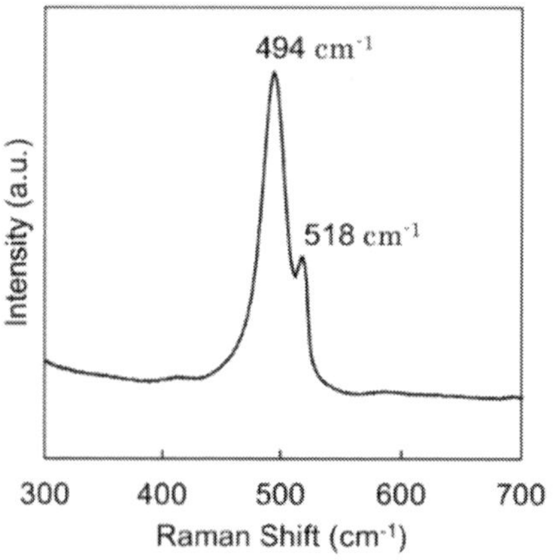

Figure 1. Raman spectrum of the silicon nanoparticles.

measurement. Therefore, the powder in the solution, which was prepared using an irradiating green laser at 0.17 J/cm^2 on a silicon wafer in ethanol, comprises silicon nanoparticles with silicon dioxide.
We investigated the effect of energy density from the irradiating laser on the morphology of the silicon nanoparticles. The irradiation time was 1 hour. Figure 2 shows TEM images of silicon nanoparticles. These images indicate that we produced silicon nanoparticles at approximately 6 nm. Moreover, the nanoparticle shapes are almost spherical. Lattice fringes were confirmed through a high magnification image. These observations indicate that the silicon nanoparticles are crystallized.

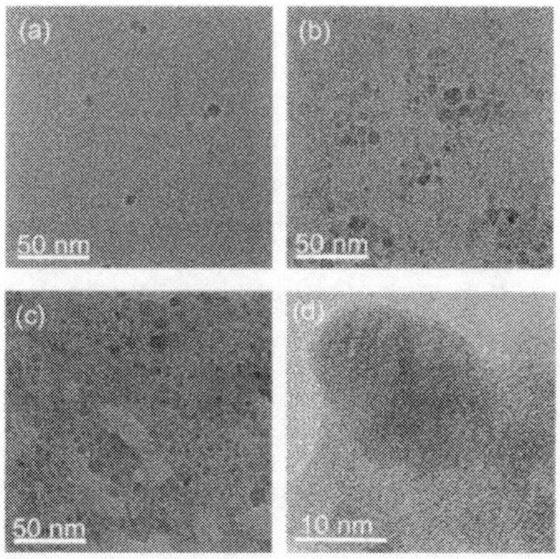

Figure 2. TEM images of the silicon nanoparticles. (Energy density: (a) 0.17 J/cm^2, (b) 0.32 J/cm^2, (c) 0.45 J/cm^2 and (d) high magnification image of 0.45 J/cm^2. Irradiation time: 1 hour.)

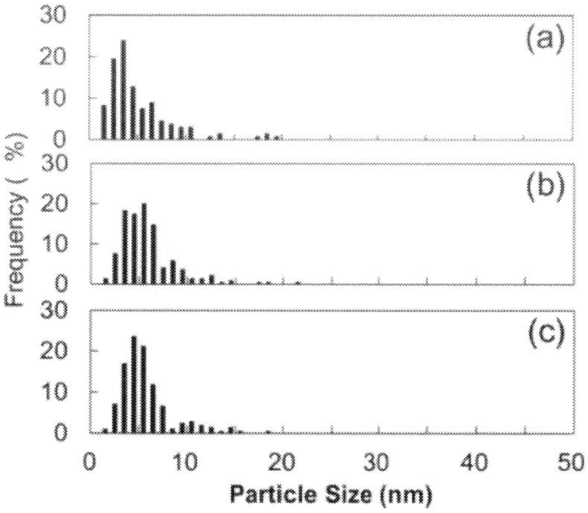

Figure 3. Particle size distributions for the silicon nanoparticles. (Energy density: (a) 0.17 J/cm^2, (b) 0.32 J/cm^2, and (c) 0.45 J/cm^2. Irradiation time: 1 hour.)

Figure 3 shows the particle size distribution of the silicon nanoparticles measured through TEM images. The average sizes were 5.6 nm, 6.3 nm and 6.2 nm at energy densities of 0.17 J/cm^2, 0.32 J/cm^2 and 0.45 J/cm^2, respectively. Silicon nanoparticles were produced with the desired size, less than the Bohr radius of exciton. Given these results, the energy density of the irradiating laser does not significantly influence the silicon nanoparticle size under such experimental conditions.
Mechanism of fabrication of crystalline Si nanoparticles in liquid might be recrystallization of liquid or vapor of silicon in cavitation bubble, which would be created by heating target using laser light. Indeed there are many reports of the relation between cavitation bubble and nanoparticle formation [22-25]. In our experiment, the particle size of silicon nanoparticles was approximately 6 nm. This particle size might be related to lifetime of bubbles, which might restrict the growth of silicon nanoparticles. The lifetime of bubbles would not be changed under our experimental conditions because pressure and temperature of ethanol were almost constant. Therefore, the particle size of silicon nanoparticles was not changed. At the same time, the recrystallization in ethanol might be possible, too.
We investigated the effect of the irradiating laser energy density on the quantity of silicon nanoparticles. Figure 4 shows an estimate of the silicon nanoparticle quantity through ICP. The threshold was approximately 0.15 J/cm^2. In general, the penetration depth as a function of energy density is saturated at a high energy density. If ablation is consistent with this rule, the ablated depth of a silicon wafer per unit energy density decreases with an increase in energy density. However, the experimental results indicate that an increase in laser energy density proportionally increases the quantity of silicon nanoparticles. This result is likely related to the ablated area of the silicon wafer as an element. The ablated area increased with an increase in energy density even in the same irradiated area. To consider the preparation efficiency of silicon nanoparticles as a function of laser power, Figure 4 was deformed. Figure 5 shows an estimate for the quantity of silicon nanoparticles per unit energy density.

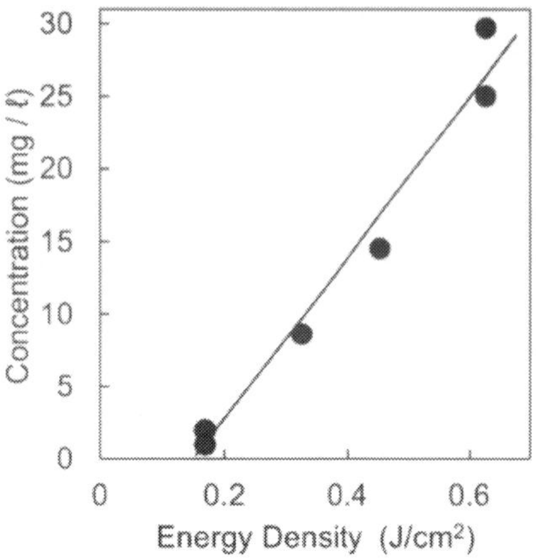

Figure 4. Silicon nanoparticle quantity as a function of energy density.

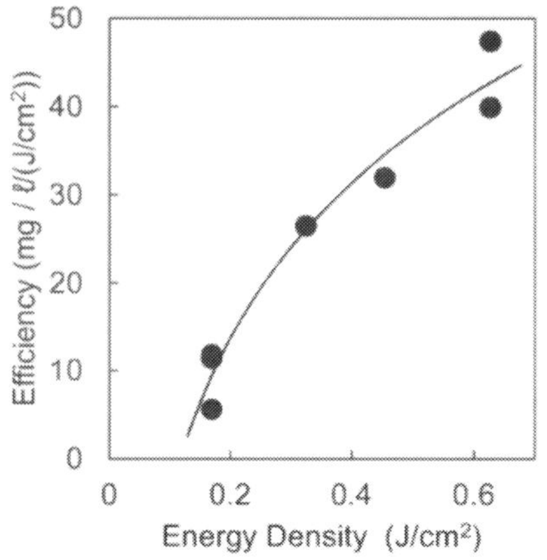

Figure 5. Silicon nanoparticle production efficiency as a function of energy density.

These concentrations in Figure 5 were calculated by dividing the original concentration with energy densities. Efficiency, which is vertical axis value in Figure 5, increased at a low energy density but was saturated at a high energy density. The monotonic increase in Figure 5 indicates that higher energy density effectively increase the quantity of silicon nanoparticles. The tendency toward saturation at a high energy density is likely due to the threshold. When the threshold was subtracted from the energy density, and the concentration was divided by this difference, these values were approximately the same at each energy density. The threshold effect was high at low energy densities and low at high energy densities.

The effect of laser ablation irradiation time on silicon nanoparticle morphology was investigated. In previous study [26], long irradiation times reduced the quantity of silicon nanoparticles because silicon nanoparticles prepared in liquid absorb and scatter the laser, which decreases the energy density at the target and efficiency of silicon nanoparticle production. To confirm the irradiation time dependency, irradiation time was varied. The laser energy density on a silicon wafer was fixed at 0.17 J/cm^2. Figure 6 shows TEM images for the silicon nanoparticles. The particle sizes of the silicon nanoparticles were desirable particle size. The nanoparticle shapes were almost spherical. Figure 7 shows the particle size distribution for the silicon nanoparticles. The average particle size was approximately 6 nm, which is desirable for quantum confinement. These results were similar under each set of experimental conditions. Therefore, irradiation time does not affect silicon nanoparticle morphology under such experimental conditions.

The effect of laser ablation irradiation time on the quantity of silicon nanoparticles was investigated. The quantity of silicon nanoparticles was estimated using ICP. This result is shown in Figure 8. The

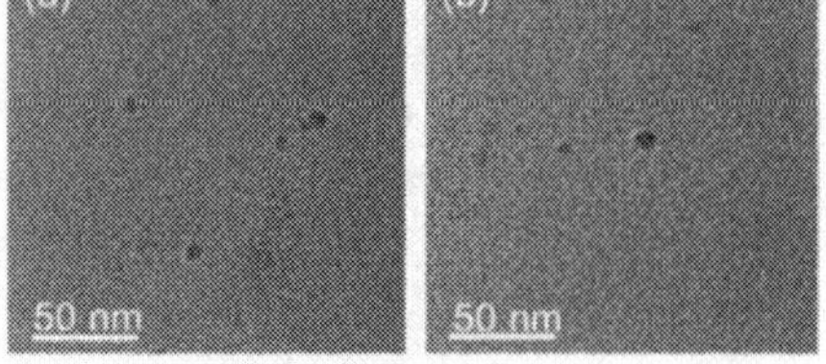

Figure 6. TEM images of silicon nanoparticles. (Irradiation time: (a) 1 hour and (b) 3 hours. Energy density: 0.17 J/cm^2.)

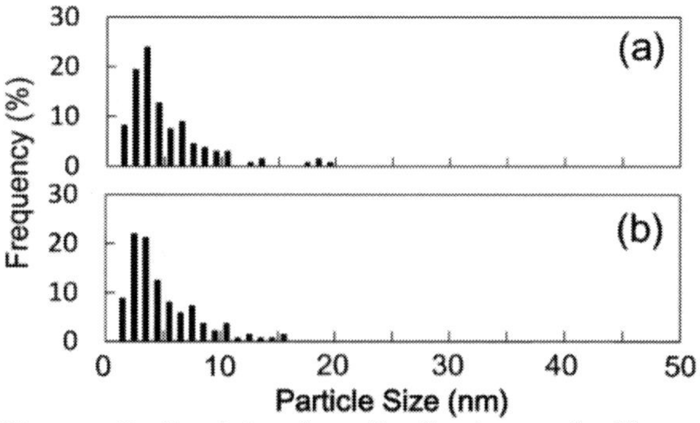

Figure 7. Particle size distributions of silicon nanoparticles. (Irradiation time: (a) 1 hour and (b) 3 hours. Energy density: 0.17 J/cm^2.)

quantity of silicon nanoparticles increased linearly with irradiation time. We observed a small threshold, which is likely related to native silicon dioxide on a target bare silicon wafer. Silicon dioxide has a higher band gap than silicon; therefore, silicon dioxide has a lower absorbance with the green laser. Further, the silicon dioxide melting point is higher than for silicon. The lower absorbance likely decreases laser ablation efficiency. In this experiment, saturation of the quantity of silicon nanoparticles prepared was not observed. Therefore, a longer irradiation time can increase the quantity of silicon nanoparticles.

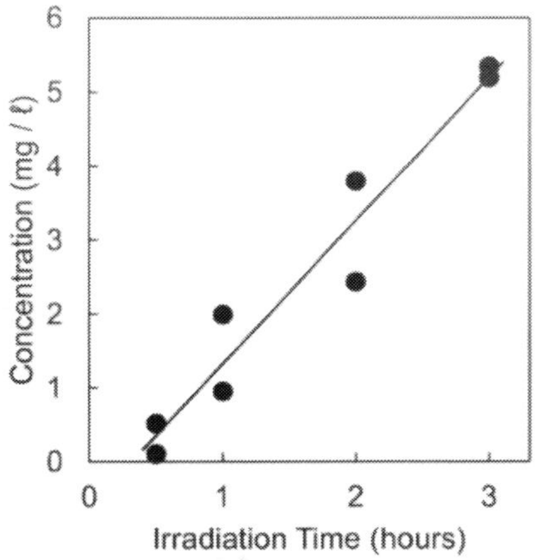

Figure 8. Silicon nanoparticle quantity as a function of irradiation time.

Figure 9 shows an absorbance spectrum for the silicon nanoparticles dispersed in the solution. Although absorption was observed at 532 nm, it has only small influence on particle generation. Therefore, the quantity of silicon nanoparticles increased linearly even at a high energy density. Figure 10 shows the square root of absorbance as a function of photon energy because silicon is indirect transition material. The intercept of horizontal axis which means bandgap E_g is 2.2 eV. Increase in bandgap by quantum confinement was observed as a result of nanosize effect.

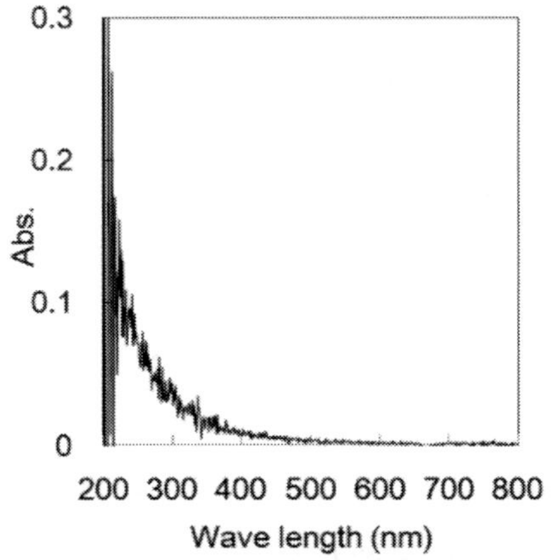

Figure 9. Absorbance spectrum for silicon nanoparticles dispersed in a solution.

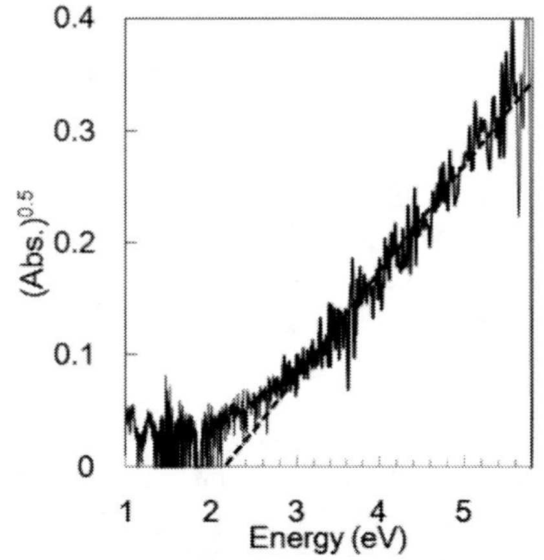

Figure 10. Absorbance spectrum for silicon nanoparticles dispersed in a solution.

4. Conclusion

Silicon nanoparticles were prepared using laser ablation in liquid. In general, the quantity of nanoparticles produced from laser ablation in liquid is low. To increase this quantity, the effect of energy density from the irradiating laser on a silicon wafer was investigated. An increase in energy density proportionally increased the quantity of silicon nanoparticles. The preparation efficiency for the silicon nanoparticles per unit laser power gradually decreased with a decrease in energy density from the laser ablation threshold. Particle size was not influenced by changing the energy density under such experimental conditions. Therefore, higher energy density compared with the laser ablation threshold is necessary for higher nanoparticle preparation efficiency. The effect of irradiation time was also studied. The quantity of silicon as a function of irradiation time was linear. The effect of silicon nanoparticle absorption in liquid was not observed under such experimental conditions. Conditions for increasing silicon nanoparticle production were determined herein. Such conditions will likely contribute to silicon nanoparticle applications, such as solar cells.

5. Acknowledgements
We would like to thank Prof. K. Nakamura (Nd:YAG laser), Ms. H. Tokimori and Mr. K. Hori (TEM), and Mr. S. Nakamura (ICP-MS) in Tokyo Tech. This work was supported by the Materials and Structure Laboratory (Tokyo Tech. collaborative research) and JSPS KAKENHI Grant Number 23119506.

6. References
[1] Švrček V, Turkevych I, Hara K, and Kondo M 2010 *Photovoltaic Specialists Conference (PVSC)*, 2010 35th IEEE 001873
[2] Švrček V, Sasaki T, Shimizu Y, and Koshizaki N 2006 *Appl. Phys. Lett.* **89** 213113
[3] Li Z F and Ruckenstein E 2004 *Nano Lett.* **4** 1463
[4] Matsui Isao 2005 *JOURNAL OF CHEMICAL ENGINEERING OF JAPAN* **38** 535
[5] Nozik A J 2002 *Physica E* **14** 115
[6] Beard M C, Knutsen K P, Yu P, Luther J M, Song Q, Metzger W K, Ellingson R J, and Nozik A J 2007 Nano Lett. 7 2506
[7] Yoffe A D, 1993 *Adv.Phys.* **42** 173
[8] Brech F and Cross L 1962 *Applied Spectroscopy* **16** 59
[9] Neddersen J Chumanov G and Cotton T 1993 *Applied Spectroscopy* **47** 1959
[10] Fojtik A and Henglein A 1993 *Berichte der Bunsen-Gesellschaft Physical chemistry, chemical physics* **97** 252
[11] Mafune F, Kohno J Y, Takeda Y, Kondow T and Sawabe H 2000 *J. Phys. Chem. B* **104** 8333
[12] Tamaki Y, Asahi T, and Masuhara H 2000 *Appl. Surf. Sci.* **168** 85
[13] Usui H, Shimizu Y, Sasaki T, and Koshizaki N 2005 *J Phys. Chem. B* **109** 120
[14] Yoshimura F, Nakamura K, Wakai F, Hara M, Yoshimoto M, Odawara O and Wada H. 2011 *Appl. Surf. Sci.* **257** 2170
[15] Yoshimura F, Ishizaki M, Wakai F, Hara M, Odawara O. and Wada H. 2012 *Adv. Opt. Technol.* **2012** 814745
[16] Ishizaki M, Katagiri T, Sasagawa T, Kitamoto Y, Odawara O. and Wada H. 2012 *J. Nanotechnol.* **2012** 435205
[17] Nunokawa T, Onodera Y, Hara M, Kitamoto Y, Odawara O and Wada H *Appl. Surf. Sci.* in press.
[18] Yang Shikuan, Cai Weiping, Liu Guangqiang, Zeng Haibo, and Liu Peisheng 2009 *J. Phys. Chem. C* **113** 6480
[19] Yang G W 2007 *Progress in Materials Science* **52** 648
[20] Faraci G et al. 2006 *Physical Review B* **73** 33307
[21] Brinker C J, Tallant D R, Roth E P and Ashley C S 1986 *J. non-cryst. solids* **82** 117
[22] Takada N, Nakano T, and Sasaki K 2009 *Appl. Surf. Sci* **255** 9572
[23] Soliman W, Takada N, and Sasaki K, 2010 *Appl. Phys. Express* **3** 035201
[24] Soliman W, Nakano T, Takada N, and Sasaki K 2010 *Jpn. J. Appl. Phys.* **49** 116202
[25] Sasaki K, Takada N Pure 2010 *Appl. Chem.* **82** 1317
[26] Tsuji Takeshi, Iryo Kenzo, Watanabe Norihisa and Tsuji Masaharu 2002 *Applied Surface Science* **202** 80

Journal of Physics: Conference Series **441** (2013) 012036
doi:10.1088/1742-6596/441/1/012036

Synthesis of Antimony Doped Amorphous Carbon Films

H Okuyama, M Takashima, H Akasaka, N Ohtake
Department of Mechanical Sciences and Engineering, Tokyo Institute of Technology,
2-12-1 O-Okayama, Meguro-ku, Tokyo, 152-8225, Japan

E-mail: akasaka@mech.titech.ac.jp

Abstract. We report the effects of antimony (Sb) doping on the electrical and optical properties of amorphous carbon (*a*-C:H) films grown on silicon and copper substrates by magnetron sputtering deposition. For film deposition, the mixture targets fabricated from carbon and antimony powders was used. The atomic concentration of carbon, hydrogen, and antimony, in the film deposited from the 1.0 mol% Sb containing target were 81, 17, 2 at.%, respectively. These elements were homogeneously distributed in the film. On the structural effect, the average continuous sp^2 carbon bonding networks decreased with Sb concentration increasing, and defects in the films were increased with the Sb incorporation because atomic radius of Sb atoms is twice larger size than that of carbon. The optical gap and the electrical resistivity were carried out before and after the Sb doping. The results show that optical gap dropped from 3.15 to 3.04 eV corresponding to non-doping to Sb-doping conditions, respectively. The electrical resistivity reduced from 10.5 to 1.0 MΩm by the Sb doping. These results suggest the doping level was newly formed in the forbidden band.

1. Introduction

Silicon has been used as primary material for electronic devices. Carbon and silicon are the homologous element at 14th group and one of the crystalline carbon materials is diamond. Diamond with energy band gap of 5.5 eV is an attractive semiconductor which can be applied to next generation of electronic devices [1,2]. As amorphous silicon is used for one of the electronic material, an amorphous carbon (*a*-C:H) has potential for the material for the electronic devices. Some studies were reported about electronic properties on *a*-C:H materials. For example, Saitoh *et al.* reported field emission properties of the nitrogenated *a*-C:H films deposited on a sharp tip of ZnO[3]. And some researcher also reported devices using *a*-C:H films [4-6].
Most of such study, dopant for the *a*-C:H materials is nitrogen[6-8]. Conway *et al.* presented that doping effect of nitrogen doping on *a*-C:H films [8]. Omer *et al.* reported photovoltaic characteristics on the device of *a*-C:H films on the silicon substrate [6]. On silicon technology, nitrogen is not primary dopant to obtain *n*-type semiconductors. Phosphor (P) or antimony (Sb) has been used to doping for silicon to obtain the *n*-type semiconductor. The investigation for P doping for *a*-C:H films was already reported [9-11], but no one reported the investigation of Sb doping effect for *a*-C:H films. In this study, Sb-doped *a*-C:H films were synthesized to understand doping effect on *a*-C:H materials.

2. Experimental

Sb-doped *a*-C:H films were deposited from the carbon and antimony mixture target by sputtering with Ar plasma. Targets for the deposition were prepared by the compression moulding from powder mixtures. Sb concentrations in these powder mixtures were coordinated to 0.1, 0.3, 0.5, and 1.0 mol%. The target was fixed to a cathode connected to a high-voltage pulse power supply. The purity and the flow rate for Ar were 99.99% and 20 cm^3/min. The pressure was controlled at 1.0 Pa during deposition, and the deposition time was set at 120 min. The pulse voltage, frequency and width are -1 kV, 20 kHz and 5 μs, respectively. Substrates used were Corning #7059 glass, pure copper plate, silicon (100). The distance from the substrate to the target was approximately 10 cm.
The *a*-C:H film structure was also evaluated using an Raman spectroscope equipped with an Ar laser light source of 514.5 nm wavelength (NRS-1000: JASCO). The peaks in the Raman spectra of different *a*-C:H films were separated by curve fitting using Gaussian functions to obtain the integrated intensity ratio I_D/I_G and G peak position. The composition in each *a*-C:H film was evaluated by the Rutherford backscattering spectrometry (RBS) and the elastic recoil detection analysis (ERDA) using a He^+ ion beam (Nissin High Voltage: NT-1700HS). The elemental distribution in the thickness direction of the *a*-C:H films was estimated by the glow discharge optical emission spectroscopy (GD-OES HORIBA / JOBIN YVON: JY - 5000RF). Optical properties of the deposited *a*-C:H films were investigated using the #7059 glass substrate in the wavelength range from 300 to 1900 nm. The optical band gap (E_g) of the *a*-C:H film was derived from the extrapolation of the linear parts of curve at 0 using the Tauc relation. Electronic defects in the *a*-C:H films were investigated by electron spin resonance spectroscopy (ESR).

3. Results and Discussion

First, the composition of the Sb doped *a*-C:H film deposited from from the 1.0 mol% Sb containing target was determined by the RBS/ERDA measurement. The contents of film were 81, 17, 2 at.% for carbon, hydrogen, and antimony, respectively. Figure 1 shows the time dependence of the optical emission intensities for each element as results of GD-OES for the Sb doped *a*-C:H film deposited from 1.0 mol% Sb containing target. The distribution of Sb in the film was relatively-uniform, and the deposited film consisted of carbon and antimony.

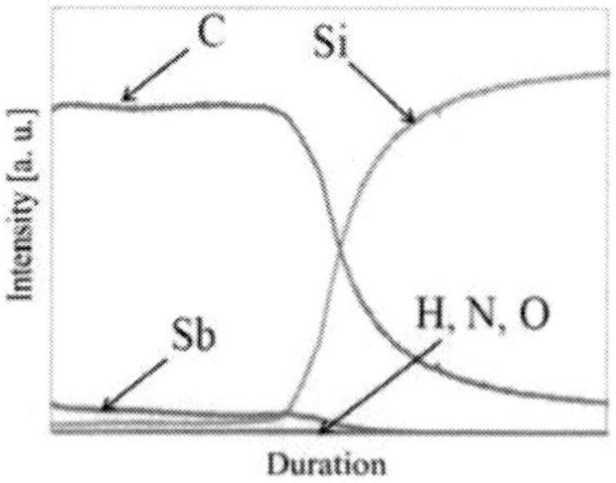

Figure 1. Elemental depth profile for the films.

Figure 2 shows the Raman spectra of the films which contained each Sb concentration. The Raman spectra of *a*-C:H films show the G (graphitic)-band around 1550 cm^{-1} and the D (Disorder)-band around 1350 cm^{-1}, indicating that films are amorphous carbon. The intensity of the G-band decreases as the Sb concentration in the target increases. These results indicate that the average size of the sp^2 cluster consisting of the continuous bonding of sp^2 carbon in *a*-C:H films depends on the Sb concentration in the target. The G-band and D-band were fit assuming a Gaussian distribution, and the ratio of peak areas on D and G band, I_D/I_G, was calculated as shown in Figure 3. I_D/I_G increases with increasing of the Sb concentration in the target. It is generally assumed that the lower the I_D/I_G ratio of the *a*-C:H is the longer average continuous sp^2 carbon bonding networks exists.

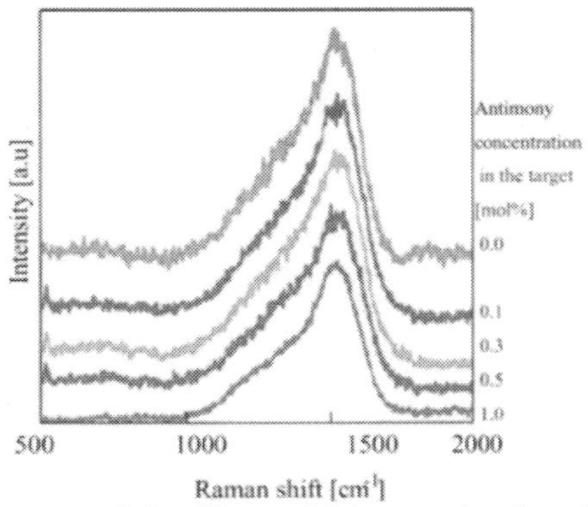

Figure 2. Raman spectra of the films which contained each Sb concentration.

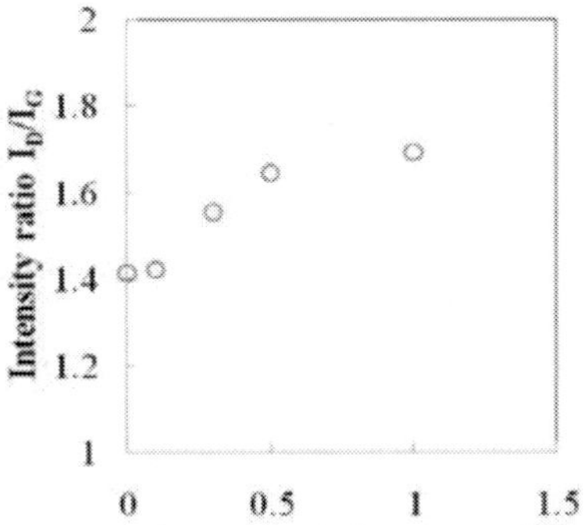

Figure 3. The integral of the fitting ratio I_D/I_G for each Sb concentration.

The variation of E_g as a function of the Sb concentration in the target increases is shown in Figure 4. E_g for the *a*-C:H films decreases with increasing of the Sb concentration. E_g is as large as 3.15 to 3.04 eV, indicating that these *a*-C:H films are semiconductors. Robertson *et al.* proposed to obtain the average sp^2 carbon cluster size in *a*-C:H films from E_g using the equation: E_g=0.77/L_a, where L_a is the sp^2 bonding cluster size [12]. As this relationship using Figure 4, the sp^2 cluster size in the films for films deposited from 0 and 1.0 mol.% Sb doped target was approximately 0.24 and 0.25 nm, respectively. This trend is not matched with the dependence of I_D/I_G calculated from results of Raman measurements. From results of Raman measurements, the sp^2 cluster size decreasing with the Sb concentration increasing suggests that the size of sp^2 cluster was limited by doped Sb. Raman measurements can only measure carbon atom networks, not estimate Sb. Hence that mismatch suggests that doped Sb make the localized state in its band gap, and appear the leading of the decreasing of band gap. It was also considered Sb doping into the 14th group element brought the formation of the new energy level in the band gap, and effected the reducing of the band gap.

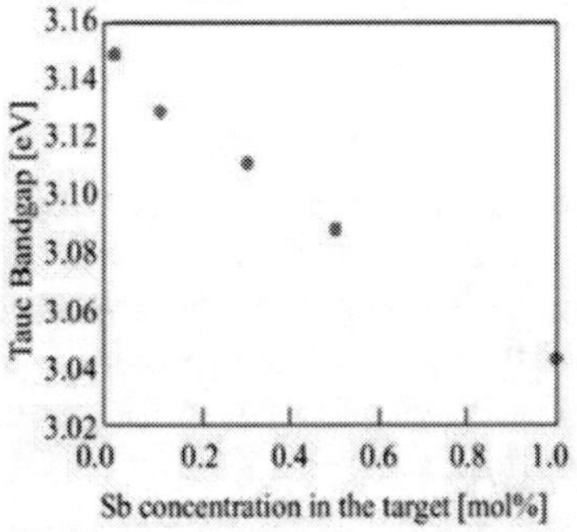

Figure 4. Band Gap of *a*-C:H films as a function of the Sb concentration in the target.

Defect densities in the *a*-C:H films increased with the increasing of Sb concentration in the target, as shown in Figure 5. Defect density was increased with the incorporation of Sb. Because atomic radius of Sb atoms is twice larger size than that of carbon, it was considered that unpaired electron of carbon atoms increased around Sb atoms.

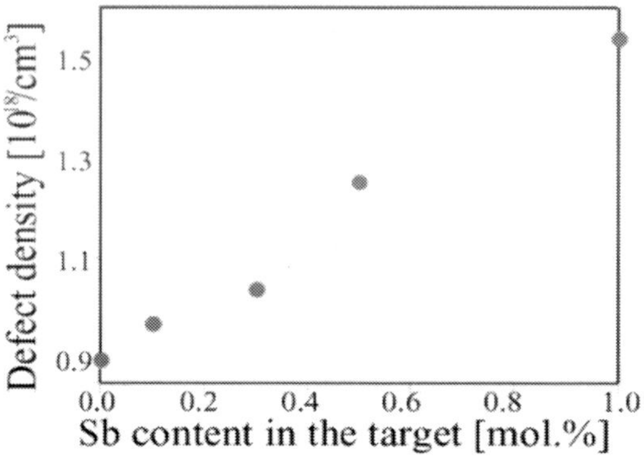

Figure 5. Defect density *vs*. Sb content in the target.

The depletion of the electrical resistivity was found when Sb concentration increased, as shown in Figure 6. The electrical resistivity of the *a*-C:H film rapidly reduced by 0.1 mol% Sb addition to the target. This rapid reducing indicated the formation of new energy level in the band gap by Sb doping. After that, the electrical resistivity was moderately decreased with the Sb concentration increase. The electrical resistivity became approximately 1.0 MΩm at the film deposited from 1.0 mol% Sb contained target. The decrease of the electrical resistivity by the incorporation of Sb was caused by this new band formation, but was also caused by the increasing of sp^2 cluster size.

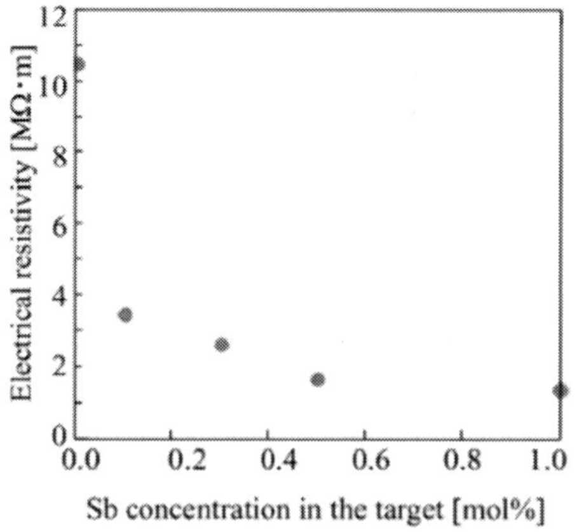

Figure 6. Electrical resistivity *vs*. Sb concentration in the target.

4. Conclusions

Sb doped *a*-C:H films were synthesized, and the effect of the Sb doping on the electrical and optical properties was investigated. On the Sb doped *a*-C:H deposited film from the 1.0 mol% Sb containing target, the atomic concentration of carbon, hydrogen, and antimony were homogeneously distributed at 81, 17, 2 at.%, respectively. By the Sb incorporation, the average of network size of continuous sp^2 carbon bonding decreased and defect densities were increased. Sb doping reduced the optical gap from 3.15 to 3.04 eV, and the electrical resistivity from 10.5 to 1 MΩm. These results suggested the doping level was newly formed in the forbidden band.

5. Acknowledgments
The author would like to acknowledge Prof. Akira Yamada, Tokyo Institute of Technology, for his help with ESR measurements. The author would like to thank Prof. Tsuneo Suzuki at Nagaoka University of Technology, for his help with RBS/ ERDA measurements and exciting discussion.

6. References
[1] Koide Y 2004 *Dia. Relat. Mater.* **13** 1963
[2] Hirama K, Miyamoto S, Matsudaira H, Yamada K, Kawarada H, Chikyo T, Koinuma H, Hasegawa K and Umezawa H 2006 *Appl. Phys. Lett.* **88** 112117
[3] Saitoh H, Akasaka H, Washio T, Ohkawara Y, Ohshio S and Ito H 2002 *Jpn. J. Appl. Phys.* **41** 6169
[4] Miyajima Y, Shkunov M and Silva S R P 2009 *Appl. Phys. Lett.* **95** 102102
[5] Silva S R P and Carey J D 2003 *Dia. Relat. Mater.* **12** 151
[6] Omer A M M, Rusop M, Adhikari S, Adhikary S, Uchida H and Umeno M 2005 *Dia. Relat. Mater.* **14** 1084
[7] Bhattacharyya S, Walzer K, Hietschold M and Richter F 2001 *J. Appl. Phys.* **89** 1619
[8] Conway N M J, Milne W I and Robertson J 1998 *Dia. Relat. Mater.* **7** 1477
[9] Pearce S R J, Filik J, May P W, Wild R K, Hallam K R and Heard P J 2003 *Dia. Relat. Mater.* **12** 979
[10] Kuo M T, May P W, Gunn A, Ashfold M N R and Wild R K 2000 *Dia. Relat. Mater.* **9** 1222
[11] Regan E M, Uney J B, Dick A D, Zhang Y, Yanez J N, McGeehan J P, Claeyssens F and Kelly S 2010 *Biomaterials* **31** 207
[12] Robertson J 1992 *Surf. Coat. Tech.* **50** 185

Prompt response and durability of polymer ablation from synthetic fibers irradiated by thermal plasmas for arc resistant clothes

Masahiro Ishida[1], Naoki Shinsei[1], Yasunori Tanaka[1,2], Yoshihiko Uesugi[1,2], Tatsuo Ishijima[2], Wataru Mio[3], Hiroyasu Hagi[3] and Keita Uchibori[3]

[1]Faculty of Electrical and Computer Engineering, Kanazawa University, Kakuma, Kanazawa 920-1192, Japan

[2]Research Center for Sustainable Energy and Technology, Kanazawa University, Kakuma, Kanazawa 920-1192, Japan

[3]Kaneka Corporation, 3-2-4, Nakanoshima, Kita-ku, Osaka 530-8288, Japan

E-mail: tanaka@ec.t.kanazawa-u.ac.jp

Abstract. Interactions between thermal plasmas and synthetic fibers such as polyamide, polyester, phenol and aramid were investigated by thermal plasma irradiation technique. Understanding the above interactions is crucial to design effective flame retardant synthetic fiber clothes with arc resistance to protect a human from arc flash accidents. To investigate the interactions, an Ar inductively coupled thermal plasma (ICTP) was used instead of the arc discharge because the ICTP has high controllability and no contamination. The ICTP irradiation raises polymer ablation in case of polyamide and polyester. Two features of the polymer ablation such as prompt response and durability were fundamentally investigated from viewpoint of shielding the heat flux. It was found that polyamide fiber has both a high prompt response and a long durability.

1. Introduction

According to statistics compiled by CapSchell Inc., Chicago, there have been 5–10 arc flash incidents in a day in the United States [1]. The arc plasma is established around the accidental portion between high voltage conductors. This arc plasma usually has extremely high temperature more than 5000 K. It also generates brilliant flash of light involving a large noise. Thus, such arc flash can result in considerable damage to electrical equipments and serious injuries to nearby personnel. To protect personnel workers, it is greatly important to develop protective clothing with a sufficient arc resistant ability.

The National Fire Protection Association (NFPA) has developed a leading internationally recognized safety standard for electrical safety in workplace as NFPA 70E [2]. This standard defines a set of safe requirements for personnel working on electrical equipment. To comply with the standard, employers must carry out a hazard risk assessment and ensure that all employees working in a potential arc-flash hazard zone use appropriate equipment and wear the right protective clothing. The NFPA 70E standard also defines the arc thermal performance value (ATPV) for clothing. This value represents the maximum capability for arc-flash protection of a particular garment. This rating also applies to fabrics. It is presented in cal/cm^2 or kW/m^2 through the specified arc exposure test. This arc exposure test needs the arc plasma to create heat flux rates from 84 to 25120 kW/m^2 exposed to the test clothing. However, this arc exposure test generally requires large test equipments and is also very expensive.

On the other hand, the authors continue to study the interactions between the bulk polymers and thermal plasma, which involves the polymer ablation. This study aims the enhancement in the arc interruption ability of low-voltage circuit breakers by using polymer ablation phenomena [3]. The polymer ablation occurs inevitably by contacting arc plasmas between the electrodes with the polymer wall during the high current interruption performance in a circuit breaker. The polymer ablation creates pressure rise, and then gas flow to enhance the arc interruption ability of the circuit breaker. The polymer ablation is also used for heat protection of a re-entry capsule in aerospace fields [4]. The re-entry capsule with a polymer ablator is exposed to high-temperature air plasma by the aerodynamic effects during re-entry process. The polymer ablator is effectively ablated to generate a heat shield vapor preventing the capsule from being exposed to too high heat flux.

One goal of the present work is to develop a high performance synthetic fibers with arc-resistance ability. The contribution of this paper contains two proposals: (i) to adopt polymer ablation phenomena of the synthetic fiber to protect from the arc plasma heat flux, and (ii) to use an induction thermal plasma irradiation technique instead of the arc exposure test. The contribution also includes fundamental experimental results on the thermal plasma irradiation test to four kinds of synthetic fiber specimens. From the experiments, strong polymer ablations were found for three of the above specimens. To use the polymer ablation for arc protection, two important features such as prompt response and durability of the polymer ablation were fundamentally investigated. The temperature of the ablated vapor was also estimated to obtain the heat shielding ability due to polymer ablation. It was found that polyamide fiber has both a high prompt response and a long durability.

2. Experimental

2.1. Test synthetic fibers

The present work targets the following four kinds of polymer synthetic fibers: polyamide-66, polyester, phenol and para-aramid. The polyamide 66 (PA66) fiber is well known as nylon-66 made of repeating units linked by amide bonds. Its heat resistant ability is relatively low, because it has a lower pyrolysis temperature. The pyrolysis temperature of the test PA66 fiber was acctually measured to be 704 K with the thermo gravimetry, the differential thermal analysis and the differential scanning calorimetry (TG-DTA/DSC) method. Thus, it can be easily ablated by the heat flux according to our previous work. This ablation phenomena is expected to be adopted to the heat shielding from arc/thermal plasmas in the present study. The polyester fiber is widely used for textile, and it is also known as the polyethylene terephthalate (PET) fiber. However, in the context of textile applications, PET is referred to rather by its common name, polyester because it contains the ester functional group in its main chain. The pyrolysis temperature of the test PET fiber was measured to be 700 K. From this lower pyrolysis temperature, PET is also expected to be ablated by heat flux. The phenol fiber has high mechanical strength, high wear, heat and flame resistance, and morphological stability. This material fiber is widely used for fire retardant clothes. For this phenol fiber, the pyrolysis temperature could not be measured clearly with TG-DTA/DTG method. The para-aramid (p-aramid) fiber has amide group, and it has various performances similar to a phenol fiber. It is categorized in heat-resistant and strong synthetic fibers. Its applications are in aerospace and military fields for ballistic rated body armor fabric, etc. It is mechanically and thermally stronger than the PA66, because amide bond is combined with benzene ring formula instead of straight chain alkane structures. The pyrolysis temperature of the test p-aramid fiber was measured to be 850 K.

2.2. Experimental arrangements

Figure 1 depicts the inductively coupled thermal plasma (ICTP) torch used in the experiments as a thermal plasma source. The plasma torch is composed of two coaxial quartz tubes. The inner diameter of the interior quartz tube is 70 mm; its length is 330 mm. Ar gas was supplied as a sheath gas in axial direction and swirl direction along the inner wall of the inner tube. An eight-turns coil is located around the quartz tubes. The coil was connected with an rf power source to supply the coil current. The coil current produces electromagnetic fields to form an ICTP in the plasma torch.

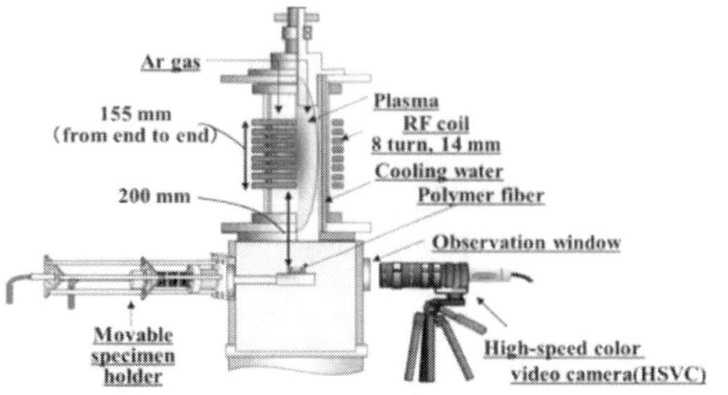

Figure 1. Plasma torch and chamber.

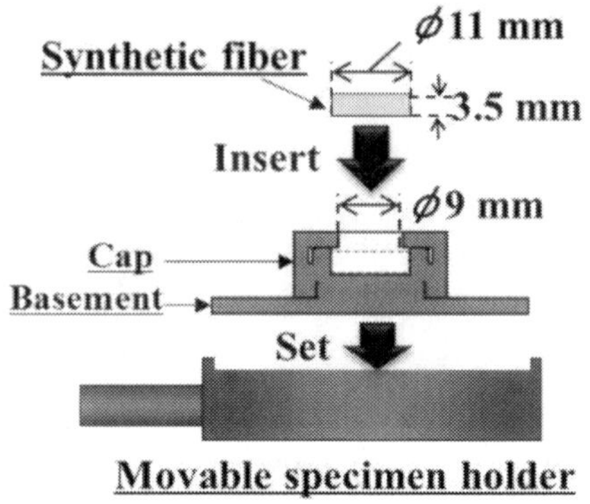

Figure 2. Specimen, specimen mount, and specimen holder.

Downstream of the plasma torch, a water-cooled specimen holder was located in the chamber. On this specimen holder, a specimen in a mount is placed to be directly irradiated by the ICTP from the plasma torch. Figure 2 shows the schematic of the specimen fiber, the mount and the water-cooled specimen holder. Specimens used in the present experiment were the polymer synthetic fibers. They are compressed to form a cylindrical shape with a diameter of 11 mm and a thickness of 3.5 mm. The specimen mount is used to fix a fiber specimen. This mount is comprised of two parts: a basement and a cap made of stainless steel. The cap has a hole of 9 mm in diameter. Through this hole, a thermal plasma is irradiated to specimen surface directly. A specimen in the mount was located on the water-cooled specimen holder for thermal plasma irradiation test.

The thermal plasma irradiation to a polymer synthetic fiber specimen may lead to the polymer ablation. The polymer ablation was observed by a color high-speed video camera (HSVC). At the same time, spectroscopic observation was carried out at 2 mm above the surface of the specimen. The optical system is constituted of a quartz lens, an optical fiber bundle, a polychromator and an ICCD detector. The spatial resolution of this observation is about 2.5 mm in diameter. Irradiation was initiated by inserting the movable specimen holder to the position just under the plasma torch.

2.3. Experimental conditions

The total flow rate of Ar sheath gas was fixed at 30.0 L min^{-1}. The pressure inside the chamber was controlled at 760 Torr (=101 kPa) with an automatic feedback control for pressure. The distance between the coil end to the specimen surface was set to 200 mm. The input power was fixed at 8.5 kW to the inverter power supply. This input power corresponds to the irradiation heat flux of about 550 kW/m^2 from the ICTP to the specimen, according to our numerical calculation result. This heat flux is in the range of the standard test method for determine the arc rating of materials for clothing [5]. The HSVC was used to capture ablated vapor from the specimen. The frame rate was 1000 frames per second.

3. Results for Ar thermal plasma irradiation

3.1. Color high speed video camera observation

Figures 3 shows the irradiating synthetic fiber situation taken by the HSVC. Panel (a) indicates the schematic explanatory view of the video capture pictures; panels (b)–(e) are the images for synthetic fibers irradiated by the ICTP. As seen in this figure, the polymer ablation strongly occurs for PA66, PET and phenol fibers irradiated by the ICTP. For these three materials, the ablated vapor emitting blue-white light was observed over the specimen mount. In particular, the PA66 emitted much more ablated vapor

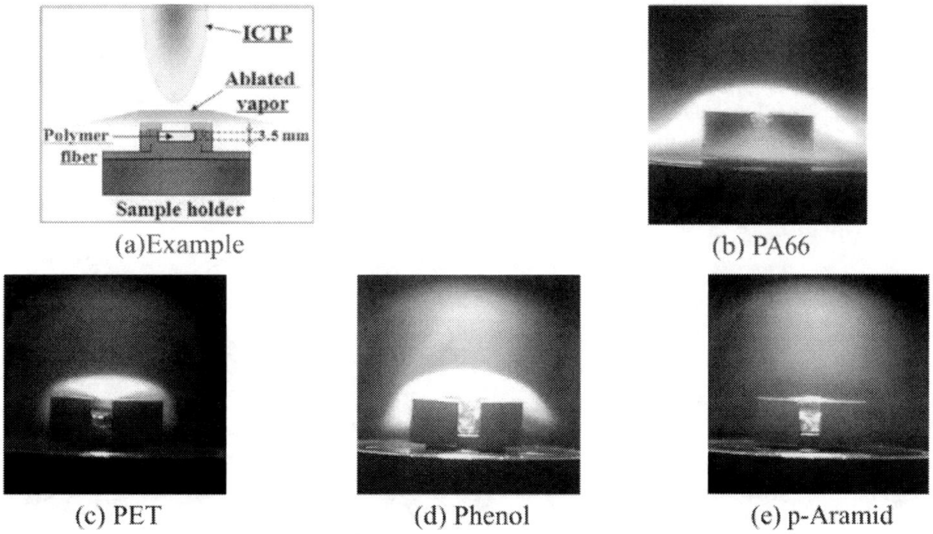

Figure 3. High speed video images for different synthetic fibers irradiated by Ar induction thermal plasmas.

Table 1. Prompt response and durability of synthetic fiber specimens.

	PA66	PET	Phenol	p-Aramid
Ablated vapor	yes	yes	yes	no
Start time of ablation [s]	0.1	0.5	0	no
End time of ablation [s]	20	20	0.4	no
Prompt response	yes	no	yes	no
Durability	yes	yes	no	no

than the others. Whereas, the p-aramid specimen irradiated by the ICTP hardly generated ablated vapor. This hard generation may be due to its high heat resistance ability.

We are considering the ablated vapor could be rather useful to shield heat flux from thermal plasmas on the polymer surface. For the shielding heat flux by the ablation phenomena, there are two important points that polymer synthetic fibers should have: a high prompt response of the ablation and a long durability of the ablation. From the measurement of the HSVC, these two points were summarized in Table 1 for test four kinds of fibers. The prompt response was judged as 'yes' if the ablation initiates in 0.5 s after the ICTP irradiation. The durability was 'yes' if the ablation is kept over 10 s.

The PA66 fiber emitted much more ablated vapor. The PA66 ablation was initiated quickly at 0.1 s after the ICTP irradiation, and then the ablated vapor continued to cover the specimen surface after 20 s from the irradiation. The PET fiber required about 0.5 s to initiate the ablation, but continued to generate the ablated vapor. The amount of the ablated vapor was less than the PA66. The phenol fiber also had a quick response to be ablated. However, the phenol finished to emit ablation vapor until 0.4 s after the ICTP irradiation. On the other hand, p-aramid fiber was scarcely ablated by the heat flux although it was damaged after the ICTP irradiation. According to these results, it was found that PA66 has both prompt response and durability of polymer ablation, which is expected to shield heat flux of the arc plasmas.

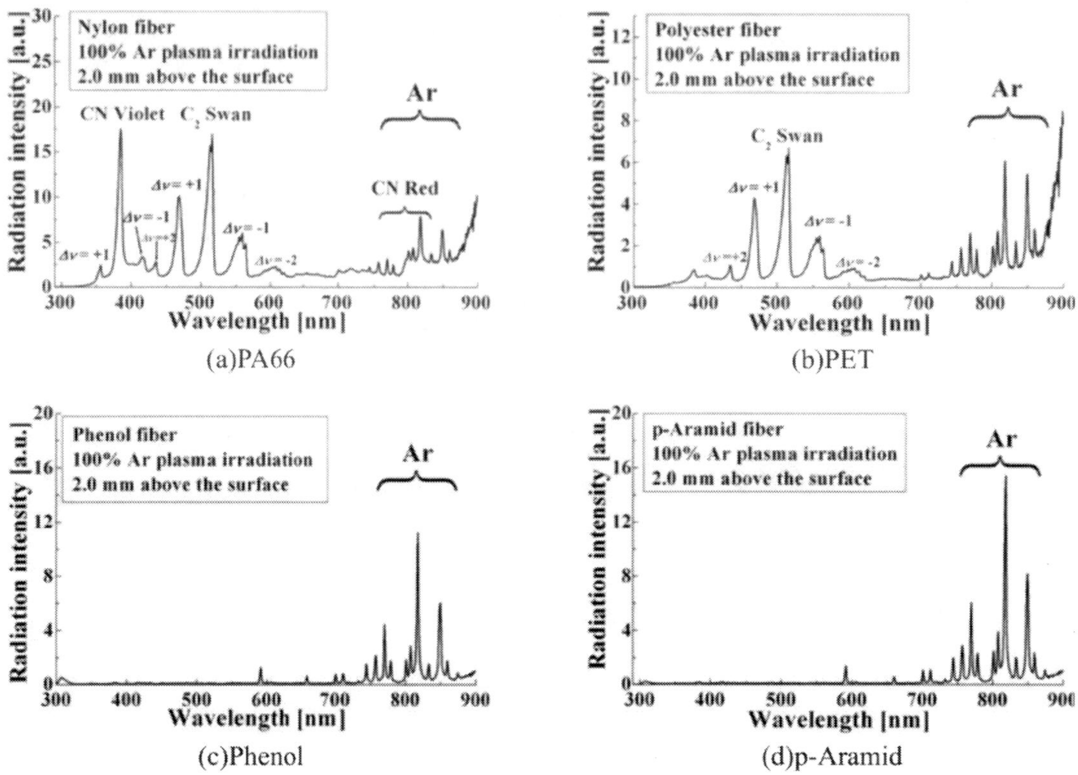

(a)PA66 (b)PET

(c)Phenol (d)p-Aramid

Figure 4. Measured emission spectra from different synthetic fibers irradiated by Ar thermal plasmas.

3.2. Spectroscopic observation and estimation of vapor temperature

The ablated vapors were observed from some of specimens in the ICTP irradiation experiments. The spectroscopic observation was conducted to analyze the composition of the vapor and then the vapor temperature. Figure 4 shows the emission spectra observed at 2 mm above the specimen surface of four kinds. These spectra were measured at 20 s after the ICTP irradiation. The p-aramid fiber is scarcely ablated, and thus almost only Ar spectral lines were observed as indicated in panel (d). The phenol fiber has a lower durability of the ablation, and thus the spectral intensity from the ablated vapor at 20 s after the irradiation is also weak as depicted in panel (c). On the other hand, strong molecular spectra from C_2 Swan system can be detected in ablation vapor from the PA66 and PET fibers. Especially, the PA66 fiber vapor has a stronger intensity from C_2 compared to the PET fiber. In addition, the vapor from PA66 also has strong intensity from CN Violet system because the PA66 contains nitrogen atoms in it. This implies that the vapor measured is indeed the ablated one from the specimen, and that bluish light from the vapor is mainly due to C_2 molecular spectra.

Use of the molecular spectra enables the estimation of the vibrational temperature and the rotational temperature of molecules. First, the emission coefficient of C_2 Swan system was theoretically calculated as functions of the vibrational temperature $T_{\rm vib}^{\rm C_2}$ and the rotation temperature $T_{\rm rot}^{\rm C_2}$ of C_2 assuming that population of exited molecules follows the Boltzmann distribution. After that, fitting the theoretically calculated emission coefficient to the measured spectra yields the both $T_{\rm vib}^{\rm C_2}$ and $T_{\rm rot}^{\rm C_2}$ simultaneously. The $T_{\rm vib}^{\rm C_2}$ and $T_{\rm rot}^{\rm C_2}$ estimated here has a relatively low accuracy because the Abel inversion was not employed for the radiation intensities. In spite of that, the estimated $T_{\rm vib}^{\rm C_2}$ and $T_{\rm rot}^{\rm C_2}$ can be used for an indicator of the temperature with ± 500 K uncertainty at the highest intensity region.

Figure 5 represents the estimated $T_{\rm vib}^{\rm C_2}$ and $T_{\rm rot}^{\rm C_2}$ from PA66 and PET fibers. This figure also includes

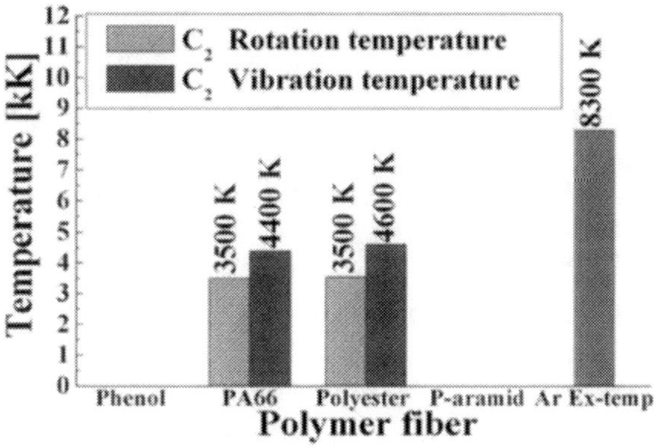

Figure 5. Rotational and vibrational temperatures of C_2 in ablated vapor. Ar excitation temperature without specimen is also added.

the Ar excitation temperature $T_{\rm ex}^{\rm Ar}$ in case of no specimen. The temperature $T_{\rm ex}^{\rm Ar}$ was estimated by the two-line method using Ar atomic lines at wavelengths 703.7 nm and 714.0 nm on the assumption of Boltzmann relation for the excitation population. In this figure, $T_{\rm vib}^{\rm C_2}$ is about 4400–4600 K and $T_{\rm rot}^{\rm C_2}$ is about 3500 K. Generally, the rotational temperature becomes close to the translational temperature of heavy particles, because the energy level difference for rotational excitation is low of the order of 0.01 eV and thus the energy transfer happens frequently by collision with many heavy particles. Whereas, $T_{\rm vib}$ is rather close to the electron temperature in a plasma because of the energy level difference for vibrational excitaion is on the order of 0.1 eV. However, the most important thing that we emphasize in Fig. 5 is the difference of $T_{\rm vib}^{\rm C_2}$ and $T_{\rm rot}^{\rm C_2}$ from $T_{\rm ex}^{\rm Ar}$. Originally, the Ar ICTP irradiates the heat flux with this $T_{\rm ex}^{\rm Ar}$ of about 8300 K. The heat flux leads to the ablation of the synthtic fibers to create cloud of the ablation vapor covering the specimen surface. The vapor temperature is of the order of $T_{\rm vib}^{\rm C_2}$ or $T_{\rm rot}^{\rm C_2}$. These temperatures are 3000–4000 K which is much lower than $T_{\rm ex}^{\rm Ar}$. This may indicate that the ablated vapor with lower temperature covers the specimen surface, reducing temperature gradient between the plasma and the specimen surface, and then reducing the actual heat flux effectively.

4. Conclusions

In this paper, interactions between thermal plasma and synthetic fiber were fundamentally investigated for effective design of the arc retardant textiles and clothes. The arc retardant clothes are desired to protect a human from the arc flash accidents. Four kinds of synthetic fibers were tested: polyamide-66 (PA66), PET, phenol, para-aramid fibers. The Ar inductively coupled thermal plasma (ICTP) was irradiated for this fundamental study instead of the conventional arc discharge because of its controllability and no contamination. It was found that the PA66 and polyester fibers were markedly ablated by the ICTP irradiation, and the ablated vapor covered the specimen surfaces. From this experiment polyamide fiber was found to have both a high prompt response and a long durability. The temperature of the ablated vapor was estimated to be 3000–4000 K much lower than the plasma temperature 8300 K. This implied that covering the specimen surface by a lower temperature vapor may reduce the temperature gradient, resulting in the heat flux reduction on the specimen surface.

References

[1] Markisan Naso 2004 *Safety & Health* **3** 42–47
[2] National Safety Council 2004 *Today's Supervisor* **9** 3
[3] Sakuyama T, Tanaka Y, Uesugi Y, Kaneko S, Okabe S 2009 *J. Plasma & Fusion Res. Series* **8** 728–31
[4] Yamada T and Abe T 2006 *J. Plasma & Fusion Res.* **82** 368–74
[5] ASTM International 2008 Standard Test Method for Determing the Arc Rating of Materials for Clothing

Journal of Physics: Conference Series **441** (2013) 012038 doi:10.1088/1742-6596/441/1/012038

Characteristics of TiO_2 Surfaces Etched by Capacitively Coupled Radio Frequency N_2 and He Plasmas

Retsuo Kawakami[1], Masahito Niibe[2], Yoshitaka Nakano[3], Masashi Konishi[1], Yuta Mori[1], Hideo Takeuchi[4], Tatsuo Shirahama[1], Tetsuya Yamada[1], and Kikuo Tominaga[1]

[1]The University of Tokushima, 2-1 Minami-Josanjima, Tokushima 770-8506, Japan
[2]University of Hyogo, 3-1-2 Kouto, Kamigoori, Ako-gun, Hyogo 678-1205,Japan
[3]Chubu University, 1200 Matsumoto, Kasugai, Aichi 487-8501, Japan
[4]The University of Shiga Prefecture, 2500 Hassaka, Hikone, Shiga 522-8533, Japan

E-mail: retsuo@ee.tokushima-u.ac.jp

Abstract. We have investigated characteristics of anatase TiO_2 thin film surfaces etched by a capacitively coupled radio frequency N_2 plasma from the viewpoint of both an experiment and a simulation. The result obtained with use of the N_2 plasma is compared with that obtained with use of the He plasma. The experimental O/Ti ratio at the surface etched by the N_2 plasma increases as a function of etching time. The increase in the experimental O/Ti ratio is independent of a change in gas pressure. The above-mentioned experimental results agree with the simulation result that the Ti atoms at the surface are preferentially removed by N_2^+ ions. The experimental O/Ti ratio at the surface etched by the He plasma also increases, which does not agree with the simulation result that the O atoms are preferentially removed by He^+ ions. The O-1s XPS spectra of the surface etched by the He plasma show the peak associated with oxygen and water adsorbed by the surface. In the case where the component of the adsorbed oxygen and water in each spectrum is subtracted, the experimental O/Ti ratio at the surface etched by the He plasma shows a decrease. This is consistent with the simulation result. The adsorption of oxygen and water seems to be caused by ambient air. Morphology of the surface etched by the N_2 plasma is almost similar to that of the as-grown surface regardless of the changes both in etching time and in gas pressure. In contrast, morphology of the surface etched by the He plasma changes as the etching time increases. This result is independent of the gas pressure.

1. Introduction

TiO_2 has been attracting much attention for applications to dye-sensitized solar cells [1,2] and environmental cleaning [3-5]. The attention to TiO_2 originates from the fact that TiO_2 has excellent photo-catalytic properties, non-toxicity, and chemical stability. In order to prepare TiO_2 thin films, sputtering is one of the promising deposition techniques for uniform and large area coating [6-8]. However, plasma-induced damage of TiO_2, which degrades the original characteristics, is a crucial problem because plasmas are used to sputter atoms from a target surface [9]. TiO_2 has also been expected for a candidate of gate dielectric material in metal-oxide-semiconductor transistors [10-12].

The expectation arises from the characteristic that TiO_2 has a high dielectric constant, which results in suppressing of the gate leakage current. However, the damage of TiO_2 in the plasma etching process is also a crucial problem because TiO_2 is a binary compound [13]. It is, therefore, necessary to study the above-mentioned issue in order to improve the characteristics of TiO_2 surfaces exposed to or etched by various types of plasmas.

So far, Kawakami *et al.* have studied He plasma-induced damage of TiO_2 [14]. Their study indicates that morphology of the etched surface changes at the gas pressure range from 10 to 100 mTorr as the etching time increases to 60 min. It is considered that the change in the surface morphology results from synergy effect of He^+ ions and ultraviolet (UV) light emitted from the plasma, i.e., UV irradiation effect [14]. The mechanism of the change in the surface morphology, however, has not been clarified in detail. Furthermore, it is not clear whether the surface morphology can be changed by other types of plasma.

In the present work, characteristics of TiO_2 thin film surfaces etched by a N_2 plasma have been studied using a capacitively coupled radio frequency (RF) plasma reactor. The surface morphology was observed, and O/Ti ratio at the surface was analyzed. A change in O/Ti ratio at the surface etched physically by N_2^+ ions was also calculated with a simulation model of plasma-surface interactions developed by the authors. The result obtained with use of the N_2 plasma is compared with that obtained with use of the He plasma. We discuss a difference between the surface characteristics induced by the N_2 and He plasmas.

2. Experiment and Simulation Procedures

For the present study, we employed a capacitively coupled RF plasma reactor with an asymmetric electrode system, named CPA: the area of anode electrode was much larger than that of the cathode one [14]. The cathode was powered with a 13.56 MHz RF generator (Peal Kogyo RP-500C). The anode, which is located 4 cm away from the cathode, was electrically grounded. The RF maximum voltage V_{RF} was set to 200 V. The gas pressure was varied from 10 to 100 mTorr while the gas flow rate was kept to be 15 ml/min. The RF power of the N_2 plasma increased from 4 to 8 W with increasing gas pressure from 10 to 100 mTorr. The RF power for the He plasma was also enhanced from 5 to 10 W. The self-bias voltage V_{DC} generated at the cathode in each plasma discharge was almost equal to $-V_{RF}$ regardless of the increase in the gas pressure: $V_{DC} \approx V_{RF} = -200$ V. Spectra of UV light emitted from the plasmas were measured using a spectroscope (Ocean Optics USB2000).

The specimen used was a thin film of TiO_2 deposited on unheated glass (Corning 1737) using the DC facing target planner magnetron sputtering technique [6]. The specimen was ~1 μm in thickness, and it showed the anatase (101) peak in the XRD pattern. The specimen was cut into the samples. After the plasma etching, the surface morphologies were observed using a scanning electron microscope (SEM, Hitachi S-4700). The O/Ti ratios at the etched surfaces were analysed with the O-1s and Ti-2p spectra taken using X-ray photoelectron spectroscopy (XPS, Shimadzu ESCA-1000).

In order to clarify the N_2 plasma etching of TiO_2, a simulation model for plasma-surface interactions, named PIS, has been developed. The developed simulation is composed of two particle models. One is a model for simulating the plasma discharge including reactions of charged particles with gas, which is based on PIC/MCS [15,16]. The other is a model for simulating the physical etching, which is based on BCA/MCS [17]. Both the self-consistent N_2 plasma behaviour and the physical etching by N_2^+ ions impinging on the surface of TiO_2 were analysed using PIS. In the model of plasma discharge, movement of charge particles, e^- and N_2^+, which is dominated by the Newton and Poisson laws, was simulated [15]. The simulation takes account of the elastic, excitation, and ionizing reactions of e^- with N_2 gas [16,18], as well as the elastic and charge transfer reactions of N_2^+ with N_2 gas [16,19]. As a result, a change in energy of the impinging N_2^+ ions with gas pressure was calculated. In the model of physical etching, the elastic and inelastic reactions of moving particles (N_2, Ti, and O) with target particles (Ti and O) and with electrons in the solid were simulated based on the same scheme as that used in BCA/MCS [17]. The surface binding energy, which is indispensable for emitting the target atoms from the surface, was assumed to be the TiO_2 binding energy, 5.5 eV [20].

Before the simulation starts, O/Ti ratio at the surface is set to be 2. Eventually, a change in the O/Ti ratio with an etching time was calculated. For the simulation of He plasma etching, the details are described in ref. 14.

3. Results and Discussion

Figure 1 shows energy of N_2^+ ions impinging on the cathode (the TiO_2 surface) as a function of gas pressure, which is calculated using the PIS simulation. The energy of N_2^+ decreases with increasing gas pressure. The decrease is due to an increase in the energy losses of N_2^+ resulting from its elastic and charge exchange reactions with N_2 [21]. The result for N_2^+ is almost similar to that for He^+ as shown in Fig. 1. The similarity between the energies of N_2^+ and He^+ originates from the fact that the self-bias voltages generated by the N_2 and He plasmas are almost the same. In contrast, the flux of N_2^+ increases with increasing gas pressure as shown in Fig. 1. The flux is roughly derived from experimental data of the RF plasma discharge [14]. The increase in the flux of N_2^+ almost resembles that of He^+. The flux of N_2^+ at each gas pressure, however, is lower than that of He^+. The lower flux of N_2^+ is probably because the N_2 plasma density, which is proportional to the flux, is lower than the He plasma density. The lower density of N_2 plasma is attributed to the phenomenon that reaction of vibration excitation of N_2 by electron impact is more pronounced than that of the ionization [18].

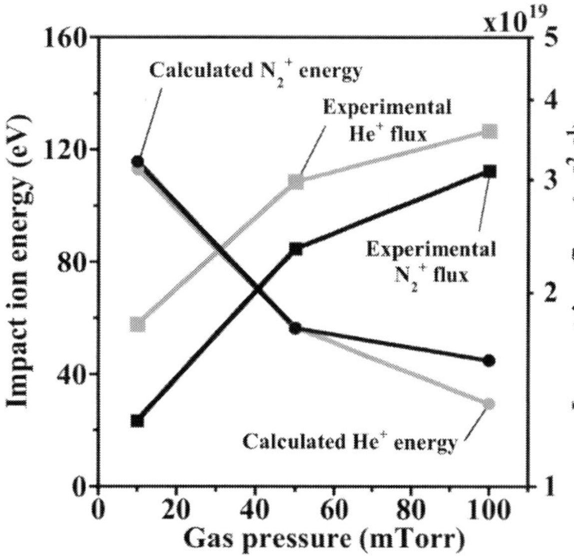

Figure 1. Energies and fluxes of N_2^+ and He^+ ions impinging on the cathode (the TiO_2 surface) as a function of gas pressure.

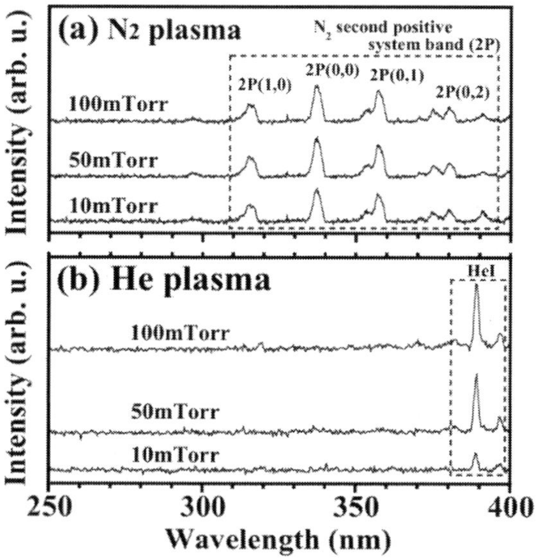

Figure 2. Spectra of UV light emitted from the plasmas generated at various gas pressures.

Figure 2 shows the spectra of UV light emitted from the N_2 and the He plasmas at various gas pressures. As shown in Fig. 2(a), at the low gas pressure of 10 mTorr, several peaks appear in the wavelength range from 310 and 390 nm. The observed peaks are assigned to N_2 second positive system band, 2P(v',v''): $C^3\Pi_u(v') \rightarrow B^3\Pi_g(v'')$, where v is the vibrational state [22]. The result of the measurement on the UV spectra indicates the significant contribution of vibrationally excited N_2 to the emitted UV light. The peak positions do not change with increasing gas pressure, but their intensities slightly increase. In the emission spectra of the He plasma generated at the low gas pressure, few peaks appear between the wavelengths of 388 and 398 nm as shown in Fig. 2(b). The peaks are due to HeI transition lines, which indicates the contribution of excited He to the UV light [14]. The intensities of the peaks are enhanced with increasing gas pressure, and the peaks positions do not change. The spectra of UV light emitted from the N_2 and He plasma are, therefore, independent of gas pressure.

Figure 3 shows the experimental and calculated O/Ti ratios at the surfaces at various gas pressures. As shown in Figs. 3(a), 3(c), and 3(e), the experimental O/Ti ratio at each gas pressure increases with an increase in the etching time. The increase in the experimental O/Ti ratio agrees with the simulation that Ti atoms are preferentially removed by the physical etching effect resulting from the N_2^+ impact. The preferential etch of Ti atoms is attributed to the phenomenon that Ti atoms move more easily than O atoms by N_2^+ impact because, on the basis of a binary collision model, the maximum transferable elastic energy of N_2^+ to Ti is higher than that of N_2^+ to O [17]. The agreement between the experimental and simulated results indicates that the physical etching effect produces Ti vacancies on the etched surface regardless of the gas pressure. In general, it is considered that the Ti vacancies form acceptor levels above the valence band [23]. Thus, the etched surface has a possibility of exhibiting p-type conductivity.

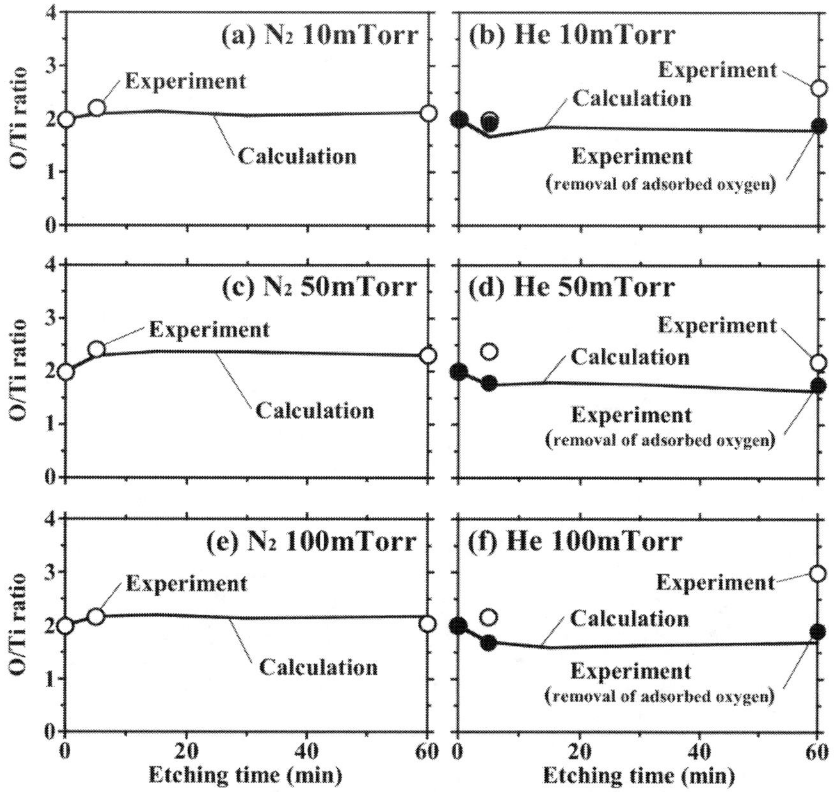

Figure 3. Experimental and calculated O/Ti ratios at the surfaces at various gas pressures, as a function of etch time.

As shown in Figs. 3(b), 3(d), and 3(f), the experimental O/Ti ratio for the He plasma (open circles) also increases with an increase in the etching time. The increase in the experimental O/Ti ratio is irrespective of gas pressure. The experimental O/Ti ratio does not agree with the simulated result that the calculated O/Ti ratio decreases with an increase in the etching time. The decrease in the calculated O/Ti ratio indicates that O atoms are preferentially removed by He^+ ions and that, as a result, the etched surface becomes a Ti-rich surface. The preferential etch of O atoms results from the phenomenon that O atoms move more easily than Ti atoms by He^+ impact, based on a binary collision model result that the maximum transferable elastic energy of He^+ to O is higher than that of He^+ to Ti [17]. In general, the O vacancies are considered to be donor-like defects in the near-surface region [24]. The simulation predicts that the etched surface has a possibility of exhibiting n-type conductivity.

The O-1s XPS spectra for the He plasma show the peak associated with oxygen bonding to titanium, as well as the peak related to oxygen and water adsorbed in the surface (not shown here), according to ref. 25. In the case where the component of adsorbed oxygen and water in each spectrum is subtracted, the experimental O/Ti ratio (solid circles) decreases. This is consistent with the calculated result as

Journal of Physics: Conference Series **441** (2013) 012038 doi:10.1088/1742-6596/441/1/012038

shown in Figs. 3(b), 3(d), and 3(f). The adsorption of oxygen and water can be explained in terms of the following fact. The etched surface becomes a Ti-rich surface, as predicted from the simulation. The Ti-rich surface easily adsorbs oxygen and water in ambient air as the samples are taken out of the vacuum chamber. The agreement between the experimental and simulated results, therefore, indicates that the preferential etch of O atoms is independent of the gas pressure.

Figure 4 shows SEM images of the surfaces etched by the N_2 and the He plasmas at various gas pressures. As shown in Figs. 4(a)-4(c), morphology of the surface etched at each gas pressure is almost similar to that of the as-grown surface even as the etching time increases to 60 min. The result of the SEM measurement indicates that the surface morphology for the N_2 plasma is independent of the gas pressure and etching time, though the surface is irradiated with the emitted UV light. As shown in Figs. 4(d)-4(f), the surface morphology for the He plasma changes as the etching time increases to 60 min, as described in ref. 14, which indicates that the change in the surface morphology caused by the He plasma is dependent on the etching time and gas pressure.

The difference between the surface morphologies caused by the two kinds of plasma cannot be fully explained only in terms of the UV radiation effect [14]. The reason is indicated by the fact that the two kinds of plasma emit the UV light as shown in Fig. 2. The above-mentioned difference can be related to the sizes of atoms removed preferentially from the surface. It is known that the ionic radius of Ti^{4+} (64 pm) is smaller than that of O^{2-} (124 pm) [26]. Since Ti atoms are preferentially removed by the N_2 plasma as described above, pits with radii comparable to the radius of Ti are formed on the surface. It is difficult to observe the formed pits because their sizes are too small. The formed pits also disappear because of the structural relaxation to the stable crystalline configuration. As a result, it is observed that the surface morphology for the N_2 plasma is almost similar to that of the as-grown sample. In contrast, since O atoms are preferentially removed by the He plasma as described above, pits with radii comparable to the radius of O are formed on the surface. The formed pits remain because their radii are too large to be recovered by the structural relaxation. As a result, the surface morphology for the He plasma probably changes in association with the UV radiation effect [14]. Detailed study for the difference between the surface morphologies by the two kinds of plasma will be necessary in the future.

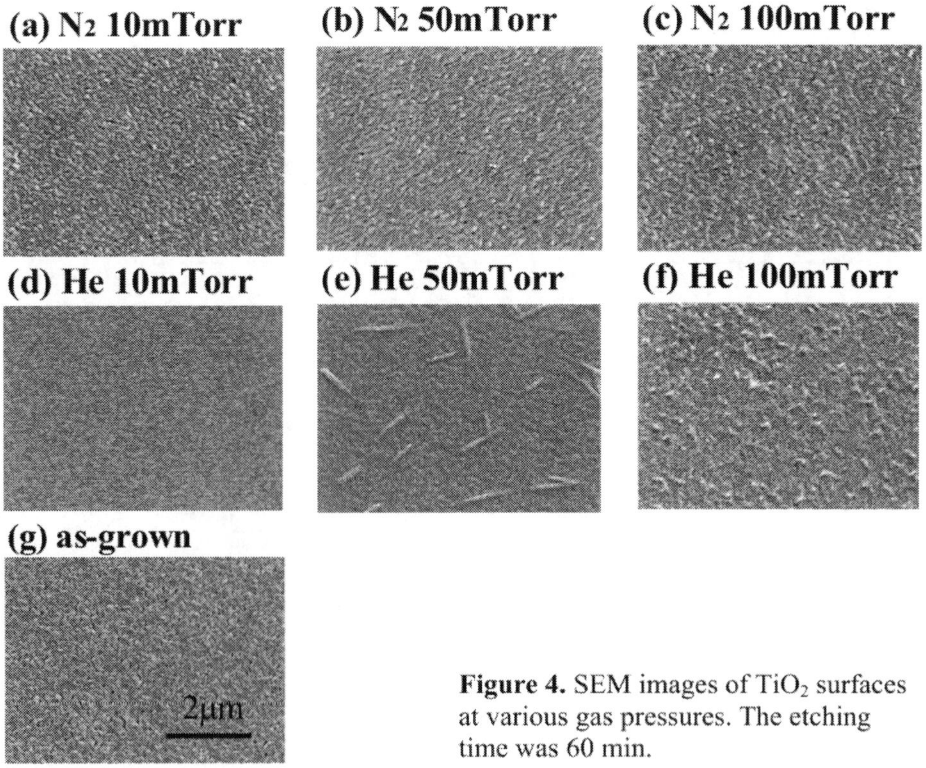

Figure 4. SEM images of TiO_2 surfaces at various gas pressures. The etching time was 60 min.

4. Conclusion

From the viewpoint of both the experiment and a simulation, we have been studied characteristics of TiO_2 surfaces etched by the N_2 plasma. The result obtained with use of the N_2 plasma was compared with that of the He plasma. The experimental O/Ti ratio at the surface etched by the N_2 plasma irrespectively increases with an increase in the gas pressure, which agrees with the simulation that Ti atoms are preferentially removed by N_2^+ ions impinging on the surface. The experimental O/Ti ratio at the surface etched by the He plasma also increases, which does not agree with the simulation that O atoms are preferentially removed by He^+ ions. The O-1s XPS spectra for the He plasma show the peak associated with oxygen and water adsorbed by the surface. As the component of the adsorbed oxygen and water in each spectrum is subtracted, the experimental O/Ti ratio for the He plasma decreases, which is consistent with the simulated one. The adsorption of oxygen and water seems to be caused by ambient air. The morphology of the surface etched with use of the N_2 plasma is almost similar to that of the as-grown surface. This is independent of the etching time and gas pressure. In contrast, the morphology of surface etched with use of the He plasma changes as the etching time increases. This is independent of gas pressure.

5. References

[1] Saxena V, Veerender P, Chauhan A K, Jha P, Aswal D K, and Gupta S K. 2012 *Appl. Phys. Lett.* **100** 133303.
[2] Chen R S, Chen C A, Tsai H Y, Wang W C, and Huang Y S 2012 *Appl. Phys. Lett.* **100** 123108.
[3] Liu L Z, Xu W, Wu X L, Zhang Y Y, Chen T H, and Chu P K 2012 *Appl. Phys. Lett.* **100** 121904.
[4] Lu J, Dai Y, Guo M, Yu L, Lai K, and Haung B 2012 *Appl. Phys. Lett.* **100** 102114.
[5] Lin Y, Jiang Z, Hu X, Zhang, and Fan J 2012 *Appl. Phys. Lett.* **100** 102105.
[6] Tominaga K, Okada K, Miyamoto Y, Ohkura S, Shiraishi K, Kusaka K, and Hanabusa T 2009 *e-J Surf. Sci. Nanotech.* **7** 290.
[7] Yu X and Shen Z 2011 *Vacuum* **85** 1026.
[8] Yildirim G, Bal S, Gulen M, Varilci A, Budak E, and Akdogan M 2012 *Cryst. Res. Technol.* **47** 195.
[9] Ushiki T, Yu M C, Kawai K, Shinohara T, Ino K, Morita M, and Ohmi T 1999 *Microelectronics Reliability* **39** 327.
[10] Woo J C, Chun Y S, Joo Y H, and Kim C 2012 *Appl. Phys. Lett.* **100** 081101.
[11] Kim S B, Brown S L, Rossnagel S M, Bruley J, Copel M, Hopstaken M J P, Narayanan V, and Frank M M. 2010 *J. Appl. Phys.* **107** 054102.
[12] Chowdhury P, Barshilia H C, Selvakumar N, Deepthi B, Rajam K S, Chaudhuri A R, and Krupanidhi S B 2008 *Physica B* **403** 3718.
[13] Park J B, Lim W S, Park S D, Park B J, Yeom G Y 2009 *J. Korean Phys. Soc.* **54** 976.
[14] Kawakami R, Takeichi A, Niibe M, Inaoka T, and Tominaga K. 2011 *Jpn. J. Appl. Phys.* **50** 08KD01.
[15] Birdsall C K 1991 *IEEE Trans. Plasma Sci.* **19** 65.
[16] Vahedi V, and Surendra M 1995 *Comput. Phys. Commun.* **87** 179.
[17] Möller W, Eckstein W, and Biersack J P 1988 *Comput. Phys. Commun.* **51** 355.
[18] Ohmori Y, Shimozuma M, and Tagashira H 1988 *J. Phys. D: Appl. Phys.* **21** 724.
[19] Yu W, Zhang L Z, Wang J L, Han L, Fu G S 2001 *J. Phys. D: Appl. Phys.* **34** 3349.
[20] Chambers S A 2007 *Hyomen Kagaku* **28** 561.
[21] Manenschijn A, and Goedheer W J 1991 *J. Appl. Phys.* **69** 2923.
[22] Gilmore F R, Laher R R, and Espy P J 1992 *J. Phys. Chem. Ref. Data* **21** 1005.
[23] Nowotny J, Bak T, Nowotny M K, and Sheppard L R 2006 *J. Phys. Chem. B* **110** 18492.
[24] Meng F and Sun Z 2009 *Appl. Surf. Sci.* **255** 6715.
[25] Vesel A, Mozetic M, Kovac J, and Zalar A 2006 *Appl. Phys. Sci.* **253** 2941.
[26] Su Y L, Xiao Y T, Du Y X, and Fu S 2010 *Chinese Sci. Bull.* **55** 2136.

Fabrication of Amorphous Silicon Carbide Films from Decomposition of Tetramethylsilane using ECR plasma of Ar

H. Ito[1], S. Onitsuka[1], R. Gappa[1], H. Saitoh[1], R. Roacho[2], K. H. Pannell[2], T. Suzuki[3], M. Niibe[4], K. Kanda[4]

[1]Department of Chemistry, Nagaoka University of Technology, Japan

[2]Department of Chemistry, University of Texas at El Paso, USA

[3]Extreme-Energy Density Institute, Nagaoka University of Technology, Japan

[4]Laboratory of Advanced Science and Technology Institute, University of Hyogo, Japan

E-mail: bu7dd8@nagaokaut.ac.jp

Abstract. Mechanically-hard hydrogenated amorphous silicon carbide (a-SiC_x:H) films were formed from the decomposition of $Si(CH_3)_4$ using the electron-cyclotron resonance plasma flow of Ar. An external radio-frequency (RF) voltage was applied to the substrate with the negative self-bias voltage ($-V_{RF}$) of 0-100 V. Compositional analysis was made with a combination of Rutherford backscattering and elastic recoil detection analysis. The C/Si ratios of films were 2.2-2.7. Film hardness was measured with a nano-indentation testing equipment. Chemical bonding was analyzed using carbon-K near edge X-ray absorption fine structure (C-K NEXAFS) spectroscopy using an accelerator NewSUBARU. The peak-fitting analysis of the C-K NEXAFS spectra yielded the $sp^2/(sp^2+sp^3)$ ratios, being fully correlated with film hardness. With supported by the IR and Raman spectroscopic measurements, the change of the chemical structure induced by $-V_{RF}$ was discussed.

1. Introduction

Thin films of amorphous silicon carbides (a-SiC_x) and hydrogenated a-SiC_x (a-SiC_x:H) have attracted much attention as potential candidates for mechanical, electronic, and biomedical applications [1]. For the synthesis of these materials, plasma decomposition of tetramethylsilane ($Si(CH_3)_4$, TMS) has been frequently used. The method of decomposition of TMS applied so far has been mainly the radio-frequency (RF) discharge where a DC bias voltage is applied to the substrate in order to obtain high mechanical hardness [2-4]. Mechanically-hard a-SiC_x:H films have also been synthesized using the dissociation reaction of TMS with microwave (MW) discharge flow of Ar where an external RF bias voltage is applied [5,6]. When an RF voltage is applied to the substrate, a negative self-bias voltage, $-V_{RF}$, is generated according to a high mobility of free electrons. Positive ions are attracted by $-V_{RF}$, causing ion bombardment of the film surface to change the bonding nature [7]. The present study reports on the fabrication of a-SiC_x:H films using the decomposition of TMS in the electron-cyclotron-resonance (ECR) plasma flow of Ar and on the structural change of films by the application of an RF bias voltage. The hardness of films was measured as a function of $-V_{RF}$. As for the mechanical hardness of the carbon-related thin films such as a-SiC_x:H, the fraction of the sp^3-hybridized state of C atoms is the chemical structure of the central importance [7]. The most precise and direct method to evaluate this fraction is the carbon-K near edge X-ray absorption fine structure (C-K NEXAFS) analysis using synchrotron radiation [8,9]. Atomic compositions including hydrogen content were evaluated by the combination of Rutherford backscattering (RBS) and the elastic recoil detection analysis (ERDA). The bonding nature of films was studied on the basis of infrared (IR) absorption and Raman spectroscopic methods. The film properties obtained in this study were compared with those reported previously using MW-discharge flow system [5,6].

2. Experiments

Figure 1 shows a schematic of the ECR-plasma flow apparatus used in the present study. It consisted of a stainless-steel chamber and an ECR discharge source (Irie Koken RBG-152L) which were evacuated to 10^{-3} Pa using turbomolecular (80 L/s) and oil-rotary (180 L/min) pumps. Ar (99.9999%) was introduced into the discharge section through a needle valve. Microwaves (2.45 GHz, 120 W) were introduced through a coaxial cable and an antenna (A). An external magnetic field was applied using permanent magnets arranged in the vicinity of the upper part of the ECR-discharge source. A stainless-steel nozzle (N) with the inner diameter of 1 mm was set downstream of the discharge section from which TMS (> 95% purity) was introduced through a Teflon needle valve. The partial pressures of Ar and TMS were 0.40 Pa and 0.13 Pa, respectively. A substrate holder (S) insulated from the wall of the chamber was set ≈5 mm downstream of the nozzle tip. Pressure was monitored using an ionization gauge. Both Ar and TMS were passed through desiccants (P_2O_5). H_2O molecules adsorbed on the substrate surface and the wall of the chamber were removed by discharging Ar for 1.5 h. An external bias voltage was applied to S using an RF generator (13.56 MHz) through a matching box (MB), and was monitored through a filter circuit (F). Films were formed onto Si substrates of ≈10 mm^2 set on S. The deposition time was in the range of 45-90 min.

The measurements of RBS/ERDA were carried out using a Tandem-type accelerator (Nisshin High Voltage, NT-1700HS) equipped at Extreme Energy-Density Research Institute of Nagaoka University of Technology. In order to make precise measurements of ERDA, the thickness of films were set ≈300 nm by adjusting the deposition time appropriately. Films were irradiated by a He^+ beam accelerated to

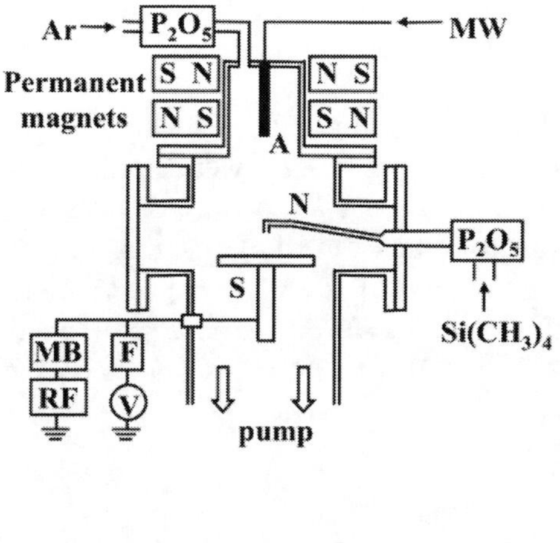

Figure 1 ECR-plasma flow apparatus used in the present study.

25 MeV. In the RBS measurement, scattered He^+ beam was detected by a solid-state detector (SSD). The irradiation and detection angles of He^+ beam were 18° and 78°, respectively, from the film surface. In the ERDA measurement, recoiled H atoms were detected with SSD positioned at an angle of 168° from the film surface.

The a-SiC_x:H film hardness was measured using a nano-indentation testing equipment (Fischer MZT-500) with Vickers indenter. The maximum load and the loading step were set to 1 mN and ~0.05 mN s^{-1}, respectively. The hardness of Si substrate was observed as (6.9±0.3) GPa under this condition.

The C-K NEXAFS measurements were carried out at the beamline-9 of NewSUBARU located at Laboratory of Advanced Science and Technology for Industry, University of Hyogo [10,11]. Synchrotron radiation provided from the 10.8-m long undulator was extracted using a Hettrick-Underwood mount type monochromator with a varied line spacing plane grating of 1200 lines/mm. The energy resolution was estimated to be less than 0.5 eV full width at half maximum (FWHM). The reading of the monochromator was calibrated against the pre-edge resonance corresponding to the $\pi^*\leftarrow$1s transition of carbon atoms that appeared at 285.38 eV in the NEXAFS spectrum of graphite [12]. The monochromatized beam in the range of 275-340 eV irradiated the sample at the incident angle of 54.7°, and the absorption spectrum was obtained in the total electron yield mode. The signals from sample films were calibrated against the incident beam intensity measured with a gold grid.

The structural analysis of films was performed using a Raman spectrometer (Jobin Yvon, Model Labram Infinity) and a Fourier Transform Infrared (FTIR) Spectrophotometer (Nicolet IMPACT-410). Raman spectra were obtained in the backscattering mode using an Ar^+ laser operating at a wavelength of 514.5 nm and a power of 100 mW. The wavelength resolution was 1 cm^{-1}, and the spatial resolution was ≤1 μm. Scattering signal was detected by using a CCD detector. IR spectra were measured in the Right AEM-Main mode with a MCT/A detector and a resolution of 4 cm^{-1}.

3. Results and discussions

Table 1 lists the atomic compositions of films measured using RBS/ERDA. As expected, Si, C, and H atoms were detected. The [C]/[Si] ratios (2.2-2.7) and the H atomic density were observed to be independent of $-V_{RF}$, because the atomic compositions measured in RBS/ERDA may have uncertainties of ≈20%.

Table 1 Atomic composition of films.

$-V_{RF}$ [V]	[Si] (at.%)	[C] (at.%)	[H] (at.%)	[C]/[Si]
0	19	52	29	2.7
50	22	48	30	2.2

Figure 2 shows the indentation hardness of films. The maximum indentation depths of the present measurements were <0.06 μm, being <10% of the film thickness (0.6-2.0 μm). Therefore, the measured hardness was free from the influence from the Si substrates. Under the condition of $-V_{RF}$=0 V, films showed mechanical hardness of 5.5±0.5 GPa, and the hardness was increased slowly to 7.8±0.6 GPa as $-V_{RF}$ increased to 50 V. This observed tendency of mechanical hardness has contrast to our previous study using MW-discharge flow where the film hardness increased rapidly from 2.4±1.1 GPa at $-V_{RF}$=0 V to 8.0±0.6 GPa at $-V_{RF}$=20 V [5,6].

The C-K NEXAFS spectra of a-SiC_x:H films are shown in Figure 3 where the NEXAFS spectrum of graphite is also shown. The narrow peaks appearing in 285.3 eV were assigned to the transitions from C1s core level to the unoccupied π^* levels of sp^2 (C=C) and sp (C≡C) sites [8]. The broad feature observed in the range of >288 eV was assigned to the $\sigma^*\leftarrow$C1s transitions of sp^2(C=C) and sp^3(C-C) bonds [12,13] and the $\sigma^*\leftarrow$C1s transitions of Si-C bond [14,15] overlapped with the direct ionization.

In order to evaluate the intensities of the $\pi^*\leftarrow$C1s peaks, the simulation analysis of the observed NEXAFS spectra was made as follows [9]. The profile of a $\pi^*\leftarrow$C1s peak was assumed to be a single

Journal of Physics: Conference Series **441** (2013) 012039 doi:10.1088/1742-6596/441/1/012039

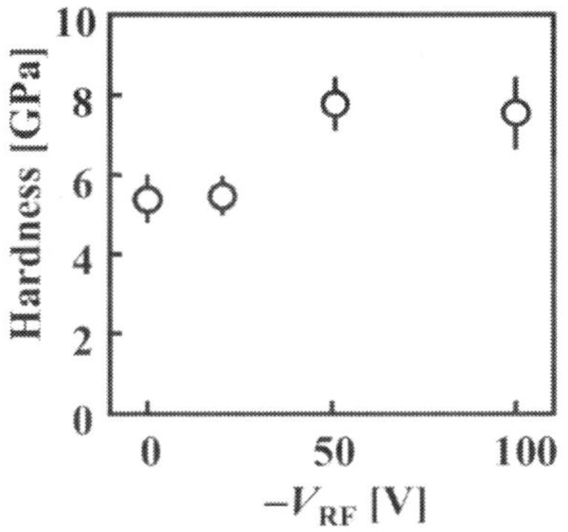

Figure 2 Hardness of a-SiC$_x$:H films.

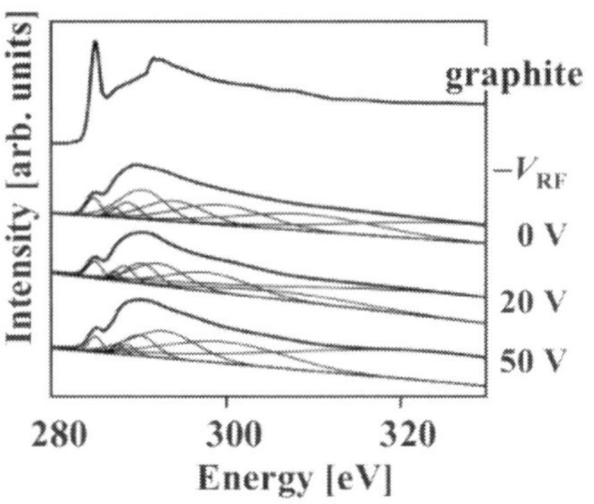

Figure 3 C-K NEXAFS spectra and their analysis.

Gaussian function. The broad feature in the region of >288 eV was represented by the superposition of six Gaussian functions as listed in Table 2 and additional Gaussian functions in the high-energy side to reproduce the profile due to direct ionization [16]. In addition, a linear baseline was assumed. The observed C-K NEXAFS spectra were fitted with the peak heights and widths of the above Gaussian functions as the parameters. Figure 3 shows the result of this analysis. Table 2 lists the height, H, and the width, W, for each Gaussian profile. From this analysis, the sp^2/(sp^2+sp^3) ratio of a-SiC$_x$:H films were estimated as

$$\mathrm{sp}^2/(\mathrm{sp}^2+\mathrm{sp}^3)=(I_\pi^{\mathrm{SiC}}/I_{\mathrm{all}}^{\mathrm{SiC}})/(I_\pi^{\mathrm{graphite}}/I_{\mathrm{all}}^{\mathrm{graphite}})\,. \quad (1)$$

In eq. (1), I_π^{SiC} and $I_\pi^{\mathrm{graphite}}$ represent the areal intensities of the π*←C1s peaks of a-SiC$_x$:H films and graphite, respectively, and $I_{\mathrm{all}}^{\mathrm{SiC}}$ and $I_{\mathrm{all}}^{\mathrm{graphite}}$ represent the areas of the total profiles of the NEXAFS spectra of a-SiC$_x$:H films and graphite, respectively, evaluated from the numerical integrations of the observed spectra. The sp^2/(sp^2+sp^3) ratios obtained from the above procedure are shown in Figure 4 as a function of $-V_{\mathrm{RF}}$. Figure 5 shows the correlation between the sp^2/(sp^2+sp^3) ratios and the hardness of films where films fabricated in this study (open circles) and those obtained in our previous study (closed circles) [6] are compared. The decline for the present films is steeper than that for the previous films. This difference will be discussed later in this section.

Figure 6 shows the IR spectra of films, where the normalization with their film thickness was not carried out. Following vibrational bands were observed; 670-760 cm^{-1} (Si-C stretching), ≈1000 cm^{-1} (bending vibration modes of Si-CH_3 and Si-$(CH_2)_n$-Si), ≈1250 cm^{-1} (Si-CH_3 symmetric bending), 2100 cm^{-1} (Si-H stretching), and 2700-3000 cm^{-1} (C-H stretching) [17]. As inferred from Figure 6, the intensities of the Si-CH_3, Si-$(CH_2)_n$-Si, Si-H, and C-H vibration modes remain unquenched with the application of $-V_{\mathrm{RF}}$. This tendency is contrast to the previous result of a-SiC$_x$:H films fabricated in the

Table 2 Analysis of C-K NEXAFS spectra.[a)]

		$-V_{\mathrm{RF}}$=0 V		$-V_{\mathrm{RF}}$=20 V		$-V_{\mathrm{RF}}$=50 V	
Energy [eV]	Assignment	H	W	H	W	H	W
285.3	π*←1s(C=C)	0.827	1.91	0.359	1.94	0.226	1.75
286.8	π*←1s (C=O)					0.047	0.78
287.8	σ*←1s (C-H)	0.453	1.84	0.200	1.58	0.138	1.59
288.8	σ*←1s (e. p.)[b)]	0.718	3.07	0.335	2.48	0.142	2.30.
290.2	σ*←1s (C-Si)	1.34	6.27	0.440	4.06	0.295	4.47
293	σ*←1s (C-C)	0.947	8.25	0.479	6.96	0.384	3.99
300	σ*←1s (C-C)	0.920	11.9	0.405	11.3	0.306	16.3
309		0.755	18.7	0.179	18.9		
327		0.828	29.4	0.563	47.6	0.424	39.8

a) H and W are the spectral height in arbitrary units and the width in units of eV, respectively.
b) Abbreviation of excitonic process.

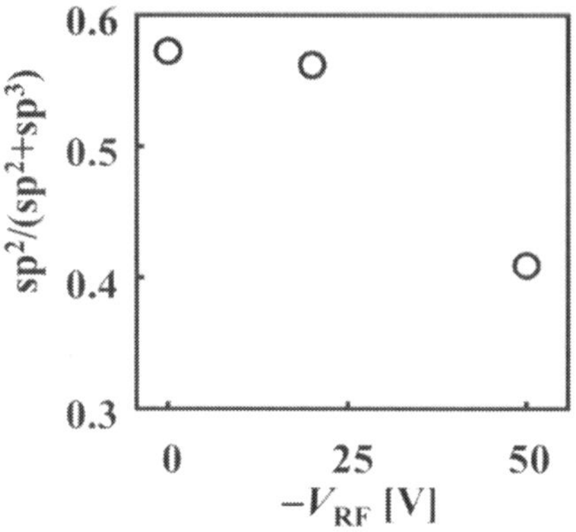

Figure 4 $sp^2/(sp^2+sp^3)$ ratios of films.

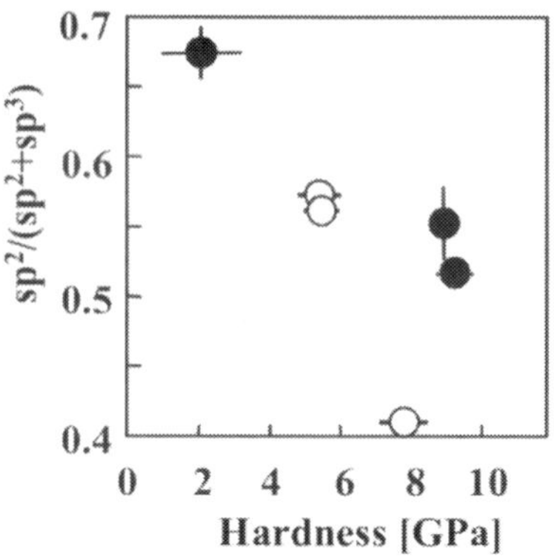

Figure 5 Correlation between the $sp^2/(sp^2+sp^3)$ ratios and the hardness of films.

MW-discharge flow system where the IR intensities of the above absorption bands quenched effectively with the application of $-V_{RF}$ [5].

Figure 7 shows the Raman spectra of films. G (graphitic) band (≈1450 cm^{-1}) was observed accompanied by weak D (disordered) band (≈1300 cm^{-1}). These bands were very weak at $-V_{RF}$=0 V, whereas they became prominent as $-V_{RF}$ increased. The observed Raman shifts of the G- and D-bands were 50-100 cm^{-1} lower than those of amorphous carbons [18], being attributable to the lowering of the C=C vibrational frequency induced by the substitution of Si atoms [19]. In addition, background signals toward higher Raman shift were observed which is attributable to the fluorescence due to the one-dimensional conjugated structures of carbons containing hydrogen terminations [20].

In the following, the difference of the film-hardening mechanism between the present films and those fabricated in the previous MW-discharge flow system is discussed. In general, film hardness may have positive correlations with the fractions of the C atoms in the sp^3-hybridized state, whereas have negative correlations with the densities of the hydrogen terminations [21]. In the case of the MW-discharge flow, films fabricated without RF bias include hydrogen terminations and Si-Si bonds [6]; they decrease under the RF-biased conditions, and instead, the C-Si bonds increase to cause the increase of the film hardness. This mechanism may be consistent with the literature [4]. The structural change observed in the present films may be somewhat different from the case of MW-discharge flow. The relative density of the C atoms in the sp^3 state has a positive dependence on $-V_{RF}$ from the analysis of the C-K NEXAFS spectra. The [C]/[Si] ratios of the present films (≈2.5) are higher than those in the previous study (≈1.0) [5,6]. Therefore, the decline shown in Fig. 5 may be caused by the increase of the density of the sp^3-hybridized C-C bonds as well as the Si-C bonds. Films fabricated under the condition of $-V_{RF}$=0 V showed lower $sp^2/(sp^2+sp^3)$ ratio than those under the same condition in the MW-discharge flow system as shown in Figure 5, and both films contain hydrogen terminations. These observations are consistent with higher mechanical hardness of the present films of $-V_{RF}$=0 V. On the other hand, films formed under the RF-biased conditions include more significant amount of hydrogen terminations than those fabricated previously. Therefore, hardness of the present films may

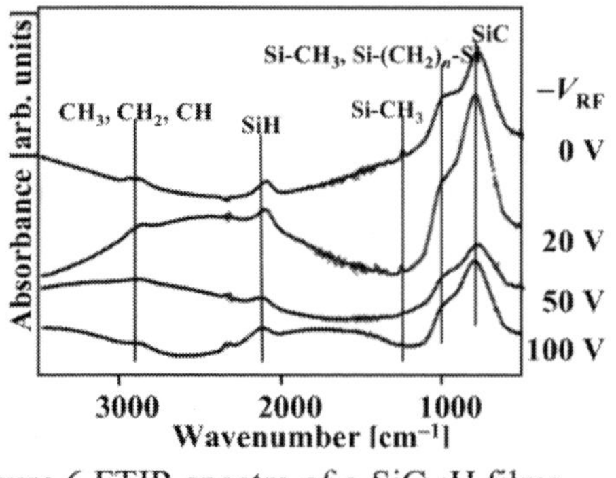

Figure 6 FTIR spectra of a-SiC_x:H films.

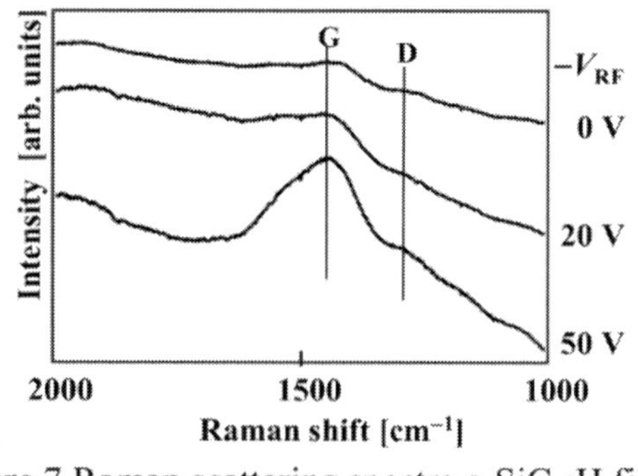

Figure 7 Raman scattering spectra a-SiC_x:H films.

be suppressed to be lower than the previous films even under the same $sp^2/(sp^2+sp^3)$ ratio (see Fig. 5).

4. Conclusions

Mechanically-hard a-SiC_x:H films were fabricated using the decomposition of TMS in the ECR plasma flow of Ar. An external RF voltage was applied to the substrate. C-K NEXAFS spectra were observed, and the $sp^2/(sp^2+sp^3)$ ratios were obtained. These ratios were found to be correlated well with the film hardness. The C/Si ratios were 2.2-2.7. Significant increase of the density of sp^3(C-C) bonds may be induced by the bombardment of positive ions (mainly Ar^+) induced by $-V_{RF}$. Mechanical hardness of the present films is explicable in terms of the fraction of the sp^3 component of C atoms and the amount of hydrogen terminations.

Acknowledgements

R. G. was supported by the SUPER-VOS program of Nagaoka University of Technology. R. R. acknowledged the short visit and short stay program of JASSO. This work was supported by JSPS KAKENHI Grant Number 22560020.

References

[1] Snead L L, Nozawa T, Katoh Y, Byun T -S, Kondo S and Petti D A 2007 *J. Nucl. Mat.* **371** 329
[2] Yoshino M, Shimozuma M, Date H, Itoh H and Tagashira H 2005 *Thin Solid Films*, **492** 207
[3] Soum-Claude A, Thomas L, Tomasella E, Badie J M, and Berjoan R 2005 *Surf. & Coat. Tech.* **200** 855
[4] Zhao J F, Lemoine P, Liu Z H, Quinn J P, Maguire P and McLaughlin J A 2001 *Diam. Relat. Mat.* **10** 1070
[5] Ito H, Shinohara A and Saitoh H 2009 *J. Plasma and Fusion Res. Ser.* **8** 1439
[6] Wada A, Ogaki T, Niibe M, Tagawa M, Saitoh H, Kanda K and Ito H 2011 *Diam. Relat. Mat.* **20** 364
[7] Robertson J 2002 *Mater. Sci. Eng.* **R37** 129
[8] Lenardi C, Piseri P, Briois V, Bottani C E, Li Bassi A and Milani P 1999 *J. Appl. Phys.* **85** 7159
[9] Kanda K, Kitagawa T, Shimizugawa Y, Haruyama Y, Matsui S, Terasawa M, Tsubakino H, Yamada I, Gejo T, Kamada M 2002 *Jpn. J. Appl. Phys.* **41** 4295
[10] Niibe M, Mukai M, Miyamoto S, Shoji Y, Hashimoto S, Ando A, Tanaka T, Miyai M and Kitamura H 2004 *8-th Int. Conf. Am. Inst. of Phys. Conf. Proc.* **705** 576
[11] Niibe M, Mukai M, Kimura H and Shoji Y 2004 *8-th Int. Conf. Am. Inst. of Phys. Conf. Proc.* **705** 243
[12] Batson P E 1993 *Phys. Rev. B* **48** 2608
[13] Lee M-S, Bent S F 2000 *J. Appl. Phys.* **87** 4600
[14] Monterio O R, D-Ogletree M –P 2003 *Surf. Coat. Tech.* **163** 144
[15] Jung H –S, Park H –H 2004 *J. Appl. Phys.* **96** 1013
[16] Wada A, Koshimura K, Niibe M, Saitoh H, Kanda K and Ito H 2012 *J. Non-Cryst. Solids*, **358** 124
[17] Soum-Claude A, Thomas L, Tomasella E, Badie J M, Berjoan R 2006 *Surf. & Coat. Tech.* **201** 174
[18] Park M, Teng C W, Sakhrani V, McLaurin M B, Kolbas R M, Sanwald R C, Nemanich R J, Hren J J and Cuomo J J 2001 *J. Appl. Phys.* **89** 1130
[19] Racine B, Ferrari A C, Morrison N A, Hutchings I, Milne W I and Robertson J 2001 *J. Appl. Phys.* **90** 5002
[20] Yang W J 2005 *Surf. & Coat. Tech.* **194** 128
[21] Saikubo A, Yamada N, Kanda K, Matsui S, Suzuki T, Niihara K and Saitoh H 2008 *Diam. Relat. Mat.* **17** 1743

Journal of Physics: Conference Series **441** (2013) 012040 doi:10.1088/1742-6596/441/1/012040

Fabrication of silicon and carbon based wide-gap semiconductor thin films for high conversion efficiency

Kohsuke Yoshinaga, Hiroshi Naragino, Akira Nakahara and Kensuke Honda*

Graduate school of Science and Engineering, Yamaguchi University, 1677-1, Yoshida, Yamaguchi-shi, Yamaguchi, 753-8512, Japan

*khonda@yamaguchi-u.ac.jp

Abstract. Nitrogen doped amorphous silicon carbide (N-doped a-SiC) thin films were fabricated by radio frequency plasma enhanced chemical vapor deposition (RF-PeCVD) method using mixed solution of tetramethylsilane (TES) and 1,1,1,3,3,3-hexamethyldisilazane (HMDS) as a liquid source. Chemical composition of N-doped a-SiC thin film was Si:C = 1:4 and sp^2-bonded carbon ratio was 0.75. N-doped DLC were multi-phase structure including a-SiC phase, sp^2 clusters and a-Si clusters. Optical gap and resistivity of the film were 1.68 eV and 4.32×10^4 Ω cm, respectively. From photocurrent measurement under UV exposure, it was clarified that the film functioned as n-type semiconductor materials with 4.87 % of quantum yield, which was on the same level as that obtained at anatase-type titanium oxide prepared by sol-gel method. To apply these films to solar cells, further improvements of optical gap and conductivity are necessary.

1. Introduction

The widespread use of a low-cost (7 yen/kWh) and high efficiency (40%) solar cell can contribute to reduce CO_2 emission from thermal power station and the amount of man-made carbon dioxide. Tandem-type solar cell composed of stacked layers of silicon and carbon-based semiconductors can achieve high efficiency solar cell at lower cost. To realize the solar cell, it is necessary to develop p- and n-type silicon and carbon-based semiconductors with band gap larger than 2.5 eV and high semiconducting properties applicable to electronic devices.

In the process of device fabrication, semiconductor layers should be synthesized on a conducting glass at a temperature below the melting point of the substrate. In this study, plasma-enhanced chemical vapor deposition (PeCVD) method was employed because amorphous thin film could be fabricated at low temperature. It has been reported that n-type semiconductor could be synthesized by incorporating phosphorous atoms in amorphous silicon carbide (a-SiC)[1]. This film, however, exhibited quite higher resistivity more than 10^4 Ω cm because the concentration of P atoms in a-SiC was not so high. The resistivity of a-SiC should be reduced less than 10^1 Ω cm to utilize the film for electronic devices.

To realize n-type semiconductive a-SiC film, nitrogen atoms were tried to be incorporated in a-SiC (N-doped a-SiC). Impurity level of N atoms in 4H-SiC has been reported to be 0.0653 eV under the conduction band of 4H-SiC. By incorporating N atoms, electrons in N impurity level can be easily activated and can function as carriers of n-type conduction at room temperature, which results in lower resistivity. In addition, a liquid source composed of hydrocarbons including impurity (nitrogen) atoms and Si was used as source material of PeCVD. The atomic ratios of Si and C atoms in a-SiC can be easily controlled by changing the composition (Si:C) of the liquid source. By increasing Si/C ratio in a-SiC films, the ratio of sp^3/sp^2 hybrids can be easily enhanced. It results in widening the optical gap of a-SiC films. Moreover, impurity (nitrogen) atoms remained single bound to silicon atoms and were introduced in a-SiC film since source material was not completely decomposed in partially-ionized plasma. Therefore, higher carrier density and lower resistivity compared to those of a-SiC synthesized with N_2, PH_3 and SiH_4 gases can be achieved, for nitrogen atoms generated from N_2 gas in plasma tend to be preferentially formed quite stable multiple bonding such as =C=N- and -C≡N, which does not contribute to conductivities.

The purpose of this study is to realize n-type a-SiC semiconductor with optical gap of 2.5 eV and volume resistivity of 10^1 Ω cm.

2. Experimental

Nitrogen-doped amorphous silicon carbide films (N-doped a-SiC) were fabricated by radio frequency plasma enhanced chemical vapor deposition system (RF-PeCVD) (13.56 MHz, SAMCO Co., Ltd. Model BPD-1). N-doped a-SiC films were deposited on highly Sb-doped silicon (111) substrate (0.02 Ω cm of resistivity) and insulating glass plate. Before the deposition, substrate on PeCVD equipment was heated to 100°C using 200 W of Ar plasma (50 sccm of Ar flow). As a liquid source, mixed solution of tetramethylsilane (TES) (Si and C sources) and 1,1,1,3,3,3-hexamethyldisilazane (HMDS) (Si, C and N sources) at a ratio of [HMDS]/[HMDS]+[TES] = 0.01 was used. After the substrate was heated, vaporized liquid source was introduced into a reaction chamber by heating at 80°C. The source gas at a flow rate of 1 sccm and H_2 gas at a rate of 150 sccm were introduced simultaneously. N-doped a-SiC thin films were deposited using plasma at various RF powers (75, 100 and 150W) to make clear the correlation between the structure and properties of semiconductor. Deposition time was 40 minutes. During depositions, total pressure in the chamber was kept at 110 Pa. Typical film thickness of the deposited thin films was 2.70 μm that was observed by SEM measurement.

Chemical composition of N-doped a-SiC was examined by XPS measurement. Transmission spectra of N-doped a-SiC deposited on glass substrate were examined by UV-Vis spectrophotometer (JASCO Co., Ltd. V-670). Absorption coefficient α was calculated using equation (1).

$$\alpha = -\ln\{(T/100)/d\times 10^8\} \qquad (1)$$

T is transmittance (%) and d is film thickness (Å) of the sample film. Optical gap (E_{og}) was estimated from Tauc's plot using Tauc equation (2), which is the result of the conversion.

$$(\alpha h\nu)^{1/2} = B(E_{og} - h\nu) \qquad (2)$$

hν is energy of incident light, B is constant, and E_{og} is optical gap of the sample film. Properties of semiconductor (carrier density, carrier mobility, resistivity and conductive type) were determined by Hall measurement with van der Pauw method (Ecopia Co., Ltd. HMS-3000). Ohmic contacts were formed by 1 mm^2 nickel vapor deposited with resistance heating deposition system on the four corners of N-doped a-SiC thin film at 1 cm interval. These four electrodes were used for Hall measurement. Hall measurement was performed at room temperature. Potentio/galvanostat (Hokuto Denko Co., Ltd. HZ-3000 system) was used for photo-electrochemical measurement. Photocurrent measurement was performed with linear sweep voltammetry (LSV) using three electrode electrochemical cells. 0.2 M NaH_2PO_4 was used as electrolyte solution. The geometric area of working electrodes was 0.1 cm^2. Sweeping rate was 5 mV sec^{-1}. Hg-Xe lamp (USHIO Co., Ltd. SP9-250DV) was used as UV sources.

3. Results and Discussion

3.1. Chemical composition

N-doped a-SiC thin film deposited on glass substrate at 100 W of RF power was highly transparent. The surface of N-doped a-SiC thin film was flat, smooth, and yellow in colour. Chemical composition of this N-doped a-SiC film was estimated by XPS measurement. The measurement within the range of 0 eV to 1100 eV at scan rate of 10 eV min^{-1} was performed. Peaks of Si 2p, C 1s, N 1s and O 1s were confirmed at 99 eV, 285 eV, 398 eV and 531 eV, respectively (Figure 1). From this result, it was clarified that the resulting film was consisted mainly of Si and C at a ratio of Si:C =1:4. It suggests that the deposited film contained relatively higher concentration of C atoms (Table 1). Nitrogen atoms that act as donor were confirmed to be 1.1 atom%.

Si 2p peak was measured at scan rate of 1 eV min^{-1} within the range of 96 eV to 108 eV. The peak attributable to Si bonded Carbon (Si-C) and Si bonded to Si (Si-Si) were observed at 100.65 and 99.709 eV by waveform separation of the measured Si 2p peak in Figure 2. This result indicated that the resulting films include a-SiC phase and amorphous silicon (a-Si) clusters. Atom % of Si atoms in N-doped a-SiC was decreased from 18.1 to 13.3 atom% and C atoms was increased from 65.2 to 68.8 atom % with increasing in RF power during CVD synthesis (75 to 150 W), as shown in Table 1. Elevating plasma power might induce decomposition of source materials and increase of the rate of methyl cation formation, which resulted that large amounts of C atoms were introduced in a-SiC films.

C 1s peak for N-doped a-SiC film at 100 W was measured at scan rate of 1 eV min^{-1} within the range of 283.8 eV to 284.6 eV. The peaks attributable to C-C (sp^3) and C=C (sp^2) were observed at 283.8 eV and 284.6 eV by altering waveform separation. The ratios of sp^2 and sp^3 bonded carbon were calculated to be 0.75 and 0.25 from the ratio of peak areas of these carbons. This result indicates that N-doped a-SiC contains sp^2-bonded carbon clusters.

The existence of sp^2 clusters in N-doped a-SiC fabricated at 75, 100, and 150W was examined by Raman spectroscopy. Broad peaks around 1500 cm^{-1} were observed in all Raman spectra and the peaks were composed by G and D peak that typically observed at a-C films (Figure 4). It indicated that sp^2-bonded carbons in a-SiC films form clusters. Size of cluster that can be estimated from the ratio of peak intensity of D and G peak (I_D/I_G) was gradually decreased with plasma power (8.47 μm at 75W, 6.13 μm at 150 W). Since C/Si ratio was increased with RF power in XPS spectra, the number of sp^2 cluster should be increased. The peaks attributable to a-Si (ca. 480 cm^{-1}) and a-SiC (ca. 750 cm^{-1}) could not be obtained in Raman spectra. It indicated that the proportion of a-Si clusters and a-SiC phase was relatively lower than sp^2 cluster. The position of G peak was sifted to lower wave number side with increasing Si atom % (lower RF power). This sift was caused by formation of Si-C bonding at sp^2 clusters (C=C-Si). Therefore, a-SiC phase and sp^2 clusters were connected and clear boundary between them might not be formed.

Figure 5 shows FTIR spectra for a-SiC with different RF power. In all spectra, peaks at ca. 2100, 1100 and 800 cm^{-1} were observed. The peak at 2100 cm^{-1} was attributable to Si-Hn stretching mode and the peak at 1100 cm^{-1} was C-Hn wagging mode. The peak at ca 800 cm^{-1} was Si-CH_3 stretching/wagging mode and the intensity of this peak was increased with increasing Si content (lowering of RF power). This observation supports that N-doped a-SiC includes a-SiC phase. The peak intensity (0.0335) at 2100 cm^{-1} of N-doped a-SiC fabricated at 150 W was quite lower than those of N-doped a-SiC at 75 and 100W (0.0742 and 0.0702, respectively). It indicates that N-doped a-SiC at 150 W include small amount of a-Si cluster than that at 75 and 100 W because of lower Si atom%.

Figure 6 shows XRD spectra of a-SiC film. Broad peak centred at 43 degree, which was attributable to sp^2 cluster, was observed. It indicates that there is no long range ordering in a-SiC film.

Therefore, these spectroscopic data suggests that resulting a-SiC films were multi-phase structure including a-SiC phase, sp^2 clusters and a-Si clusters. C/Si ratios and the number of sp^2 clusters in N-doped a-SiC films were increased with RF power of CV synthesis.

Table 1 Chemical composition of N-doped a-SiC by XPS measurement

Plasma power(W)	Si[atom%]	C[atom%]	N[atom%]	O[atom%]	C sp^2
75	18.1	65.2	1.1	16.8	0.840
100	17.6	68.5	1.1	13.9	0.750
150	13.3	68.8	1.1	17.8	0.786

Journal of Physics: Conference Series **441** (2013) 012040
doi:10.1088/1742-6596/441/1/012040

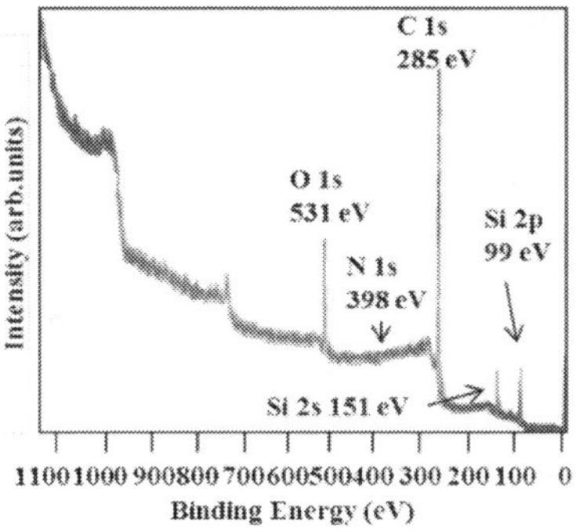

Figure 1 XPS spectrum of N-doped a-SiC

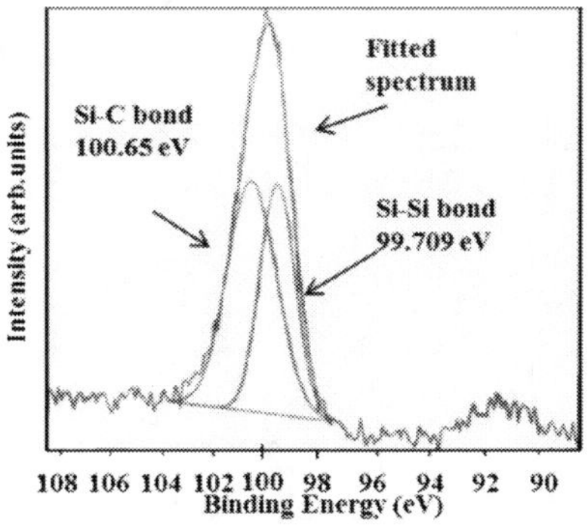

Figure 2 XPS Si 2p spectrum for N-doped a-SiC, scan rate, 1eV min^{-1}

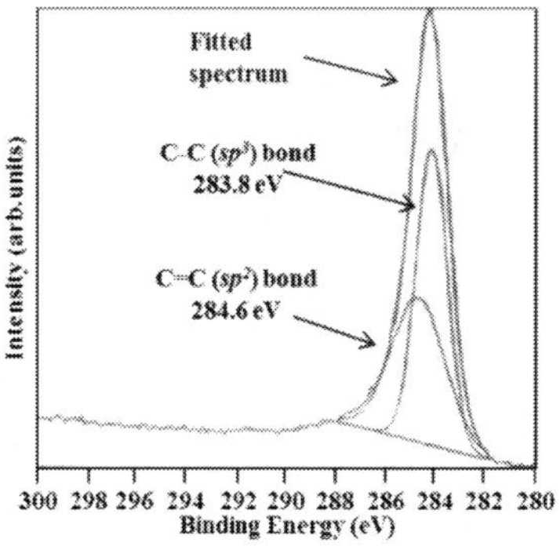

Figure 3 XPS C 1s spectrum for N-doped a-SiC, scan rate, 1eV min^{-1}

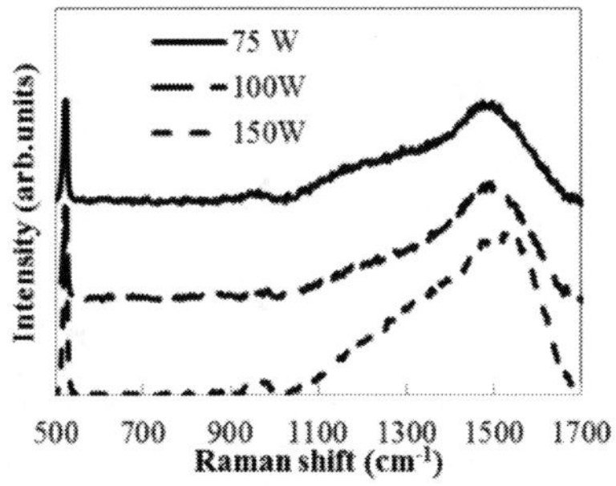

Figure 4 Raman spectra of N-doped a-SiC

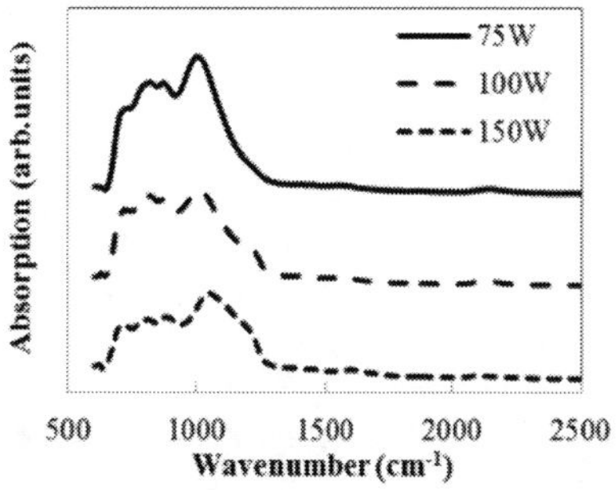

Figure 5 FTIR spectra of N-doped a-SiC

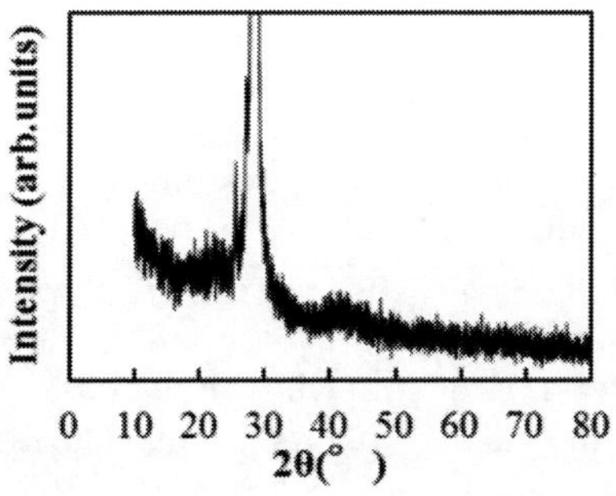

Figure 6 XRD spectrum of N-doped a-SiC

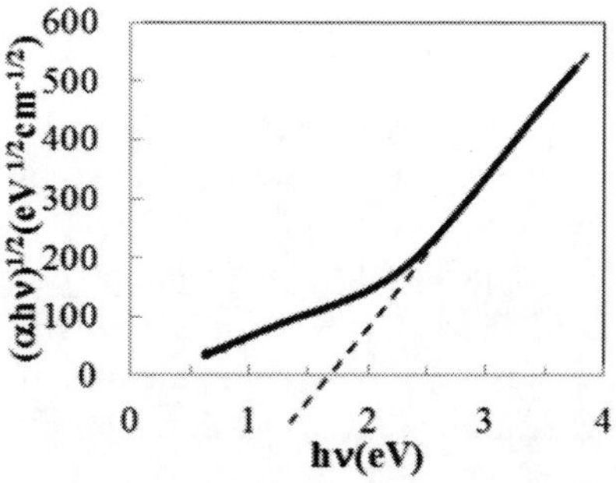

Figure 7 Tauc plot of N-doped a-SiC

3.2. Semiconductor properties

Semiconductor properties were examined using N-doped a-SiC deposited on glass substrate (insulator) by Hall measurement. Carrier density, carrier mobility, and resistivity of N-doped a-SiC synthesized with various RF power were shown in Table 2. The resistivity of N-doped a-SiC at 100W was relatively low and on the same level as the reported value of P-doped a-SiC fabricated using PeCVD, 10^4 Ω cm [1].

There are two possibilities to explain the causes of low resistivity; (1) Increase of carrier density (decrease of resistivity) by introducing N atom as a donor, (2) Enhancement of carrier mobility by forming (carbon) nanocrystal in the film. Carrier density and resistivity of non-doped a-SiC were 3.49×10^{14} cm^{-3} and 5.17×10^4 Ω cm, respectively. By incorporating N atoms in a-SiC, carrier density was increased by 100-fold (4.79×10^{14} cm^{-3}) and resistivity was decreased to 4.32×10^4 Ω cm. It was, therefore, clarified that N atom might act as n-type donor (cause 1). Carrier mobility of N-doped a-SiC was within the range of 10^{-1} Ω cm that was fairly lower than that of bulk SiC. In Raman spectra, peaks attributable to sp^2 clusters were observed. Hence, π-electron on sp^2-bonded carbon might be related closely to the conductivity of N-doped a-SiC (sp^2 clusters contribute to shorten the length of hopping conduction) (cause 2). This assumption was supported by the fact that the lowering of the resistivity at a-SiC fabricated at 150 W of RF power (higher C contents) in Table2.

Table 2 Optical gap, carrier density, carrier mobility, resistivity and conduction typefor N-doped a-SiC

Plasma Power (W)	Optical gap (eV)	Carrier density (cm^{-3})	Carrier mobility ($cm^2V^{-1}s^{-1}$)	Resistivity (Ω cm)	Conduction type
75 (N-doped)	1.69	3.05×10^{13}	2.71×10^0	7.57×10^4	n
100 (Undoped)	1.68	3.49×10^{12}	1.11×10^1	5.17×10^4	n
100 (N-doped)	1.68	4.79×10^{14}	2.98×10^{-1}	4.32×10^4	n
150 (N-doped)	0.585	6.75×10^{14}	7.85×10^{-1}	1.18×10^4	n

3.3. Optical properties

Optical gap of N-doped a-SiC at 100 W of RF power was estimated to be 1.68 eV in Figure 7. This value was 0.67 times lower than the target value, 2.5 eV and 0.52 times lower than 3.26 eV of crystalline 4H-SiC band gap [2]. Optical gap of N-doped a-SiC was related closely to sp^2 clusters in N-doped a-SiC, which were conformed in Raman spectra.

The colour of N-doped a-SiC changed from yellow to dark brown by increasing RF power from 100 W to 150 W employed in PeCVD synthesis. C/Si ratios were increased with RF power, as mentioned above. Optical gap was monotonically decreased from 1.68 eV to 0.59 eV with RF power (C/Si ratio). It has been reported that optical gap of amorphous carbon was inversely proportional to the average number of six-membered ring that was included in sp^2 clusters. a-C including sp^2 clusters with 10 rings exhibited 1.68 eV of optical gap and a-C that has sp^2 clusters with 120 rings was 0.59 eV. The number and average size (number of rings) of sp^2 cluster might affect strongly the value of optical gap. Similar tendency was confirmed in literature relating a-SiC prepared using PeCVD method [3].

3.4. Photo-electrochemical properties

The performance of photo-electron conversion at N-doped a-SiC was examined by photo-electrochemical method (Figure 8) to check applicability for photo voltaic system. The value of the current with exposure to Hg-Xe lamp (photocurrent) at N-doped a-SiC (the ratio of sp^2 carbon = 75 atom%) fabricated at 100 W of RF power was achieved to be 153 μA cm^{-2} (Figure 8). This value was on the same level as that obtained at anatase-type titanium dioxide thin films prepared by sol-gel method. Quantum yield of the thin film was calculated using the value of photocurrent. The quantum yield of photo-excited electron at N-doped a-SiC was calculated to be 4.87 % using 365 nm of wavelength and 11.8 W cm^{-2} of intensity of Hg-Xe lamp (photons were completely absorbed in the film). This value was lower than half of the values reported in [4] and [5] (15 % at anatase-type titanium dioxide prepared by sol-gel method [4] and 10 % at a-SiC by PeCVD [5]). It is necessary to enhance quantum yield of N-doped a-SiC.

The a-SiC film reported in literature has shown higher resistivity (~10^4 Ω cm). In this study, relatively lower resistivity of N-doped a-SiC was obtained by controlling the amount of sp^2 carbon content (higher C/Si ratio > 1.0), resulting in the higher photocurrents at N-doped a-SiC film. Both currents with and without exposure to Hg-Xe lamp were monotonically increased with increasing RF power (=sp^2 carbon content) from 75 W to 150 W, as shown in Figure 9. The increase of both currents might be caused by lowering of resistivity of a-SiC films with increase in sp^2 content. These results support that the increasing in sp^2 content (the lowering resistivity) plays important role for higher photo-electron conversion efficiency. However, by increasing RF power up to 200 W, photocurrent values were monotonically decreased. This is because excess sp^2 carbons in N-doped a-SiC worked as recombination center of photo excited carries. Therefore, the amount of sp^2 content in N-doped a-SiC film should be moderately controlled.

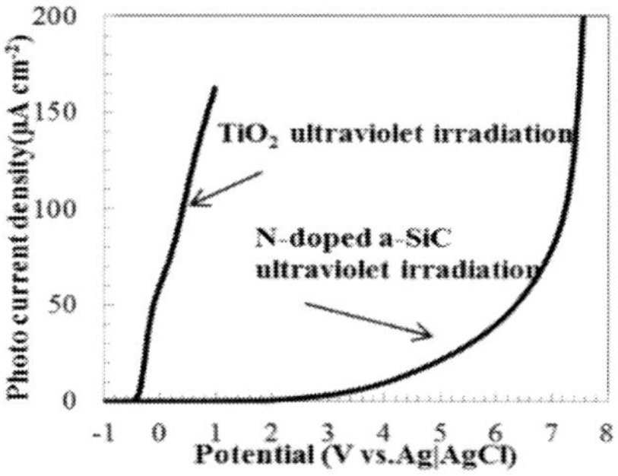

Figure 8 Current-voltage curves with exposure to Hg-Xe lamp of N-doped a-SiC

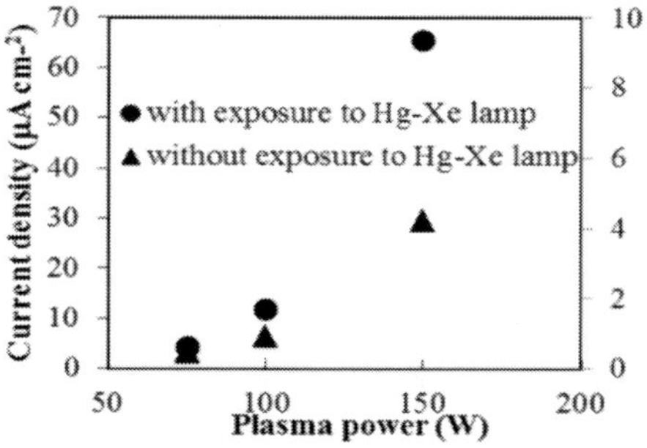

Figure 9 Current-voltage curves with, without exposure to Hg-Xe lamp of N-doped a-SiC

4. Conclusion

n-type a-SiC film with high performance, which can be utilized to electronic device, coulde be fabricated by optimizing the size and number of sp^2 clusters that was included in multi-phase structure of the film. The resulting N-doped a-SiC with 10 six-membered rings in sp^2 clusters shows 1.68 eV of optical gap and 153 μA cm^{-2} of photo-current. Quantum yield was on the same level as that obtained at anatase-type titanium oxide prepared by sol-gel method. It is, however, necessary to increase optical gap and conductivity in order to make the utilization possible. The size enlargement of sp^2 cluster brings higher conductivity but induces a lower optical gap. It suggests that another approach should be employed to develop N-doped a-SiC semiconductor with specific performance. Improvement of optical gap will be attempted by incorporating many Si atoms to increase a-SiC phase in N-doped a-SiC films. Moreover higher conductivity will be attempted by incorporating other impurity atoms like S that do not form insulating components such as SiN.

References

[1] Oliveira A.R, Carreno N.N.P 2006 *Mater. Sci. Eng.* V **128** 44
[2] Matunami H 2003 *Technology of Semiconductor SiC and Its Application* (Nihonbashikoamicho: Nikkan Kogyo Shimbun) p 11
[3] Wang L, Xu J, Ma T, Li W, Huang X, Chen K 1999 *J. Alloys. Compd.* **290** 273
[4] Kiisk V, Reedo V, Sild O, Sildos I 2009 *Optical. Materials.* **31** 1376
[5] Chhowalla M, Robertson J, Chen C. W, Silva S. R. P, Davis C. A, Amaratunga G. A. J, Milne W. I 1997 *J. Appl. Phys.* **81** 139

Journal of Physics: Conference Series **441** (2013) 012041 doi:10.1088/1742-6596/441/1/012041

Influence of molecular structure on plasma carbonization of organic semiconductor molecules

Toshihiro Shimada[1,2], Naoki Muraya[3], Jun Tomita[3], Takashi Yanase[4], Taro Nagahama[1]

[1] Division of Materials Chemistry, Faculty of Engineering, Hokkaido University, Kita-ku, Sapporo, Hokkaido 060-8628, Japan
[2] CREST-JST, Gobancho, Chiyoda-ku, Tokyo 100-0013, Japan
[3] Graduate School of Chemical Science and Engineering, Hokkaido University, Kita-ku, Sapporo,
Hokkaido 060-8628, Japan
[4] Frontier Chemistry Center, Faculty of Engineering, Hokkaido University, Kita-ku, Sapporo, Hokkaido 060-8628, Japan

Abstract

The plasma treatment of organic semiconductor molecules includes issues related with nanoscale plasma interfaces. The process was examined by optical emission spectroscopy. It was found that molecules are sublimed by the heat and decomposed by the interaction with the plasma. The product carbon solids were examined by Raman spectroscopy. The raman spectra are strongly dependent on the molecular structures. The carbon solids which does not show graphitic Raman spectra have nanostructures originating from initial molecules.

1. Introduction

Carbon materials with controlled nanostructure are gathering much attention recently for the application in electrode materials for rechargeable batteries[1], carbon-based catalysis[2] and new semiconductors such as carbon nanotubes (CNTs)[3] and graphene[4]. However, the strategy for finely control the nanostructure is rather limited so far, for example calcination of polymers with various functional moieties[5], diameter control of the CNTs by controlling the size of the catalysts[6], and doping of other elements during the growth[7]. In order to explore the possibility of producing nanostructured carbon from organic molecules, we have studied the high temperature treatment of the molecules by vacuum deposition at 1800 ^{o}C[8]. However, the amount of the product was too small to allow detailed systematic examination. We here report the study on the possibility of synthesizing carbon materials with well-defined nanostructures by treating organic semiconductor molecules with plasma. Recent advances in the organic electronics, various molecules with wide p-electron systems have been synthesized. By exposing these molecules to plasma, radicals and ions of these molecules will be formed and they will polymerize to make a solid. It is expected that the product may contain the structure of the initial organic semiconductor molecules. In this paper, we examined the effect of microwave plasma by optical emission spectroscopy of the plasma and Raman spectroscopy of the product solids.

2. Experiment

The organic semiconductor molecules used as the raw materials in this study are shown in Fig. 1. They are solid powders at room temperature and used as purchased. 1~2 mg of the powder was sealed in a quartz tube (4mm OD, 2.5mm ID, 4 cm length) and the gas inside the tube was replaced with 99.9% N_2. It was pumped by a rotary pump to the pressure of 100 Pa and then sealed by fusing an oxygen burner. It was observed that the molecules are sublimed in the tube when the tube was heated to 100 ~ 300 °C (the temperatures are dependent on the molecules). The tube was set to a surfatron reactor shown in Fig. 2. Microwave (~ 500 W) from magnetron was irradiated and the

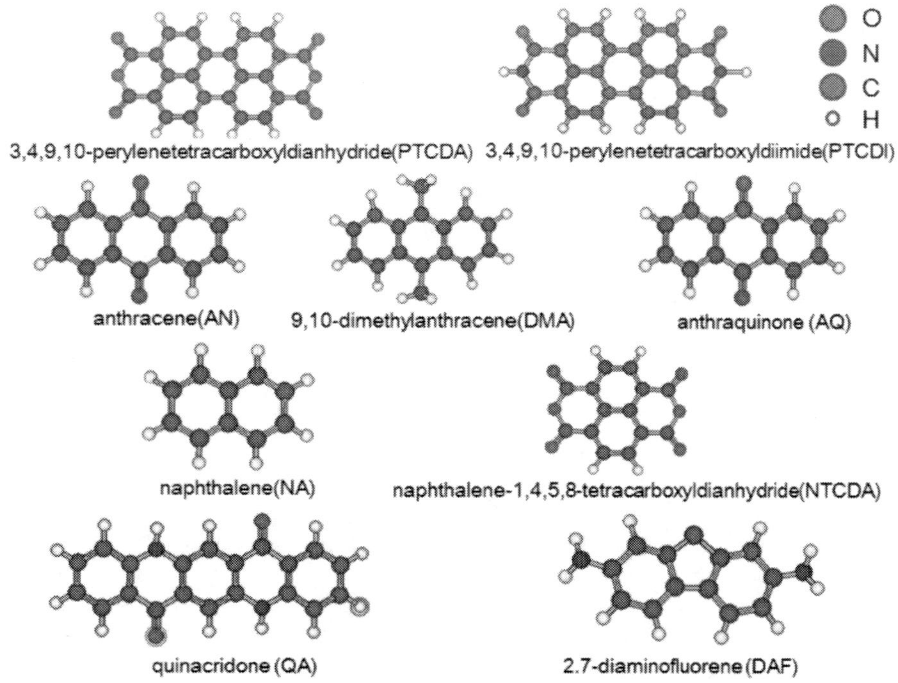

Fig. 1: Structure of the molecules used as source materials.

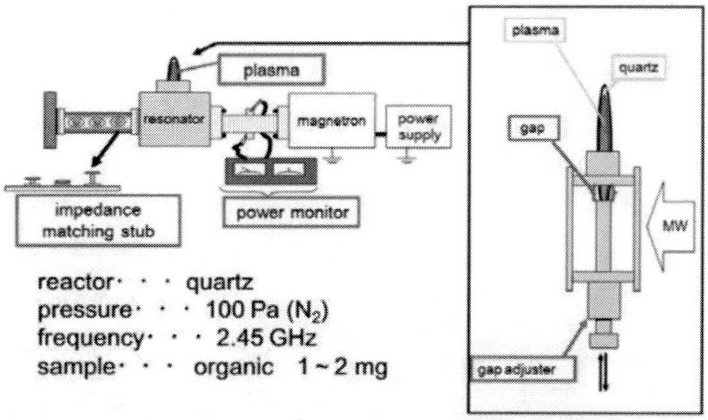

Fig. 2: Schematic drawing of the surfatron reactor.

plasma was excited. The colour of the plasma changed as the time elapsed and the plasma ceased eventually. The optical emission spectra were monitored by a fibre optic spectrometer (Ocean Optics

USB-2000) which was calibrated by emission wavelengths of neon. After the plasma treatment, the inside of the tube turned brown or black colour. The obtained solids were examined by Raman spectroscopy using Renishaw Invia spectrometer using 532 nm or 780 nm laser excitation. The laser wavelength was chosen to achieve high resolution spectra while avoiding fluorescence.

3. Results and discussion

The typical time evolution of optical emission spectra during the plasma treatment is shown in Fig. 3. The molecule was PTCDA. After the plasma ignition, emission peaks arising from the N_2 gas plasma

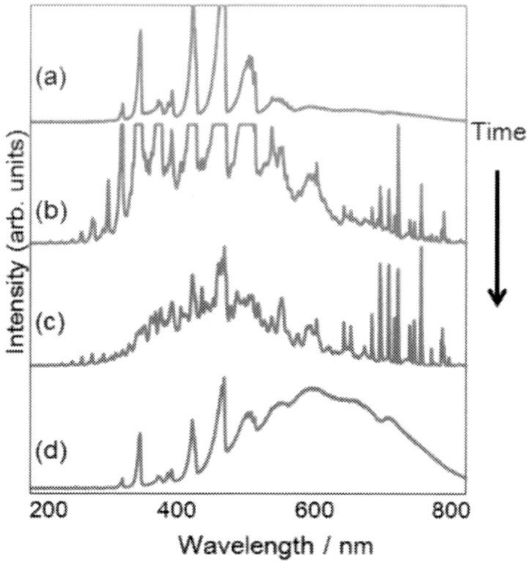

Fig. 3: Optical emission spectra from PTCDA- containing tube during the plasma treatment.

was observed (Fig. 3(a)). After several second, the quartz tube was heated by microwave and the plasma, and molecules are vaporized. Corresponding emission was appears as peaks in visible region (Fig.3(b)). When the molecule contains oxygen, as in the case of PTCDA, peaks around 700-800 nm appears. It shows the decomposition at the oxygen containing part of the molecule. After several seconds, the N-originated peaks diminished and the many small peaks were observed in the visible region(Fig.3(c)). Finally a broad peak appeared from infrared region and shifted toward the visible region(Fig.3(d)). At this moment, the quartz tube glowed and sometimes it deformed. We consider that the broad peak was the blackbody radiation from the quartz tube and the sample. It is supposed that the temperature was elevated to the glass transition temperature of the quartz glass (~1000 °C).

We found that the molecules are categorized into three groups. The firs group is easily graphatized molecules. Figure 4 shows the Raman spectra of the product obtained from DAF and PTCDA. D peak (~1350 cm^{-1}) and G peak (~1580cm^{-1}) are observed clearly and no other peaks are noticeable. It is the signature of the formation of graphite-like carbon. The broadness of the D and G peaks and their ratio indicates that the size of the graphite is very small, probably not far beyond the size of the molecules [9]. It is noticed that PTCDA shows shaper peaks then DAF, which is in accordance with the fact that PTCDA easily produces graphemes by CVD [10]. The broad background peak observed in the Raman spectra of these materials probably correspond to various ill-defined structures having various vibration frequencies. It is difficult to analyze the structure because fluorescence is also present even with infrared (780nm) excitation.

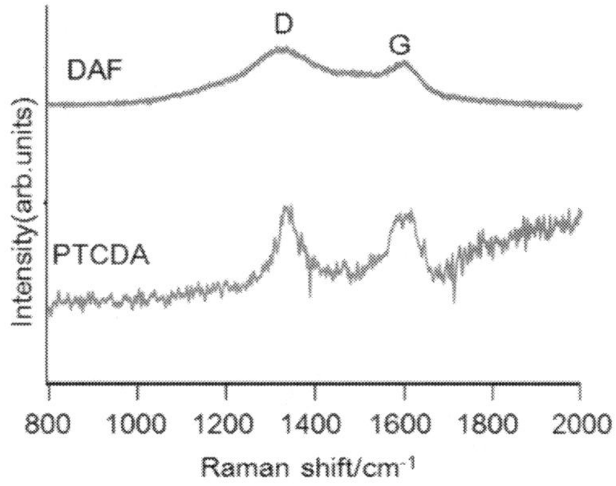

Fig. 4: Raman spectra of solids obtained from the plasma treatment of DAF and PTCDA (group 1).

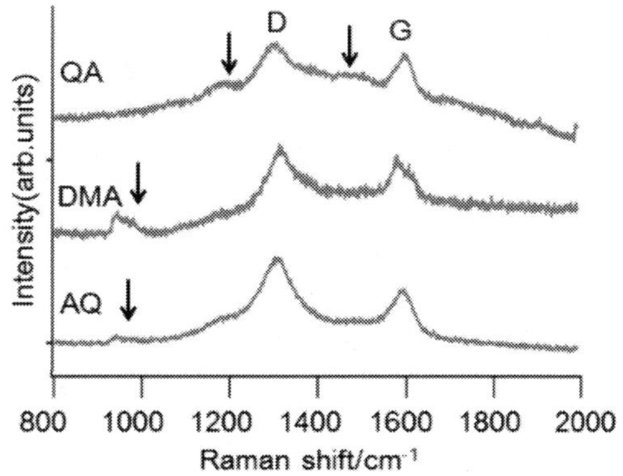

Fig. 5: Raman spectra of solids obtained from the plasma treatment of QA, DMA and AQ (group 2).

Figure 5 shows the Raman spectra of the second group, i.e., giving strong single D and G peaks with additional peaks, indicated by arrows, originating from substitution groups. These molecules have linearly fused benzene rings. It is considered that these molecules are decomposed to give graphite-like carbons but substantial amount are not incorporated in the graphite lattice. Although the origin of the additional peaks are still under investigation, preliminary quantum chemical simulation suggests that vibration of the CH_3- group coupled with that of the fused benzene rings corresponds to the peak around 900 cm^{-1}.

The third group is shown in Fig. 6. They shows strong additional peaks with D and G peaks weaker and broader than other groups. The initial molecules are derivative of small linear aromatics. It shows that these molecules are not easily graphitized. The striking difference is the difference between NTCDA and PTCDA, both having the carboxylicdianhydride structure. It is noticed that the size of the core aromatics is very important.

4. Conclusion

We investigated the plasma treatment of organic semiconductor molecules. We found that molecules are sublimed by the heat and decomposed by the interaction with the plasma. The product carbon solids were examined by Raman spectroscopy and whether the graphitic carbon is formed or not is strongly dependent on the molecular structure. The carbon solid which does not show graphitic Raman spectra have nanostructures probably originating from initial molecules and its functions as an electrode or catalyst are promising.

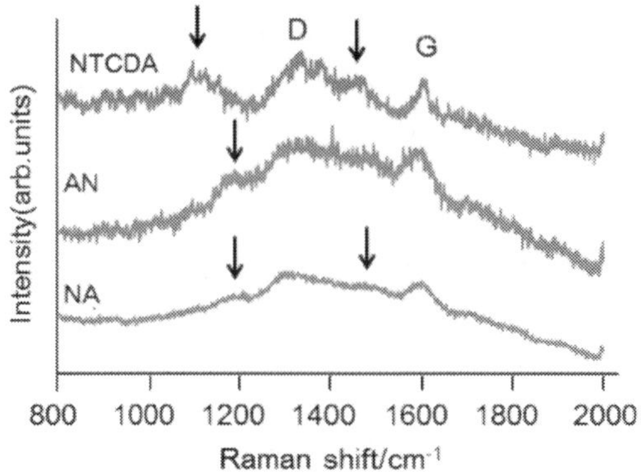

Fig. 6: Raman spectra of solids obtained from the plasma treatment of NTCDA, AN and NA (group3).

Acknowledgments

The present work has been partly supported by KAKENHI for Scientific Research on Innovative Areas "Frontier Science of Interactions between Plasmas and Nano-Interfaces". Raman spectra were taken with the facility of CRIS, Hokkaido University. The authors are grateful to the discussion with Prof. T. Hasegawa (University of Tokyo) for the valuable discussion under CREST-JST project.

References

[1] Liwen Ji , Zhan Lin, Mataz Alcoutlabi and Xiangwu Zhang 2011 *Energy Environ. Sci.* **4** 2682.
[2] Ozaki J., Tanifuji S., Kimura N., Furuuchi A. and Oya A 2006 *Carbon* **44** 1324.
[3] Baughman R.H., Zakhidov A.A., de Heer W.A. 2002 *Science* **297** 787.
[4] Novoselov K.S. et al. 2004 *Science* **306** 666.
[5] Meng Y.et al. 2006 *Chem. Mater.* **18** 4447.
[6] Terrones M et al. 1997 *Nature* **388** 52.
[7] Lin Y.C. et al. 2010 *Appl. Phys. Lett.* **96** 133110.
[8] Noguchi T.,Shimada T., Hanzawa A. and Hasegawa T. 2009 *Thin Solid Films* **518** 778.
[9] Concardo L.G. et al. 2006 *Appl. Phys. Lett.* **88** 163106.
[10] Yudasaka M. et al. 1994 *Appl.Phys.Lett.,.***64** 3237.

Enhancement of electrical conductivity and electrochemical activity of hydrogenated amorphous carbon by incorporating boron atoms

Hiroshi Naragino, Kohsuke Yoshinaga, Akira Nakahara, Sakuya Tanaka and Kensuke Honda

Graduate school of Science and Engineering, Yamaguchi University, 1677-1, Yoshida, Yamaguchi-shi, Yamaguchi 753-8512, Japan

E-mail: khonda@yamaguchi-u.ac.jp

Abstract. Conductive boron-doped hydrogenated amorphous carbon (B-DLC) thin films were successfully synthesized with RF plasma-enhanced CVD method. By incorporating boron atoms in amorphous carbon, conduction types were changed from n- to p-type, and volume resistivity was decreased from 30.4 (non-doped) to 6.36 x 10^{-2} Ω cm (B/C = 2.500 atom%). B-DLC film with $sp^2/(sp^2 + sp^3)$ carbons of 75 atom% exhibited high resistance to electrochemically-induced corrosion in strong acid solution. Furthermore, it was clarified that boron atoms in DLC could enhance kinetics of hydrogen evolution during water electrolysis at B-DLC surface. B-DLC is, therefore, a promising electrode material for hydrogen production by increasing the concentration of boron atoms in B-DLC and enhancing the reactivity of H_2 evolution.

1. Introduction

High stability for electrochemically-induced corrosion and availability at lower production costs are necessary for the electrode materials used in electrolytic industries that produce chemicals such as sodium hydroxide, hydrogen and chlorine, which are essential in industrialized world.

Incorporation of impurity atoms has been commonly used to enhance the conductivity of semiconductor. Our research group has developed conductive hydrogenated amorphous carbon thin films by incorporating nitrogen atoms (N-DLC) that could be synthesized at low cost [1]. N-DLC films showed high chemical stability, and could be used as a polarizable electrode material in electroanalytical assay as well as boron-doped diamond electrodes. It, however, is necessary to reduce power loss resulting from electrode resistance and overpotential for hydrogen evolution in order to apply the film practically to electrolytic industries that produce pure material by passing larger current on electrodes.

N-DLC film contained a large number of nitrogen atoms (> 8 atom%, 7.0 x 10^{21} cm^{-3}). Nevertheless, carrier density (7.5 x 10^{18} cm^{-3}) of N-DLC film was relatively lower than atom% of nitrogen atoms in DLC. Therefore, most of nitrogen atoms in DLC did not contribute to increase carrier density of n-type conduction. In this study, boron atoms were employed as impurity atoms. The conductivity of boron-doped hydrogenated amorphous carbon (B-DLC) is expected to be higher than that of N-DLC because boron atoms do not form multiple bonding with carbon atoms.

In this study, amorphous carbon electrodes with high conductivity and high chemical stability were fabricated by doping boron atoms and by controlling the amount of sp^2 bonded carbons. Moreover, the improvement of electrochemical activity of hydrogen evolution was attempted by controlling the concentration of boron atoms.

2. Experimental

B-DLC thin films were prepared using plasma-enhanced chemical vapor deposition (CVD) with RF-self bias system (model BP-10, SAMCO Co., Ltd.). Films were deposited on n-Si (111) substrates (< 0.02 Ω cm, SUMCO Co., Ltd.) after ultrasonic cleaning with 2-propanol for 15 minutes and in-situ sputter cleaning with argon plasma (200 W for 15 min.). Vapor at a flow rate of 5 mL min^{-1} was introduced into an evacuated reaction chamber from a degassed mixed solution of *n*-hexane and trimethyl borate ($B(OCH_3)_3$, TMOB) (carbon and boron sources) kept at room temperature. Argon was simultaneously introduced into the reaction chamber with a source gas. The atom% of boron in DLC was adjusted by the concentration of TMOB in *n*-hexane. B-DLC Films with a thickness of 0.68 μm were grown by applying RF power (350 W, 13.56 MHz) to cathode that held Si substrates on a quartz liner. The atom% of $sp^2/(sp^2 + sp^3)$ carbons in B-DLC was controlled by the temperature of a substrate during CVD synthesis. The temperature of the substrate used for films with sp^2 content of 82 % was set at 300 °C, and for films with sp^2 content of 75 % was set at 275 °C. The pressure in the reaction chamber was adjusted at 40 Pa. The deposition time was 40 minutes.

The resulting B-DLC films were characterized by Raman spectroscopy, X-ray photoelectron spectroscopy (XPS) and Hall-effect measurement. The electrochemical measurements were carried out using B-DLC films (geometric area: 0.1 cm^2) as a working electrode, Ag|AgCl electrode as a reference electrode, and Pt coiled wire as a counter electrode in a three-electrode cell.

3. Results and Discussion

3.1. Film Characterization

TMOB used as boron source is resistant to decomposition in partially-ionized plasma of CVD synthesis because B-O bonding with higher bond strength (5.4 eV, C-C; 3.6 eV) is included in TMOB molecule. To enhance TMOB decomposition by the collision with Ar* matastable atoms and/or Ar^+ ions, argon gas was introduced in the reaction chamber during deposition [2]. By adding Ar gas, conduction types in B-DLC were changed from n- to p-type. And then, higher values of carrier density and carrier mobility of B-DLC could be achieved than those of N-DLC as shown in **Table 1**.

The atomic concentration ratio of B-DLC was examined by XPS spectra. B 1s peak was clearly observed. The change of conduction type from n- to p-type shown in **Table 1** suggests that boron atoms are included in DLC and act as acceptors. The resistivity of B-DLC reached 6.36 x 10^{-2} Ω cm (at 100 sccm of Ar flow rate). This value was more than 20 times lower than that of N-DLC. By

Table 1 Hall measurement results of non-doped DLC, B-DLC (B/C = 2.500 atom% in the source gas) and N-DLC

Sample	Ar flow rate (sccm)	Carrier density (10^{18} cm^{-3})	Mobility (cm^2 V^{-1} s^{-1})	Resistivity (Ω cm)	Hall coefficient (cm^3 C^{-1})	Type
non-doped DLC	100	0.12	1.73	30.4	-52.57	n
B-DLC	0	2.93	3.60	5.93	-21.30	n
B-DLC	**100**	**12.90**	**7.59**	**6.36 x 10^{-2}**	**+0.48**	**p**
N-DLC	-	0.57	7.52	1.44	-10.85	n

using B-DLC films as electrode materials in electrolytic industries, power loss caused by electrode resistance can be reduced to one twenty-second. The concentration of impurity atoms (dopant atoms) in N-DLC films (8.0 atom%) is larger than that of B-DLC films (1.3 atom%). However, carrier density of N-DLC was lower than that of B-DLC. This lower carrier density resulted in the resistivity of N-DLC higher than that of B-DLC. The incorporation of boron atoms was found to be more efficient for enhancing conductivity than nitrogen atom incorporation. A similar result has been reported by C. Ronning *et al.* [3]. In order for impurity atoms in DLC to function as donors or acceptors, four single bonds with carbon atoms should be formed. The calculation results that boron atoms in amorphous carbon are preferentially bonded to sp^3-bonded carbons have been reported [4]. On the other hand, it has been reported that nitrogen atoms are preferentially formed quite stable multiple bonding such as =C=N- and -C≡N, which does not contribute to conductivities [5]. Consequently, boron atoms are more effective to increase carrier density than nitrogen atoms. The carrier density higher than that at N-DLC can be achieved at B-DLC although dopant concentration of B-DLC is lower than that of N-DLC.

3.2. Susceptibility to surface corrosion in strong acid media

To evaluate the inertness of B-DLC films toward electrochemically induced corrosion, the potential cycle treatment was carried out in 1.0 M HNO_3 + 0.1 M KF from -0.65 V to +1.3 V at 25°C for 2 hours [5]. **Fig. 1** shows optical micrographs of B-DLC and glassy carbon (GC) surfaces before and after potential cycling. At the surfaces of B-DLC with $sp^2/(sp^2 + sp^3)$ carbons of 82 atom% and GC, pits (30 – 100 μm in diameter) were formed (**Fig. 1-A** and **C**). The electrode surface corrosion might occur primarily by carbon gasification producing CO_2 and/or CO. On the contrary, B-DLC surfaces with $sp^2/(sp^2 + sp^3)$ below 75 atom% were identical after cycling (**Fig. 1-B**).

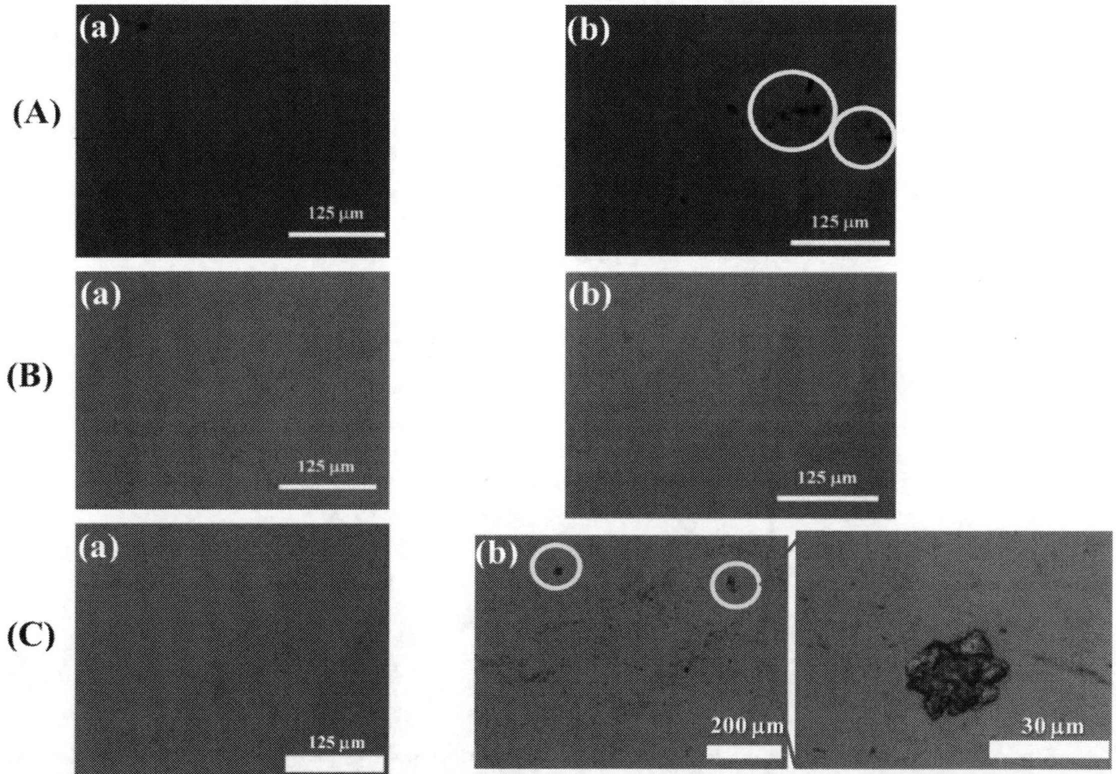

Fig. 1 Optical micrographs of (A) GC, (B) B-DLC (sp^2; 75 %) and (C) B-DLC (sp^2; 82 %) surfaces (a) before and (b) after potential cycling

B-DLC films with sp^2 of 75 atom% were fabricated under the following conditions; A temperature of substrate was set at 275 °C (< 275 °C). RF power of 350 W was applied. It has been known that sp^2 bonded carbon based electrodes show lower resistance to corrosion at anodic polarization than that of electrode materials including sp^3 bonded carbon [6]. In DLC films with sp^2-bonded carbons of 50 atom% (sp^2:sp^3 = 1:1), which were uniformly distributed, sp^2 clusters are thought to be not formed because all carbon forms sp^2-sp^3 bonding. In DLC films with sp^2 of 75 atom%, excess sp^2-bonded carbons might form sp^2-sp^2 bondings. As a result, sp^2 cluster (sp^2-sp^2-sp^2) is formed. The resistance to corrosion at DLC was lower with higher (> 75 atom%) sp^2-bonded carbons. B-DLC electrode with much higher resistance to electrochemically-induced corrosion could be fabricated by decreasing the amount of sp^2-bonded carbons. From these results, it can be concluded that highly conductive B-DLC film fabricated in this study can be applied as an electrode material with reliability in long-term electrochemical use.

3.3. Electrochemical Characterization

Fig. 2 shows the cyclic voltammograms for B-DLC electrode in 0.1 M H_2SO_4. The onset potential for hydrogen evolution (at which current densities reached -0.2 mA cm^{-2}) was positively shifted from -1.61 to -1.41 V vs. Ag|AgCl when boron concentration in the source gas was increased from 0.625 to 2.500 atom% during B-DLC synthesis. The overpotential for hydrogen evolution was decreased by approximately 200 mV. The overpotential for oxygen evolution was not changed (slightly decreased by 30 mV). The amount of boron atoms in DLC may strongly affect the rate of hydrogen evolution.

In order to clarify the dependence of the parameter of the rate of hydrogen and oxygen evolution on the concentration of boron atoms in DLC, values of exchange current density i_0 and transfer coefficient α were estimated from Tafel plot (**Fig. 3**). The values are summarized in **Table 2**. The exchange current density i_0 is the value proportional to standard rate constant k_0 and the transfer coefficient α is a measure of the energy barrier. **Fig. 3** shows Tafel plots of B-DLC synthesized under various boron concentrations in 1.0 M H_2SO_4.

For oxygen evolution, the values of transfer coefficient α stayed constant for all boron concentration in the source gas. The value of exchange current density i_0 was slightly increased with increasing boron concentration. These results indicate that boron concentration is indifferent to the kinetics of oxygen evolution. In contrast, for hydrogen evolution, the values of transfer coefficient α were not changed (0.18 ~ 0.14) and the values of exchange current density i_0 were increased by one order of magnitude (from 1.47 x 10^{-8} to 3.70 x 10^{-7} A cm^{-2}) with increasing boron concentration from 0.625 to 2.500 atom%. The kinetics of hydrogen evolution was drastically enhanced with boron atoms in DLC films.

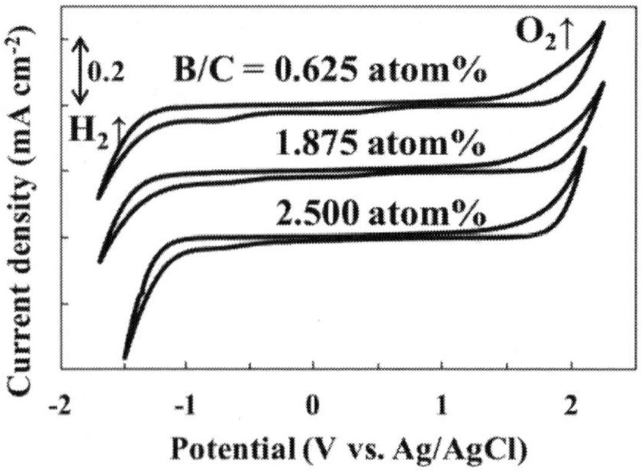

Fig. 2 Cyclic voltammograms in 0.1M H_2SO_4 at B-DLC electrodes deposited under various boron concentrations in the source gas. Sweep rate, 100 mV s^{-1}.

Table 2 The values of i_0 and α for hydrogen and oxygen evolution at B-DLC electrodes.

B/C ratio (atom %)	H_2 evolution			O_2 evolution		
	i_0 (10^{-8} A cm^{-2})	slope^{-1} (mV dec^{-1})	α	i_0 (10^{-6} A cm^{-2})	slope^{-1} (mV dec^{-1})	α
0.625	1.47	336	0.18	2.63	415	0.14
1.875	2.14	339	0.17	3.53	405	0.15
2.500	37.0	413	0.14	5.08	418	0.14

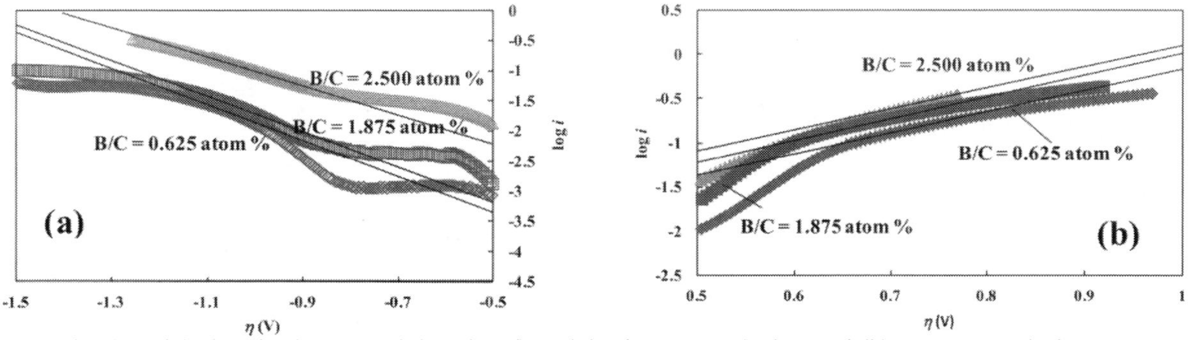

Fig. 3 Tafel plots in the potential region for (a) hydrogen evolution and (b) oxygen evolution at B-DLC electrodes. Electrolyte, 1.0 M H_2SO_4. Sweep rate, 50 mV min^{-1}.

These results indicate that the shape of potential curve of H_2 evolution was not changed but the activation energy was decreased with boron concentration. It suggests that boron atoms in DLC may act as an electrochemical catalyst for H_2 evolution. Thus, the electrochemical reactivity for hydrogen evolution at B-DLC surface could be enhanced by increasing boron concentration.

Fig. 4 shows cyclic voltammograms for B-DLC and N-DLC electrodes in 0.1 M H_2SO_4. The overpotential for hydrogen evolution at B-DLC with B/C = 2.500 atom% was lower than that at N-DLC by ca. 400 mV. The loss of electric power for H_2 production by water electrolysis could be decreased by using B-DLC electrode. However, the overpotential for hydrogen evolution at B-DLC surface is extremely higher than that at metal electrodes such as Pt. B-DLC can be applied as electrode materials with high corrosion resistance in electrolytic industries by increasing the concentration of boron atoms in DLC and enhancing the electrochemical reactivity for H_2 evolution.

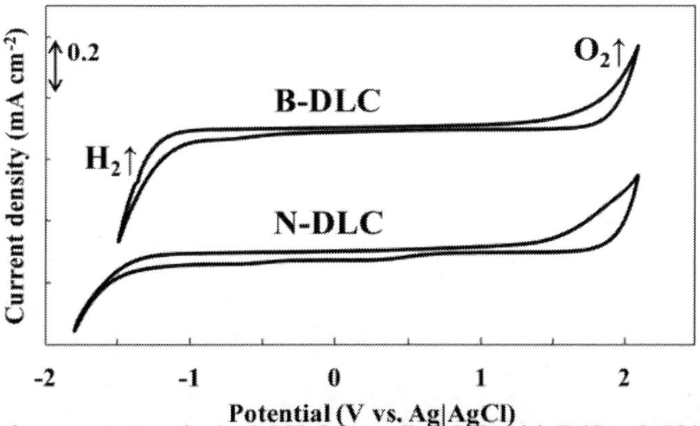

Fig. 4 Cyclic voltammograms in 0.1M H_2SO_4 at B-DLC with B/C = 2.500 atom% and N-DLC electrodes. Sweep rate, 100 mV s^{-1}

Journal of Physics: Conference Series **441** (2013) 012042 doi:10.1088/1742-6596/441/1/012042

4. Summary

B-DLC films with lower resistivity than that of N-DLC films could be fabricated by incorporating boron atoms as a dopant. Boron concentration and volume resistivity of B-DLC synthesized at B/C = 2.500 atom% in the source gas were 1.3 atom% and 6.36 x 10^{-2} Ω cm, respectively. The electric power loss resulting from electrode resistance of B-DLC was lower than that of N-DLC because the volume resistivity of B-DLC was 22 times lower than that of N-DLC.

B-DLC films with sp^2-bonded carbon of 75 atom% could be fabricated under the following deposition condition; A temperature of the substrate was < 275 °C and RF power was < 350 W. B-DLC films with sp^2 of 75 atom% show a high resistance to electrochemically-induced corrosion in strong acid solution in long-term electrochemical use.

The electrochemical reactivity of hydrogen evolution could be improved by increasing the concentration of boron atoms in DLC. Hence, the loss of electric power derived from the overpotential for hydrogen evolution at B-DLC was lower than that of N-DLC.

B-DLC is promising electrode material with higher reliability in long-term use and lower electric power loss in hydrogen production by water electrolysis.

References

[1] Tanaka Y., Furuta M., Katsuki Y., Kuriyama K., Kuwabara R., Fujishima A. and Honda K. 2011 *Electrochim. Acta.* **56** (3) 1172

[2] Shimozuma M. and Tagashira H 1991 *Hokkaido University Collection of Scholarly and Academic Papers* **4** 49

[3] C.Ronning, U. Griesmeier, M. Gross, H.C. Hofsäss, R. G. Downing and G. P. Lamaze 1995 *Diamond Relat. Mat.* **4** 666

[4] P. K. Sitch, Th. Köhler, G. Jungnickel and D. Porezag 1996 *Solid State Commun.* **100** 549

[5] C. W. Chen and J. Robertson 1999 *Carbon* **37** 839

[6] G.M. Swain 1994 *J. Electrochem. Soc.* **141** 3382

Growth Rate of Titanium Thin Film by High-Power Pulsed Sputtering (HPPS) Penning Discharge Plasma with the Inner Electrode

Kingo Azuma, Toru Higuchi, Yusuke Inoue

Department of Electrical Engineering and Computer Sciences, University of Hyogo, 2167 Shosha, Himeji, Hyogo, Japan

E-mail: easter11_azuma@mem.iee.or.jp

Abstract. High-power pulsed sputtering (HPPS) penning discharge is featured that the plasma is generated at a narrow gap consisting of a pair of cathodes as sputtering target in parallel each other. The magnetic field is provided by setting a set of permanent magnets behind the targets and the electric field is parallel to the magnetic field. The conventional HPPS penning plasma source is disadvantage in a narrow pressure range on the plasma production, and in particular, the lower limit of the pressure is 2 Pa or higher. This problem has been solved by setting an electrically-grounded electrode at the central region of the plasma source. The lower limit of plasma generation is lowered to about 0.5 Pa. In the case of titanium target, titanium films are deposited on the collector electrode for an HPPS Penning discharge in the presence of the inner electrode, and the deposition rate is 16 nm/min at argon gas pressure of 0.5 Pa, where the peak power is 15 kW at power consumption of about 310 W.

1. Introduction

The sputtering deposition is one of the technologies that cover the surface of a substrate with functional thin film. It is known as the deposition process to able to form dense thin film. Especially a magnetron sputtering process has high deposition rate. However, improvements of film quality and adhesion are horny issues for the magnetron sputtering because the ionization rate of the sputtering particles is very low. These problems may be cleared up by a high power pulsed sputtering (HPPS) discharge. HPPS glows produce high density metallic ion species, and the thin film physical vapor deposition coating using these plasmas is emerging technologies both to improve tribological characteristics and to control film structure.

There are two major categories of HPPS discharge plasma by the relation of direction of an electric field and a magnetic field. One is HPPS magnetron discharge plasma [1], the other is HPPS penning discharge plasma [2]. At the magnetron discharge, an electric field and a magnetic field intersect perpendicularly in the ion sheath. In general, HPPS discharge means HPPS magnetron discharge, because the magnetron electrode is commonly used in the sphere of the hard coatings. One of the weak points of metal coating by HPPS magnetron discharge is to generate reflected high energy neutrals. These arise from ions bombarding a surface in a low-pressure environment [3]. These have a bad influence on the film quality on workpieces in many cases [4]. HPPS penning discharge technology can control generation of particles. It is largely based on configuration of target electrodes.

HPPS penning discharge is featured that the plasma is generated at a space consisting of a pair of cathodes as sputtering target in parallel each other [2]. The chamber wall works as an anode to collect

Journal of Physics: Conference Series **441** (2013) 012043 doi:10.1088/1742-6596/441/1/012043

electrons. The electric field is parallel to the magnetic field at the narrow gap. The magnetic field is provided by setting a set of permanent magnets behind the targets. Energetic argon ions accelerated by a voltage difference between the cathode and the plasma bombard the target to sputter metallic species, which are mixed in the plasma and ionized. Second emission electrons from the target are even strongly ionized metallic particles. The reflected high energy neutrals is trapped the narrow space. Therefore these rarely reach the workpieces.

The plasma density of HPPS penning discharge plasma is as high as 10^{18} m^{-3} due to high power consumption on the order of 100 kW [5]. The fraction of titanium ions is about 70% [6]. However, the HPPS source is disadvantage in a narrow pressure range on the plasma production, and in particular, the lower limit of the pressure is 2 Pa or higher. This problem has been solved by setting an electrically-grounded rod (inner electrode) at the central region of the plasma source.

2. The experimental setup

A HPPS penning discharge plasma source is configured a pair of cathode targets, permanent magnets, U-type magnetic yoke and inner electrode. The cathode targets have the same potential. A magnetic field is created by a set of permanent magnets installed behind the targets. The magnetic field is oriented perpendicular to the targets. When a pulsed voltage is applied to the targets and initial plasma is ignited in the narrow gap, the electric field in the ion sheath of target electrodes is parallel to the magnetic field. Thus, the plasma is a kind of Penning discharge [7].

HPPS plasma was generated at the gap formed by a pair of parallel targets. The surface area of each target was 12 cm^2 (length: 60 mm and height: 20 mm). The gap length was 10 mm, and the target material was titanium. A magnetic field was oriented perpendicular to the targets, at strength of approximately 0.3 T at the target surface. The inner electrode is inserted into the center void of the magnetic yoke. It is grounded. The gap distance of the inner electrode and the yoke is 2 mm.

Figure 1 shows a schematic diagram of the experimental set up. The vacuum chamber had an inner diameter of 310 mm and a height of 300 mm. A pulsed voltage was supplied from the pulsed power source (maximum ratings: current of 100 A, output voltage of –800 V, repetition rate of 5 kHz, and pulse duration of 1-200 μs). The plasma was generated in an argon environment. The argon gas pressure was varied 0.5 Pa to 3 Pa. The applied voltage was –800 V with a repetition rate of 500 Hz. The current through the circuit was regulated by a series resistor, R_S (the maximum rated power of a unit resistor was 300 W).

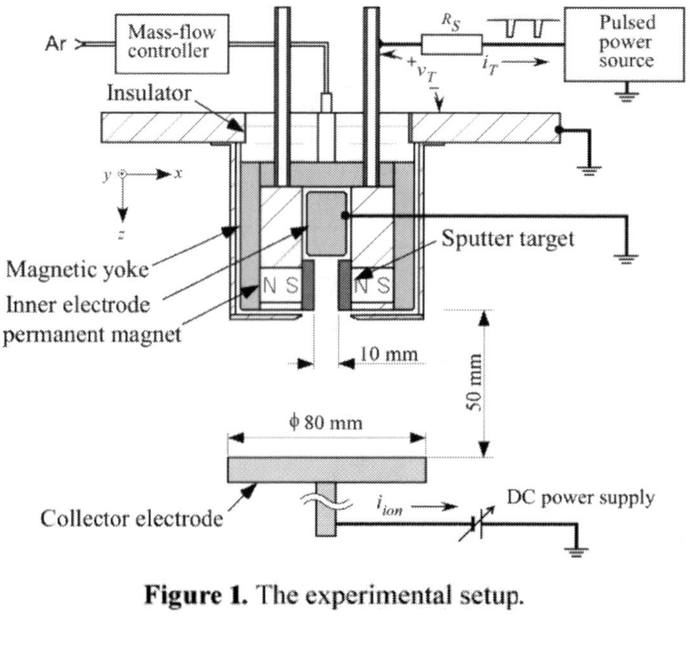

Figure 1. The experimental setup.

Journal of Physics: Conference Series **441** (2013) 012043 doi:10.1088/1742-6596/441/1/012043

The evacuation system consisted of a turbo molecular pump (TMP, 350 L/min) (as the main evacuation device), a mechanical booster pump (MBP) and a rotary pump (RP) (as a backup for the TMP). The inlet gas pressure was controlled by a mass flow controller. The argon gas was introduced from behind the plasma source through a ¼ inch-PFA (Perfluoroalkoxy) tube. The pressure inside of the vacuum chamber was measured using a ceramic capacitance manometer (ULVAC, GM-2001/CCMT-10A, range of measurement: 1.3×10^{-1} to 1.3×10^{3} Pa) and an ionization gauge (ANELVA, MIG-430/MG-2, range of measurement: 10^{-6} to 13 Pa).

The target current, i_T, through the circuit was observed using a current transformer (Pearson type 110A, sensitivity: 0.1 V/A, rise time: 18 ns, observatory bandwidth: 1 Hz-20 MHz). The target voltage, v_T, was observed using a voltage probe (Tektronix, type 5100, rated voltage: DC+AC peak of 2.5 kV, frequency bandwidth: DC-250 MHz). Waveforms were monitored with an oscilloscope (Tektronix, TDS3034B, bandwidth: 300 MHz, maximum sampling rate: 2.5 GS/s).

In order to extract ions from the HPPS plasma, a disc-shaped holder electrode with a diameter of 80 mm was positioned nearby the plasma source. As shown in Figure 3, the distance from the edge of the plasma source was 50 mm. A negative voltage was applied to the collector electrode at –50 V. The measured current is ion current, i_{ion}, because the floating potential of the collector is about –20 V.

3. Results

3.1. Temporal electrical characterisation

Figures 2 (a), (b) and (c) show waveforms of target voltage, v_T, target current, i_T, and ion current, i_{ion}. The pulse duration was 50 μs and the initial target voltage was –800 V. The series resistor, R_S, set in the plasma generation circuit had a resistance of 10 Ω. The source voltage was maintained constant during the plasma generation. Thus, the source voltage is delivered to R_S and the plasma.

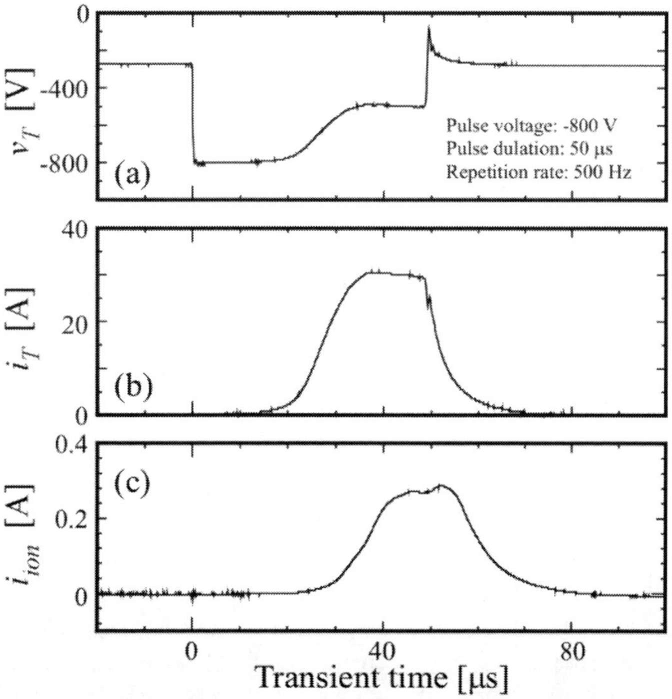

Figure 2. Typical time response of (a) target voltage V_T, (b) target current i_T and (d) ion current i_{ion} on the collector electrode under the ambient argon gas pressure of 0.5 Pa.

Just after a pulse voltage is applied to the target, the voltage shows the initial source voltage, because there is no plasma generation. The voltage begins to decrease after less than a dozen microseconds, which indicates HPPS plasma generation. The target voltage then gradually decreases, followed by a stationary state. Thus, a constant glow voltage is observed after approximately 35 μs.

It is shown in Fig. 4 that the glow voltage and current at 35 μs are approximately −500 V and 30 A, respectively, at a pulse duration of 50 μs. Thus, the instantaneous power consumption is obtained by the product of these values and is approximately 15 kW. Ion current rises later than target current about 10 μs. The peak of ion current appears about 5 μs after a pulse-off time, and is decreased after that. The peak value of ion current is about 290 mA.

3.2. Gas pressure dependence

Figure 3 shows the gas pressure dependence of average power and ion charge. The average power P_{ave} is the value which carried out the time average of the instant electric power $p_T(t) = v_T(t) \times i_T(t)$ supplied to a plasma source, and is denoted by the following formula:

$$P_{ave} = f\int_0^{1/f} p_T(t)dt \qquad (1)$$

where f is the repetition frequency of high-voltage pulses. Average power declined with the fall of gas pressure p. When p was less than 0.7 Pa, the decreasing rate became large greatly at gas pressure. Average powers ware 305 W at $p = 2$ Pa and 190 W at $p = 0.5$ Pa.

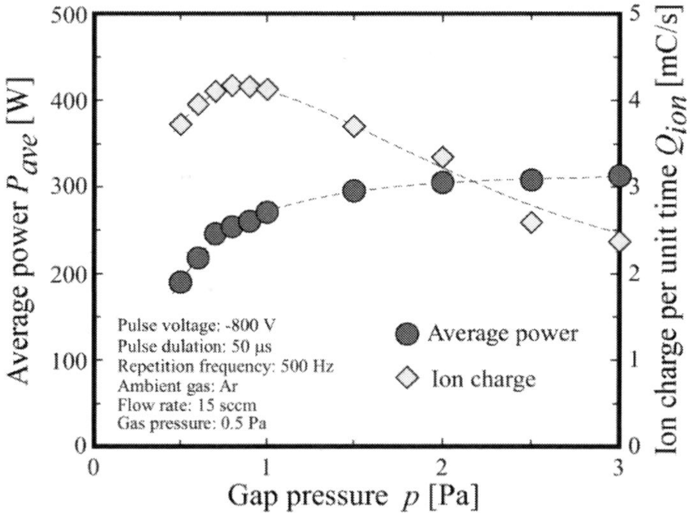

Figure 3. The gas pressure dependence of the average power and the ion charge at the collector electrode.

The ion charge quantity per unit time Q_{ion} on the collector electrode is given by the following formula:

$$Q_{ion} = f\int_0^{1/f} i_{ion}(t)dt \qquad (2)$$

Ion charge increased with the decreases of gas pressure, and became the maximum at $p \approx 0.8$ Pa. When p is less than 0.7 Pa, Ion charge Q_{ion} decreased, and Q_{ion} was 3.7 mC/s at $p = 0.5$ Pa.

While gas pressure becomes low, although average power (as it is called power consumption) decreases, ion current increases. That is, the ion charge per power consumption, Q_{ion}/P_{ave}, increases under low-pressure power. This has suggested that the rate of the metal ions contained in plasma increases.

3.3. Growth rate of titanium film

The silicon substrate was placed on the collector electrode and the titanium thin film was made to deposit for 30 minutes. The thickness of the thin film was measured using a step gauge (VEECO DEKTAK 3 ST AUTO I). Growth rate of the titanium film was determined from film thickness. The relation between the pulse duration of a high-voltage pulse and the growth rate of the titanium film is shown in Fig. 4.

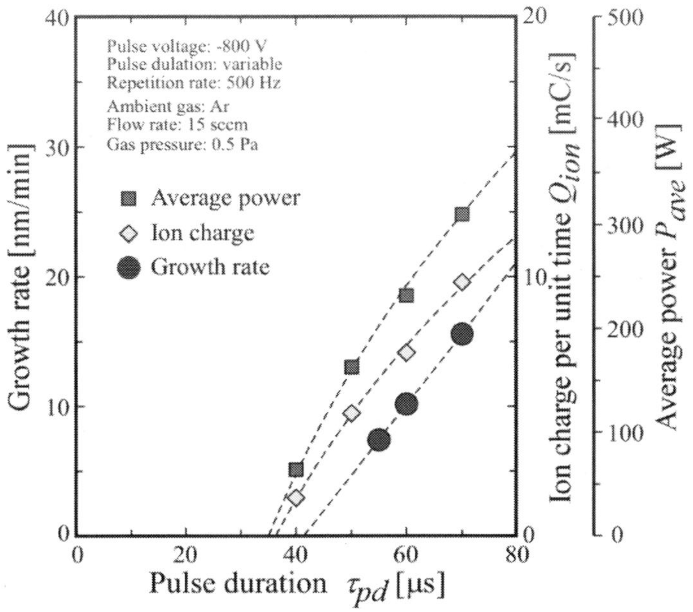

Figure 4. The pulse duration dependence of the average power, the ion charge and the growth rate.

Each of average power, ion charges, and film forming speed increased with the increase in pulse duration. When pulse duration was 70 μs, the average power was 310W, the ion charge is 9.8 mC/s and the growth rate is 16 nm/min. Also, the peak power is 15 kW and peak current is 27 A.

The growth rate increases almost linearly to pulse duration. A fitting analysis has suggested that titanium deposition occurs with the pulse duration for 41 μs or more. On the other hand, the relation between average power and pulse duration has indicated HPPS discharge occurs with the pulse duration of 35 μs or more. In HPPS penning discharge, the metal plasma is produced between a pair of target electrodes. Since the electrons are trapped by the magnetic field and the electric field between target electrodes, plasma is not diffused on the outside of a plasma source if the plasma density does not become high enough. Metal neutrals and ions reach the collector electrode with diffusion of plasma. So, it is thought that the minimum pulse duration required for deposition becomes larger than it for generating HPPS plasma.

4. Conclusion

The HPPS penning discharge plasma source was disadvantage in a narrow pressure range on the plasma production, and in particular, the lower limit of the pressure was 2 Pa or higher. This problem has been solved by setting an electrically-grounded electrode at the central region of the plasma source. According to the results, the lower limit of plasma generation was lowered to about 0.5 Pa with setting the inner electrode. Titanium films were deposited on the collector electrode for an HPPS discharge in the presence of the inner electrode. The deposition rate was about 16 nm/min at argon gas pressure of 0.5 Pa, applied voltage of −800 V, pulse duration of 70 μs and repetition rate of 500 Hz. At this time, the HPPS penning plasma source has target current of 30 A and the peak power of 15 kW and power consumption of 310 W.

References

[1] Kouznetsov V, Macák K, Schneider J M, U. Helmersson U and Petrov I 1999 A novel pulsed magnetron sputter technique utilizing very high target power densities *Surface Coating and Technology*, **122** (2–3) 290–293.

[2] Yukimura K, Mieda R, Tamagaki H and Okimoto T 2008 Generation of droplet-free high-power pulsed sputtering (HPPS) glow plasma with several tens of kilowatts *physica status solidi (a)*, **205**, (4) 949–952.

[3] Lee W Y and Daniel Oblas 1975 Argon entrapment in metal films by dc triode sputtering *Journal of Applied Physics* **46** (4) 1728–1732

[4] Mattox D M 1989 Particle bombardment effects on thin-film deposition: A review Journal of Vacuum Science & Technology A 7 (3) 1105–1114

[5] Azuma K, Mieda R, Yukimura K, Tamagaki H and Okimoto T 2009 Electrical and Optical Characteristics of High-Power Pulsed Sputtering Glow Discharge *IEEE Transactions on Plasma Science* **37** (7) 1178–1188

[6] Yukimura K, Ehiasarian A P, Ogiso H, Nakano S and Azuma K 2011 Metal Ionization in a High-Power Pulsed Sputtering Penning Discharge *IEEE Transactions on Plasma Science*, **39** (11) 3125–3132

[7] Fridman A and L. A. Kennedy L A 2004 Penning Glow discharges in Plasma Physics and Engineering, Taylor and Francis, New York 484–486

Spectroscopic analysis of a nanostructure roughness of plasma-deposited Au films using organic monolayer

M Shindo, T Sawada, K Doi, K Mukai and K Shudo*
Faculty of Engineering, Yokohama National University,
79-5 Tokiwadai, Hodogaya-ku, Yokohama 240-8501, Japan

*E-mail: ken1@ynu.ac.jp

Abstract. The morphology and roughness of nanostructure of Au film are discussed using an atomic-force microscope (AFM) and a laser Raman spectroscopy. The Au films are deposited on mica and Si substrate by means of a dc plasma sputtering technique, and a self-assembled monolayer (SAM) of benzene-thiol is formed on the Au films. The cluster size and roughness on the surface increase with substrate annealing during the deposition on mica substrate, while they decrease on Si substrate. Raman spectra show a surface-enhanced Raman scattering (SERS) occurs at the optimum roughness R_A~0.3 of the substrate.

1. Introduction

Organic thin films have been studied widely in relation with fabricating self-integrated devices, where a self-assembled monolayer (SAM) is widely used [1]. A SAM is created by the chemisorption of organic molecules onto specific metal/semiconductor surfaces. For example, a benzenethiol (BT) molecule, which is known to easily form a SAM, contains the most fundamental aromatic functional that offers a variety of applications and binds at the sulfur site to gold (Au)[2-4]. As molecular vibrations of the SAMs on the surfaces are fingerprints of adsorbed structures, the observations of those molecules clarify the molecule-substrate interaction at the nanometer scale.

The molecules in such monolayer films are detectable with surface-enhanced Raman scattering (SERS) [5,6], since SERS offers very high sensitivity spectroscopically; the enhancement can be at the order of 10^6, compared with ordinary Raman process. Although a SERS is usually used for nano-sized particles [7-11], the surface roughness will also induce SERS effect. Actually, this was confirmed by an ultrafast measurement of coherent molecular motion of adsorbed Au layers [12], on which the SAM of BT was adsorbed [13]. Detected signals of the ultrafast motion of molecules depended on the molecular array and surface morphology. Thus, a SERS measurement will give us the information on the surface roughness of Au thin films.

In this study, films of Au, which are deposited on mica and Si substrates using a dc plasma sputtering technique, are observed with an atomic-force microscope (AFM) to see their surface morphology including roughness. Then a SAM of BT is formed on the Au surface and observed by laser Raman spectroscopy to discuss the relationship between the roughness and onset of SERS. The effect of the substrate annealing during the deposition process is also discussed.

2. Experimental

Au atoms were attached to (001) surfaces of mica [14-17] after cleavage with a scotch tape and silicon substrates[18-20] at room temperature (RT) and 400°C by plasma dc sputtering using JFC1500 (JEOL).

The silicon surfaces usually have native oxide with non-uniform structures. The deposition rate was kept constant at approximately 0.5 nm/sec and the film thickness (θ) was controlled in the range of 0.3~20nm. Several authors demonstrated that the surface morphology is not as sensitive to the deposition rate as to the annealing temperature [21,22]. After the surface morphology was observed with an atomic-force microscope (AFM), substrates were soaked into BT-ethanol solution (1mM) for 24 hours at RT to make SAM. After rinse with ethanol, monolayer films were left on the surface. A laser Raman spectra of the surface rinsed with ethanol were measured using a 532nm laser with a spot size of 1μm in diameter.

3. Results and discussion

3.1 Au film deposited on mica substrates

Typical AFM images of Au thin film surfaces deposited on mica substrate are three-dimensionally shown in figure 1. The surface roughness with varied θ is summarized in table 1. Here, the diameter of the clusters designates for k_{max}^{-1} where k_{max} is the wavenumber of a peak in two-dimensional fast Fourier transformation (FFT) spectra (for example, see figure 1(c)) from AFM surface profiles. This represents the average of the clusters very well: for instance, k_{max}~0.053nm^{-1} at the peak in FFT for θ =20nm provides the cluster size k^{-1}~19nm. The height of the clusters is evaluated in terms of two indexes, R_A and R_{RMS}. One notes that R_{RMS} is 20~30% larger than R_A. The amplitude in FFT component at k_{max} is very close to these two indexes. Two times of the index (R_A, R_{RMS} or the FFT amplitude at the peak) approximates the height of clusters of the average diameter. The average diameter is increased as θ is increased. However, the roughness is increased as θ is increased except for θ=2nm.

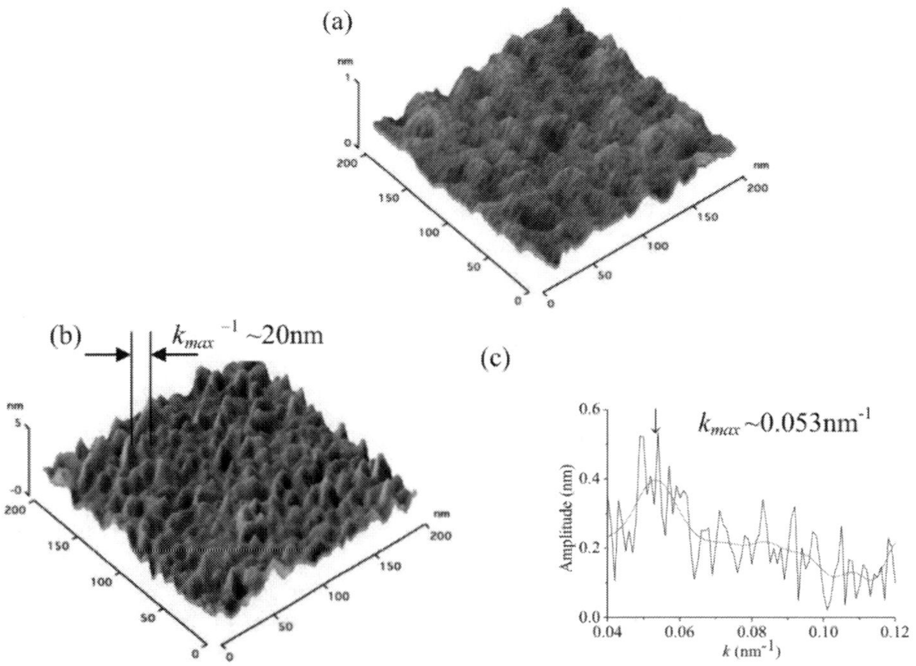

Figure 1. Topographic AMF images of Au films deposited on mica substrate at RT at (a) θ=5nm (b) θ=20nm. The cluster size is consistent with the inverse of the wave number at a peak in FFT spectrum (c) obtained with the surface profile (b). Note that the amplitude at k_{max} is ~0.4 nm. See table 1 and text for detail.

Table 1. The surface roughness of Au films estimated from AFM images. Film thickness specified with superscripts * showed the enhanced Ramana scattering (SERS) singal.

Substrate	Mica					Si					
thickness θ	2nm	5nm*	10nm*	20nm	5nm*	0.3nm	0.3nm*	1nm	1nm	3nm	3nm
Temperature	RT	RT	RT	RT	400°C	RT	400°C	RT	400°C	RT	400°C
Diameter	56.53	32.28	21.41	19.04	50.04	29.09	20.17	35.16	19.25	33.99	18.47
R_A [a]	1.24	0.076	0.269	0.395	1.04	0.69	0.34	1.09	0.595	0.793	0.66
R_{RMS} [b]	1.66	0.102	0.341	0.500	1.33	1.48	0.54	1.37	0.78	1.01	0.85

[a] Arithmetic mean value $R_A = \int |z(x,y) - z_0| dxdy \Big/ \int z(x,y) dxdy$.

[b] Root mean square (RMS) value $R_{RMS} = \left[\int \{z(x,y) - z_0\}^2 dxdy \Big/ \int z(x,y) dxdy \right]^{1/2}$,

where z_0 is an average in the height $z(x,y)$ over position (x,y) in the observed area.

This indicates that the diffusive process of Au on mica in the initial phase of the deposition is different from that on Au after the surface is enough covered with Au.

Figure 2 shows Raman spectra of the films for various θ. A clear peak of C=C bond at 1570cm^{-1} appeared only for θ =5 and 10nm in RT deposition. This suggests that the surface roughness is optimum for SERS at θ=5~10nm. The C=C bond peak shifted to the lower energy with θ increased (at the order of a few cm^{-1}), indicating that the surface roughness tilted BT orientation with respect to the surface [23,24].

3.2. Annealing effect to Au deposition on mica substrates

Au film deposition was also carried out with the substrate annealed at 400 °C for θ =5nm. An AFM image and Raman spectrum of Au film deposited on mica substrate annealed at 400°C are shown in figure 3. The annealing process during the deposition did not form a single crystal but increased the size of Au clusters and surface roughness. This is because Au on mica substrate has a large diffusion coefficient [25] and form (111) surface only at high-temperature (>473°C) regardless to deposition rate[26]. However, this fact is in contrast to a post-process annealing of Au thin films on mica substrate forming a large single crystal typically at 400~500°C [14-16]. Therefore, Au cluster seems to have a high crystallinity on the substrate after deposition at high temperature. The annealed surface yields the strong SERS as shown in figure 3(b). The diameter of RT-deposited surface with θ=2nm, whose roughness and diameter are similar to the annealed case, does not give SERS. This indicates that the larger cluster yields the stronger SERS, and that SERS needs a density of clusters.

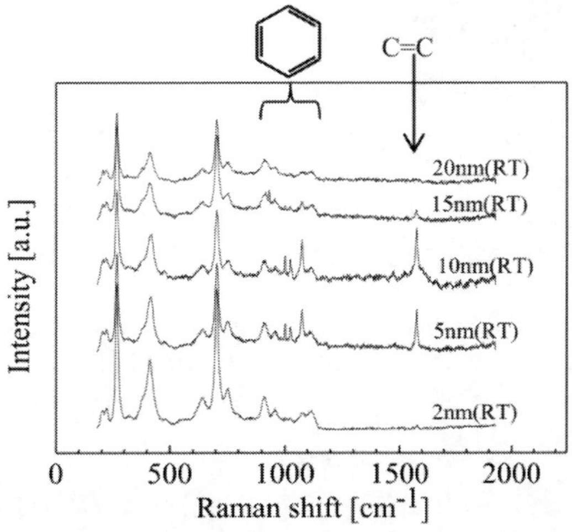

Figure 2. Raman spectra of BT on Au film deposited on mica substrate.

Typical signals of BT-derived vibrations are marked.

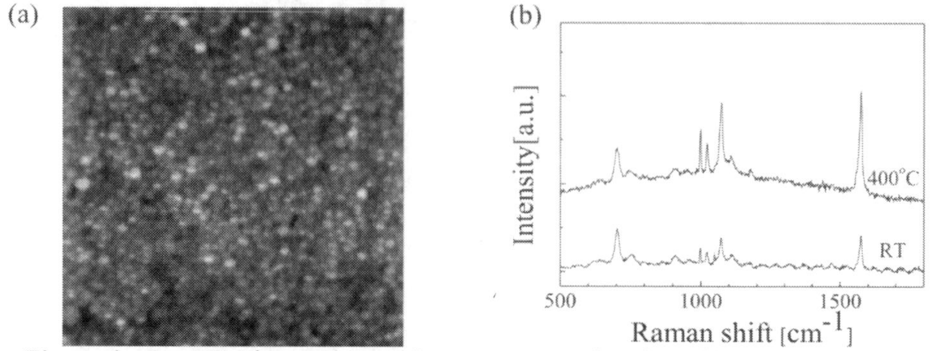

Figure 3. An AFM image (a) and Raman spectra (b) of Au film deposited on mica substrate annealed at 400°C to θ=5 nm. Raman peaks from BT are strengthened at the annealed deposition while a substrate component does not change.

3.3. *Au film deposited on Si substrate*

AFM images of Au film deposted on Si substrate with θ=0.3nm at RT and 400°C are shown in figure 4. Actually, the surface of Si was covered with native SiO_x ($x \cong 2$) layer, and it has non-uniform amorphous-like structure. Thus, one cannot identify the deposition plane in the case of Si. Annealing during plasma sputtering decreased both the size and roughness of Au clusters. On the other hand, it is reported that a post-process annealing of Au thin films on Si substrate induces grain growth [14], indicating this process is different from our high temperature deposition. In our case, the deposition at high temperature did not grow grains but enhanced the cluster size. The height of the clusters corresponds to few monolayers of Au atom. This is because the lateral diffusion of Au on SiO_2 surfaces dominates the vertical growth [27]. From Raman spectra in figure 5, the intensity of the peak was enhanced most at θ =0.3nm with annealing, at which the roughness is optimum for SERS. It is concluded that BT was adsorbed on the ultimately thin Au surface and that SERS is effective on atomic-scale islands. We note that the roughness optimum for SERS is R_A~0.3nm, which agrees with R_A for mica substrate.

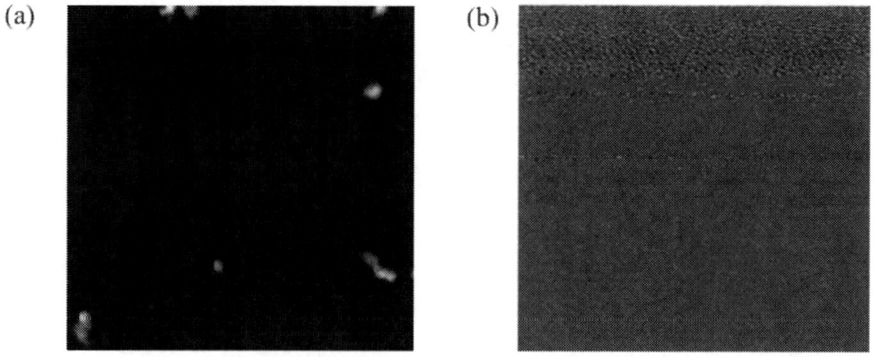

Figure 4. AFM images of Au films deposited on Si substrates with with θ= 0.3nm at (a) RT and (b) 400 °C.

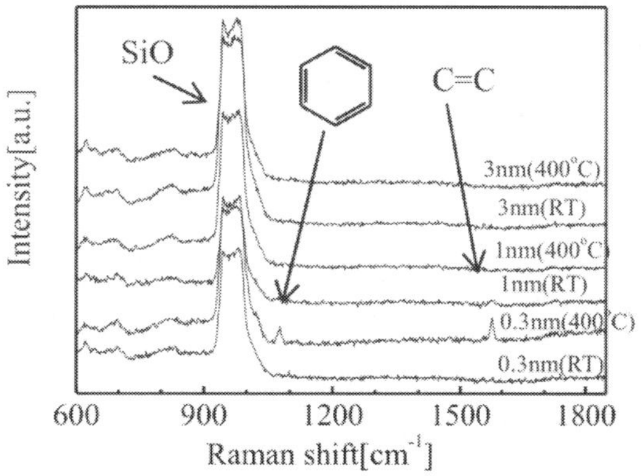

Figure 5. Raman spectra of Au films deposited on Si substrates.

4. Summary

BT-SAM on Au films deposited on mica and Si substrate were observed by means of Raman spectroscopy. Au deposited by dc plasma sputtering with and without annealing was compared. The cluster size and roughness of the surface increase on annealed mica substrate, while they decrease on Si substrate. It is found that a SERS yields strong signals at the specific diameter of Au clusters deposited with the thickness of more than a few monolayers of Au atoms and that a SERS offers a sensitive response to the surface roughness.

References

[1] Pensa E, Cortes E, Corthey G, Carro P, Vericat C, Fonticelli M H, Benitez G, Rubert A A and Salvarezza R C 2012 *Accounts of Chem. Res.* **45** 1183
[2] Price R C and Whetten R L 2005 *J. Am. Chem. Soc.* **127** 13750
[3] Kang H, Lee H, Kang Y, Hara M and Noh J 2008 *Chem. Commun.* **12** 5197
[4] Kang H, Park T, Choi I, Lee Y, Ito E, Hara M and Noh J 2009 *Ultramicroscopy* **109** 1011
[5] Zaera F 2012 *Chem. Rev.* **112** 2920
[6] Rycenga M, Cobley C M, Zeng J, Li W, Moran C H, Zhang Q, Qin D and Xia Y 2011 *Chem. Rev.* **111** 3669
[7] Hossain M K , Shibamoto K, Ishioka K, Kitajima M, Mitani T and Nakashima S 2007 *J. Luminescence* **122-123** 792
[8] Elias J, Gizawska J M , Brodard P, Widmer P, deHazan Y, Graule T, Michler J and Phlippe L 2012 *Nanotechnology* **23** 255705
[9] Wang J , Zhu T, Song J and Liu Z 1998 *Thin Solid Films* **327-329** 591
[10] Laor U and Schatz G C 1981 *Chem. Phys. Lett.* **82** 566
[11] Bell S E J and McCourt M R 2009 *Phys. Chem. Chem. Phys.* **11** 7455
[12] Katayama I, Koga S, Shudo K, Takeda J, Shimada T, Kubo A, Hishita S, Fujita D and Kitajima M 2011 *Nano Lett.* **11** 2648
[13] Shudo K, Doi K, Katayama I, Kitajima M and Takeda J 2012 *European Phys. J. Web Conf.* in press
[14] Ferrero S, Piednoir A and Henry C R 2001 *Nano Lett.* **1** 227
[15] Kowalczyk P, Kozlowski W, Kluse Z, Olejniczak W and Datta P J 2007 *App. Surf. Sci.* **253** 4715
[16] Li F, Tang L, Zhou W and Guo Q 2011 *Surf. Sci.* **605** 1016
[17] Ostendorf F, Schmitz C, Hirth S, Kuehnle A, Kolodziej J J and Reichling M 2008 *Nanotech.* **19**

305705
[18] Chan K, Aspanut Z, Goh B, Muhamad M R and Rahman S A 2011 *Thin Solid Films* **519** 4952
[19] Ruffino F, Canino A, Grimaldi M G, Giannazzo F, Roccaforte F and Raineri V 2008 *J. Appl. Phys.* **104** 024310
[20] Wang J, Zhu T, Song J and Liu Z 1998 *Thin Solid Films* **327-329** 591
[21] Liu Z H, Brown N M.D 1997 *Thin Solid Films* **300** 84
[22] Malinsky P, Slepicka P, Hnatowicz V and Svorcik V 2012 *Nanoscale Res. Lett.* **7** 241.
[23] Benkataramanan M, Skanth G, Bandyopadhyay K, Vijayamohanan K and Pradeep T 1999 *J. Colloid and Interface Sci.* **212** 553
[24] Joo S W, Han S W and Kim K 2001 *J. Colloid and Interface Sci.* **240** 391–399
[25] Ruffino F, Torrisi V, Marletta G and Grimaldi M G 2011 *Nanoscale Res. Lett.* **6** 112
[26] Higo M, Fujita K, Tanaka Y, Mitsushio M. Yoshidome T 2006 *Appl. Surf. Sci.* **252** 5083
[27] Ruffino F, Grimaldi M G, Bongiorno C, Giannazzo F, Roccaforte F, Raineri V 2008 *Superlattices and Microstructures* **44** 588

AUTHOR INDEX

AUTHOR INDEX